Responding to Environmental Issues through Adaptive Collaborative Management

Focused on forest management and governance, this book examines two decades of experience with Adaptive Collaborative Management (ACM), assessing both its uses and improvements needed to address global environmental issues.

The volume argues that the activation and the empowerment of local peoples are critical to addressing current environmental challenges and that this must be enhanced by linking and extending such stewardship to global and national policymakers and actors on a broader scale. This can be achieved by employing ACM's participatory approach, characterized by conscious efforts among stakeholders to communicate, collaborate, negotiate and seek out opportunities to learn collectively about the impacts of their action. The case studies presented here reflect decades of experience working with forest communities in three Indonesian Islands and four African countries. Researchers and practitioners who participated in CIFOR's early ACM work had the rare opportunity to return to their research sites decades later to see what has happened. These authors reflect critically on their own experience and local site conditions to glean insights that guide us in more effectively addressing climate change and other forest-related challenges. They showcase how global and regional actors will have to work more closely with smallholders, Indigenous Peoples and local communities, recognizing the key local roles in forest stewardship.

This book will be of great interest to students, scholars and practitioners working in the fields of conservation, forest management, community development, natural resource management and development studies more broadly.

Carol J. Pierce Colfer is currently a Senior Associate at the Center for International Forestry Research (CIFOR-ICRAF) and Visiting Scholar at Cornell University's Southeast Asia Program, Ithaca, New York, USA. She is the author/editor of numerous books, including *Adaptive Collaborative Management in Forest Landscapes: Villagers, Bureaucrats and Civil Society* (Routledge, 2021), *Masculinities in Forests* (Routledge, 2020), *The Earthscan Reader on Gender and Forests* (Routledge, 2017) and *Gender and Forests* (Routledge, 2016).

Ravi Prabhu is a Forester with thirty years of international experience. He began his international career with the Center for International Forestry Research (CIFOR) and has been with World Agroforestry (ICRAF) since 2012, where he is now its Director General. He won the Queen's Award for Forestry in 2005, presented by HM Queen Elizabeth II. He is the co-editor of *Adaptive Collaborative Management in Forest Landscapes: Villagers, Bureaucrats and Civil Society* (Routledge, 2021).

The Earthscan Forest Library

This series brings together a wide collection of volumes addressing diverse aspects of forests and forestry and draws on a range of disciplinary perspectives. It is aimed at undergraduate and postgraduate students, researchers, professionals, policy-makers and concerned members of civil society.

Forest Conservation and Sustainability in Indonesia
A Political Economy Study of International Governance Failure
Maxton Bernice-Lee

Masculinities in Forests
Representations of Diversity
Carol J. Pierce Colfer

Adaptive Collaborative Management in Forest Landscapes
Villagers, Bureaucrats and Civil Society
Edited by Carol J. Pierce Colfer, Ravi Prabhu, and Anne M. Larson

Sacred Forests of Asia
Spiritual Ecology and the Politics of Nature Conservation
Edited by Chris Coggins and Bixia Chen

The Cultural Value of Trees
Folk Value and Biocultural Conservation
Edited by Jeffrey Wall

Responding to Environmental Issues through Adaptive Collaborative Management
From Forest Communities to Global Actors
Edited by Carol J. Pierce Colfer and Ravi Prabhu

For more information about this series, please visit: www.routledge.com/The-Earthscan-Forest-Library/book-series/ECTEFL

Responding to Environmental Issues through Adaptive Collaborative Management

From Forest Communities to Global Actors

Edited by
Carol J. Pierce Colfer and
Ravi Prabhu

The Funder is: CIFOR

Routledge
Taylor & Francis Group
LONDON AND NEW YORK

earthscan
from Routledge

First published 2023
by Routledge
4 Park Square, Milton Park, Abingdon, Oxon OX14 4RN

and by Routledge
605 Third Avenue, New York, NY 10158

Routledge is an imprint of the Taylor & Francis Group, an informa business

British Library Cataloguing-in-Publication Data
A catalogue record for this book is available from the British Library

Library of Congress Cataloging-in-Publication Data
Names: Colfer, Carol J. Pierce, editor. | Prabhu, Ravi, editor.
Title: Responding to environmental issues through adaptive collaborative management : from forest communities to global actors / Edited by Carol J. Pierce Colfer and Ravi Prabhu
Description: New York, NY : Routledge, 2023. | Includes bibliographical references and index.
Identifiers: LCCN 2022044010 (print) | LCCN 2022044011 (ebook) | ISBN 9781032352282 (hardback) | ISBN 9781032352299 (paperback) | ISBN 9781003325932 (ebook)
Subjects: LCSH: Landscape protection. | Forest landscape management. | Forestry and community.
Classification: LCC SD387.L35 R47 2023 (print) | LCC SD387.L35 (ebook) | DDC 719/.33—dc23/eng/20220928
LC record available at https://lccn.loc.gov/2022044010
LC ebook record available at https://lccn.loc.gov/2022044011

ISBN: 978-1-032-35228-2 (hbk)
ISBN: 978-1-032-35229-9 (pbk)
ISBN: 978-1-003-32593-2 (ebk)

DOI: 10.4324/9781003325932

Typeset in Goudy
by codeMantra

We dedicate this book to our families who have graciously accepted our extra hours of work and loss of our attention throughout the writing and editing of this book:

from Ravi to Sabine, Anasuya, Maya and Kavan;
and
from Carol to Richard, Megan, Brian, Amy, and Alan.

We also thank our parents for their enduring care and affection, and for filling us with this sometimes inconvenient desire to 'contribute'.

Ravi thanks Jennifer and Mohan; and
Carol thanks Gwendolyn and Joe Eugene [Gene].

Contents

Figures

Tables

Boxes

Contributors

Hasantoha Adnan obtained a degree in anthropology from the University of Indonesia in 1999. He received his Master of Science for Social Intervention from the Faculty of Psychology, University of Indonesia in 2009. At CIFOR, his projects have involved action research for the Adaptive Collaborative Management of Forest (ACM) Project (2002–2007), Agroforestry and Forestry: Linking Knowledge to Action in Southeast and South Sulawesi (AGFOR Project, 2012–2016), Partnerships for Enhanced Engagement in Research (PEER Project, 2016–2017) and Collaborating to Operationalise Landscape Approaches for Nature, Development and Sustainability (COLANDS Project, 2021–present). He has also worked for Kemitraan/Partnerships as a project manager for a Community Based Natural Resource Management project (2016–2018) and the Peatland Care Village Project (2018–2020). In all these projects, he has had a leading role developing and implementing matters related to forestry research, sustainable landscape management, commodities and local livelihoods, as well as empowerment of forest village communities in Indonesia. Adnan has written and contributed to several papers and journal publications. He is currently a Social Learning and Knowledge Management specialist in Kemitraan/Partnership for Governance Reform, where he promotes the application of good governance principles in natural resource management in Indonesia.

Shintia Arwida started her career as a Journalist and developed expertise in communications and publishing, both for commercial and non-profit enterprises. She pursued her master's degree at Wageningen University in The Netherlands, focusing on social studies and the technical environment. Halfway through her career, while working for CERINDO, an environmental consultancy company, she cultivated her interest in research and joined Sajogyo Institute as an intern for one year. Later on, she joined CIFOR, shaping her knowledge on REDD+, gender, and the environment. She continued her career path by working for UNOPS, which sharpened her program management skills and knowledge on peatland restoration. She is currently working in the climate change mitigation and adaptation sector with specific focus on developing Nature Based Solutions in a project covering Asia and the Pacific.

Caroline Bilogui was born in 1985 in Nkolofoumbi in the Center Region of Cameroon. She obtained a master's degree in economics at the University of

Yaoundé II in September 2009. In December 2009, she began as an Executive Assistant in *Cooperative Agroforestière*, a Cameroonian organisation, where she acquired expertise in the management and development of community forests. She joined the African Model Forest Network (AMFN) Secretariat in February 2011, where she assumed the responsibilities of project assistant (2011–2015) and project manager (from 2016 onward). Bilogui was an innovator and a full part of the Model Forests team's major breakthroughs. During that time, she assisted in setting up and managing several projects and small business initiatives, including her support to two women's platforms' income-generating projects, the development of the *Model Forest Green Business* label initiative and the management of the Moringa production project and the Moringa Tea processing plant in the Campo-Ma'an Model Forest. She also coordinated other Non-Timber Forest Product value chain projects for the Secretariat, including the *Big Forests Value Chains* project preparation with the Islamic Development Bank. She was also the AMFN Focal Point for its partnership with UAE countries.

Yus Budiyono is a Researcher at the National Research and Innovation Agency (BRIN), Indonesia. He joined the Agency for the Assessment and Application of Technology (BPPT) in 1994, which recently became part of BRIN. He has led research projects focusing on flood risk assessment. Budiyono is also the coordinator of the Delta Alliance International Indonesia Wing, coordinating a team assigned to implement on-going research. He is keen to solve problems using dynamic models. Recently, he integrated land subsidence phenomena, extreme climate events, and dynamics of land use changes into flood risk assessment for northern Java. Past research was on a regional economic model of the wetland ecosystem in South Kalimantan, precipitation-induced landslides in West Java, and socio-hydrology of the tsunami in Aceh and Banten provinces. He obtained his master's degree in 2000 at the University of Melbourne, Australia. In 2018, he defended his doctoral thesis at the Amsterdam Free University. Budiyono is eager to work using Linux and free/open-source software. He uses Python programming language for iterative works and occasionally teaches science writing at various institutions in Indonesia and abroad.

Edward Chuma is an Agro-ecologist who has been working with communities, universities, and many development organisations. He is a process facilitator in PICOTEAM, Institute for People, Innovation and Change in Organisations, in Southern Africa, based in Zimbabwe.

Mariteuw Chimère Diaw was the Director General of the African Model Forest Network until July 2021, and is now the Deputy Director General and a Board Member of the Consortium d'Entreprises. He holds a PhD in Economic anthropology from Laval University, an MA in Rural Sociology from Michigan State University and a Master's in Philosophy and Sociology from the University of Dakar. He is a member of the Intergovernmental Science-Policy Platform for Biodiversity and Ecosystem Services (IPBES) Scientific Panel (MEP) and of IPBES's Management Committees for Values Assessment and the Task Force on Scenarios and Models. He is Board member emeritus of Ecoagriculture Partners and contributes to the alliance of Landscapes for People, Food and Nature (LPFN), the African Landscape Dialogue and the 1000 Landscapes initiative. Diaw has been a researcher and program manager for over 40 years, with the Senegalese Institute of Agricultural Research, the Center of Oceanographic Research, the International Institute of Tropical Agriculture, the Center for International Forestry Research, and the AMFN. His research interests and publications include African history, migrations, share system modeling and embedded commons, tenure and property rights, climate change, deforestation and food production, parks and people, adaptive collaborative management, landscape governance and transformation, Model Forests, social enterprises, participatory action research and interactive social methodologies. Diaw has lived and worked in Senegal, the United States, Canada, Indonesia and Cameroon.

Ida N. S. Djenontin is a Human-environment Geographer and Interdisciplinarily-trained Social Scientist. She is interested in the human dimensions of global environmental and climate changes affecting forest and agroecological systems and how to achieve socioecological sustainability in the pursuit of sustainable development. Djenontin examines specifically the governance and institutional challenges and the socio-cultural and economic dimensions of environmental degradation, resource conservation and restoration, and climate change adaptation and mitigation in sub-Saharan African contexts. She questions how to balance biodiversity, mitigation, and other ecological needs with natural resource-based livelihoods, food security, and development goals. Djenontin is currently an Assistant Professor at Penn State University in the Department of Geography, College of Earth and Mineral Sciences (US) and a Visiting Fellow of Grantham Research Institute on Climate Change and the Environment at the London School of Economics and Political Science (UK). She holds a dual PhD in Geography (Nature-Society Studies) and in Environmental Science and Policy from Michigan State University; a Master's in Development Practice from the University of Arizona; and MA and BSc in Agricultural Sciences from the University of Parakou (Benin).

Felicitas Egunyu is an Adjunct Professor at the School of Environment and Sustainability, University of Saskatchewan, Canada. She holds a PhD in Environment and Sustainability from that university. Her research focuses on identifying how institutions can better represent and respond to under-represented groups and how such groups can be empowered to participate more effectively. Her research is primarily conducted in Canada and Uganda. In addition to working in academia, she has also worked in government and not for profit. Egunyu is currently working with the British Columbia Ministry of Land Water and Resource Stewardship as a Land and Resource Specialist.

Paolo Ficarelli has run several programs on agricultural service delivery, and, in the past decade, on Information and Communication Technology in agriculture. He is working as an independent consultant based in Italy.

Micah Fisher is interested in engaged methodologies that enhance public participation on complex challenges of environmental change. He is a Research Fellow at the East-West Center in Honolulu, Hawaii; and serves as Affiliate Graduate Faculty at the Matsunaga Institute for Peace and Conflict Resolution.

Julie Gagoe Tchoko holds a professional master's degree in Natural Resources Management and another in Anthropology of Development and the Environment. She has more than 13 years of professional experience in the field of Monitoring and Evaluation (M&E) of development projects, gender and local development, including work for international organisations like AMFN, SNV, CIFOR, IITA, and GIZ; and she is a co-author on several articles published in scientific journals. As an expert in M&E, she has developed and implemented systems of M&E focused on gender analysis of international projects and programs. Her skills as an evaluator brought her to Ottawa, Canada in 2013 with the International Program of Development Evaluation Training – IPDET 2013. She has also contributed to the development of monitoring and evaluation training modules for civil society organisations and local communities in Congo Basin Landscapes. As a Gender Focal Point for the Central African Forest Commission (COMIFAC), she strengthened the gender capabilities of some representatives of Congo Basin organizations and also worked hard on the development of the COMIFAC gender strategy, a document validated in 2016 by COMIFAC Ministries. She has extensive experience with various stakeholders, such as women, youth, and indigenous peoples.

Jürgen Hagmann has focused on organizational development and change management over the past three decades at different levels – from communities to high political levels. He is the Team Leader of PICOTEAM, Institute for People, Innovation and Change in Organisations, where he also practices as a process facilitator.

Yayan Indriatmoko obtained a degree in anthropology from the University of Indonesia in 1999. He received a Masters in Environmental Management and Development from the Crawford School of Public Policy at the Australian National University (ANU) in 2015. Indriatmoko was a fellow in the Asian Public Intellectual Fellowship program (API) with his main activity doing research among indigenous minority peoples in Peninsular Malaysia from 2003 to 2004. Under the Center for International Forestry Research (CIFOR) for 12 years and Zoological Society of London (ZSL) for 2 years, he has had leading roles in developing and implementing projects in areas related to Sustainable Landscape Management and REDD+ projects in Indonesia. He has worked in research and practice in several sectors, including forestry research, local livelihoods, and land tenure security. Indriatmoko has authored and contributed to multiple journal papers and publications. Currently, he is a Senior Social and Environmental Specialist at Indonesia Infrastructure Finance (IIF), where he promotes the implementation of IIF's Social and Environment Principles in close cooperation with infrastructure project developers.

Kamaluddin (Kamal) works as a Facilitator to address land and resource conflict and supports village development initiatives. He heads the partnership division at Perkumpulan OASE, a local organization in Bantaeng, South Sulawesi focused on environment and development concerns for local communities in the region.

Judith F.M. Kamoto led the team that conducted ACM research in Chimaliro, Malawi in the early 2000s. She is an Associate Professor of Forestry and Rural Development at Lilongwe University of Agriculture and Natural Resources (LUANAR). She has been with the LUANAR since 1999. She is currently serving as a College Director of Bunda College of LUANAR. Her PhD is in International and Rural Development (2009) from the University of Reading in the United Kingdom; she has a Master of Science degree in Tropical Forestry (1999) from Wageningen Agricultural University in The Netherlands and a Bachelor of Science degree in Agriculture (1994) from the University of Malawi. Her PhD research was on 'Forest Governance and Rural Livelihoods'; for her master's degree, she studied 'Indigenous Silvicultural Management Practices of the Miombo Woodlands in Chimaliro Forest Reserve, Kasungu, Malawi'. Her research interests focus on community forestry, participatory forest resource management, sustainable livelihoods, bio energy, policy development and analysis, agroforestry, climate change, landscape restoration, extension and gender. She previously worked with the Forestry Department of the Ministry of Natural Resources under the Government of Malawi. She has published in refereed books and international peer reviewed journals such as *Land Use Policy; Sustainability*, and *IJAS*, among others.

Witness Kozanayi was part of the team that implemented the ACM project in Mafungautsi forest from 2001 to 2006. Currently, he is a Lecturer at Marondera University of Agricultural Sciences and Technology in Marondera, Zimbabwe. Kozanayi's recent work has been on the governance of non-timber forest products with a special interest on how the interface between customary and statutory forms affects livelihoods and ecological outcomes. A second area of interest is regenerative agriculture, in his home area of Chimanimani, Zimbabwe.

Klaus Krebs has been focused on training and personal development for the last two decades in various contexts. He presently freelances at his own company in Germany.

Trikurnianti (Yanti) Kusumanto is an Indonesian-born Research and Development Consultant, based in The Netherlands. Holding an advanced degree in tropical agriculture and forest policy from Wageningen University, she specialises in social learning and inclusivity in sustainability governance. Between 2000 and 2006, she co-led CIFOR's ACM research in Jambi, Sumatra. Over the years, she has been engaged as a development practitioner, researcher, and manager in the areas of participatory development, institutions, and socio-ecological sustainability. Before joining CGIAR (ICRAF, then CIFOR) in 1997, she worked with international development agencies operating in Indonesia. She started her own consulting company in 2013. In 2020, she co-initiated a Dutch-Indonesian project for delivering policy briefs on the flooding problem of Greater Jakarta in the context of climate change.

Ardi Labarani works as a Mediator of land and natural resource conflicts and also serves as the Head of Perkumpulan OASE, a local organization in Bantaeng, South Sulawesi, Indonesia, focused on environment and development concerns for local communities in the region.

Nining Liswanti is a Researcher in the Governance, Equity and Wellbeing (GEW) program at CIFOR-ICRAF, Indonesia. She obtained an MSc in 2004 on Environment and Natural Resources Management from the Bogor Agriculture Institute in Bogor, Indonesia. She has considerable experience in biodiversity, landscape assessment, land use planning, tenure, gender, indigenous people, governance, and multistakeholder processes. She is currently a field coordinator for CIFOR research activities in Indonesia and has conducted research herself in Sumatra, Kalimantan, Papua, Maluku and Sulawesi in collaboration with various stakeholders (government and non-government) from national to local level.

Marzoni obtained a degree in Agronomy and Agriculture from Jambi University in 1998. He was a community facilitator for Yayasan Gita Buana – a Jambi-based NGO – focused on community-based forest management in Jambi Province, Indonesia. During the implementation of CIFOR's Adaptive Collaborative Management (ACM) project (2000–2006), Marzoni played a significant role as a multi-stakeholder facilitator with the key role to help and facilitate the local community of Baru Pelepat to participate in the process of forest management within the ACM framework. He is also a GIS specialist and experienced in doing community mapping of natural and forest resources. Currently, Marzoni and his family live in Baru Pelepat and have become part of the community.

Frank Matose is an Associate Professor in the Department of Sociology and the Environmental Humanities South Centre at the University of Cape Town, South Africa. His research interests are in the political economy of commons in Africa. He just published *The Violence of Conservation in Africa* (with Maano Ramutsindela and Tafadzwa Mushonga).

Edward Missanjo is an Alumnus of Mzuzu University (MZUNI), University of Zimbabwe (UZ) and Kyushu University (KYUDAI) with academic training in education, environmental science, and forestry at the respective institutions. He is currently working as a Research Coordinator at Malawi Assemblies of God University (MAGU). His research interests include tree improvement, terrestrial assessment of forest biomass and carbon in relation to climate change, and the Forest and Other Land Uses (FOLU) subsector, education for sustainable development, participatory forest management, and interdisciplinary research. He has published numerous papers in peer reviewed journals.

Moira Moeliono, a CIFOR Senior Associate, has over 20 years' experience working on forest governance issues with a strong focus on social forestry. At CIFOR, she has been involved in research on decentralization, adaptive collaborative management, social and community-based forestry and rights-based issues. Recently, she has shifted more to work on climate change, in particular REDD+-related policy processes.

Tristam Pascal Moeliono holds a doctorate from Leiden University, The Netherlands. Currently, he is the head of the postgraduate programme at the Faculty of Law, Parahyangan Catholic University, Indonesia. He teaches legal philosophy and legal comparison in the undergraduate programme and environmental and spatial law in the master's programme. His scope of work also covers the link between human rights and the above-mentioned legal disciplines. Tristam is the co-editor of a collection of publications under the theme *Agrarian Law and Society in Indonesia*.

Agus Mulyana is a Mediator and Facilitator working on forestry and land conflicts in Indonesia. He has previously conducted research with the Center for International Forestry Research and served as a mediator for the Compliance Advisor Ombudsman of the World Bank Group.

Kudakwashe Murwira, Community facilitator (Deceased), formerly Independent Consultant in Zimbabwe. Kudakwashe was a true master of empowering community facilitation, which he practiced until his early death.

Tendayi Mutimukuru-Maravanyika is a dedicated Participatory Action Researcher who has worked in various fields, including forestry, agriculture, and fishery in many African countries. She is currently a Lecturer at Great Zimbabwe University in Zimbabwe.

Marie Françoise Roselle Ngo Baneg, born in Ekoamaen in 1980, has an advanced diploma in Soil Science, from the Faculty of Science at the Yaoundé University in Cameroon. An expert in biodiversity and integrated management of soil fertility, with an accent on biological agriculture, she was, between 2015 and 2020, the leader of Biofertilising Products in the African Model Forests Network. Between 2017 and 2022, she became a member of the Cameroonian team of the Intergovernmental Scientific and Political Platform on Biodiversity and Ecosystem Services (IPBES) and coauthor of Chapter 3 of the Cameroonian national evaluation of biodiversity. Since 2020, she has worked as a consultant on several projects on rural development and the preservation of the environment for the Study Bureau 12D. She is also the author of several works published by Harmattan and Croix du Salut, respectively.

Joachim Nguièbouri is a 3rd Degree Chief of Dibouné Village in the Campo-Ma'an Model Forest (Lokoundje); he is a Knight of the Cameroonian National Order of Merit and Knight of the National Order of Agricultural Merit (Cameroon). He has been the Senior Development Officer and Sites Coordinator at the African Model Forest Network Secretariat since 2010. He is also a specialist in participatory monitoring and evaluation, development of local criteria and indicators (C&I) sets, environmental education, conflict management, and facilitation of multi-stakeholder governance frameworks. Nguièbouri previously worked as a Research Assistant (2000–2010) at the Center for International Forestry Research where he was in charge of the Model Forests feasibility studies and was involved in the first implementation in Cameroon of Adaptive Collaborative Management. He also trained and supported local organizations, established multi-actor governance frames, and monitored timber value from uncontrolled sawing. His career in research, development and sustainable management began at the International Institute of Tropical Agriculture (IITA; 1996–1999), where he mainly worked in the characterization of IITA Benchmark Villages and the monitoring of inland artisanal fisheries. Earlier, before focusing on sustainability, he took his first professional steps teaching as a volunteer in Professional Training Centers for Young People (1988–1996).

Hlamalani Ngwenya is a Facilitator/Consultant/Writer in South Africa. She has engaged in community development and later in agricultural policy work and agricultural extension.

Mkhululi Ngwenya is a Forester and Development Practitioner currently working with Assemblies of God projects on supporting smallholder farmers to adapt to climate change through promotion of conservation agriculture, agroforestry, small livestock production and utilization of forestry resources. Ngwenya was part of the Forestry Commission Team that developed the Mafungautsi resource sharing project in Gokwe. He later went on to be part of the ACM pilot project around Mafungautsi forest.

Bruhnel Vambi N'Tambu has been the Technical Coordinator of the Model Forests Program in the Democratic Republic of Congo (DRC) since 2014. He holds a degree in Water and Forestry Engineering from the University of Kinshasa, and a master's degree in Development, Environment and Society from the Catholic University of Louvain in Belgium. Currently a PhD student in forestry sciences at Laval University, he is affiliated with the Center for International Forestry Research, the Institute of Integrative and Systems Biology, and is also a Senior Lecturer in the Department of Natural Resource Management at the University of Kinshasa in DRC. He was also an environmental consultant for the Canadian Cooperation Field Support Services Project in DRC between April 2017 and March 2022. His expertise covers the following areas: agriculture and food, agroforestry and tropical forestry, environmental sustainability, biodiversity and climate change, regional development and local dynamics.

Richard Nyirenda is a Senior Lecturer and Forest Governance Expert with Development and Training, University of Wolverhampton, UK on community forestry, forest governance, sustainable livelihoods and climate change, and REDD++. As a forester, his most recent work has been on supporting and researching capacity, strengthening initiatives for forestry law enforcement, governance and trade deliberative processes, funded mainly by the UK FCDO and EU in Central Africa, West Africa and SE Asia. Nyirenda was part of a team of researchers that pioneered research on ACM of protected forests in Zimbabwe between 1999 and 2000.

Effi Permatasari graduated from the Faculty of Agriculture, Jambi University in 2000. Permatasari joined Yayasan Gita Buana – a Jambi-based NGO – with her specialty as a gender and village women's group facilitator for CIFOR's Adaptive Collaborative Management (ACM) project in Baru Pelepat, until 2007. From 2007 to 2010, she was a women's group facilitator for the governmental National Community Development Program in Jambi. She then continued her professional work as a Jambi Provincial Government staff member as a farmer extension agent.

Henning Peter has been in forestry development at many different levels during his career. He has extensive experience in Latin America as well as in Central Asia and Africa. He is the Managing Director at PICOTEAM Germany and a Process Facilitator.

Ravi Prabhu is an accomplished Scientist who has engaged in multi-disciplinary research and action in forested landscapes for over 20 years. He earned his professional degree and doctorate in forestry from the University of Goettingen, Germany. At CIFOR, Prabhu co-led the ACM program from its inception. Ravi was previously a Senior Programme Officer, Forests and Climate Change with the United Nations Environment Programme (UNEP) in Nairobi. He led the UNEP team that contributed to the UN-REDD Programme – mainly by supporting countries to realize multiple benefits from REDD+, and to use REDD+ as a catalyst to transform to a green economy. Prior to that, he coordinated the Regional Plan for Collective Action in eastern and southern Africa, a joint initiative of CGIAR. He also worked in various capacities at CIFOR. Ravi joined World Agroforestry (ICRAF) as the Deputy Director General (Research) in 2012, becoming Director General in 2022. He has served on numerous international initiatives and committees, including the Millennium Ecosystem Assessment where he served on the review and editorial team, and the UN Millennium Projects Taskforce 6 on Environmental Sustainability. He received the Queen's Award for Forestry at Buckingham Palace in 2005.

Joe Ramaru is an Agro-ecologist and Process Facilitator, leading PICOTEAM, Institute for People, Innovation and Change in Organisations, in Southern Africa. Joe has vast experience with community facilitation.

D. Ary A. Samsura is a Senior Researcher and Lecturer at the Department of Geography, Planning, and Environment, Radboud University, The Netherlands. His research focuses on decision-making in spatial planning processes that involve different stakeholders with conflicting interests and goals. He is also interested in issues related to land management, climate mitigation and adaptation, sustainable development management, and collaborative planning. Much of his work has a global orientation, with ongoing research in Indonesia, Vietnam, China as well as European countries.

Lizwelabo Sibanda, Community member from Gokwe (Former Community Partner for the ACM team for Batanai Resource Sharing Committee), Gokwe South District, Zimbabwe. Sibanda is a resident of Batanai village, Gokwe District. As a community partner from the ACM team, he was involved in all ACM steps from inception to the time the ACM team left Gokwe District. He continues to witness residues of ACM processes in his village to this day.

Gusti Ayu Ketut Surtiari is a Human Geographer by training and is currently a Researcher at the Research Center for Population, the National Agency for Research and Innovation (BRIN) (formerly the Indonesian Institute of Sciences or LIPI). Her research interests are mainly on climate change adaptation, transformative adaptation, disaster risk reduction, community and family resilience, temporary shelter, urban poverty and informal settlements. Currently, she is in the final stage of her PhD research at the University of Bonn, studying the dynamics of vulnerability among low-income groups as a consequence of adaptation interventions in coastal Jakarta, Indonesia. Furthermore, during the COVID-19 pandemic, she has been exploring the pandemic's impact on social systems, particularly the livelihoods of vulnerable and marginal communities. She is a co-author of a paper in the UN Global Assessment Report on Disaster Risk Reduction (UNGAR 2022) titled "*Governance Strategies in Indonesia for Addressing Systemic Risks: Where do we Stand and the Future Outlook*".

Ade Tamara is a Research Consultant at CIFOR-ICRAF Indonesia. She has experience working on research on multi stakeholder forums, governance, gender, tenure, and safeguards in the Indonesian context. She holds an MSc degree in Forest and Nature Conservation from Wageningen University, The Netherlands.

Annisa Triyanti is an Assistant Professor in disaster and climate risk governance at the Copernicus Institute of Sustainable Development, Faculty of Geosciences, Utrecht University, The Netherlands. She obtained her PhD on the topic of governance of ecosystem-based disaster risk reduction (Eco-DRR) from the University of Amsterdam in 2019. From 2017 to 2019, she was appointed as a member of the Global Science and Technology Advisory Group of the UN Office for Disaster Risk Reduction (UNDRR) representing young scientists.

E. Linda Yuliani has been working as a Scientist at CIFOR since 2000, at which time she co-led ACM work in Jambi. Her research interests are ecology, adaptation and resilience of social-ecological systems, wildlife conservation, traditional knowledge and community-based conservation, and governance of protected areas. Most of Yuliani's field research has been in Indonesia with a special focus on Kapuas Hulu, West Kalimantan where she has worked on various projects since 2004. She also carried out research in Jambi and in South and Southeast Sulawesi. She obtained a degree in freshwater ecology from the Bandung Institute of Technology in 1994. She was awarded a Chevening Scholarship from the UK Government to study for a master's degree in ecology and environmental management at the University of York, which she received in 1999. She has just received a PhD in environmental science from Radboud University Nijmegen in The Netherlands.

Chris Zevenbergen is a Professor of urban (flood) resilience at the Water Engineering Department of IHE Delft and of Delta Urbanisme at the Faculty of Architecture and the Built Environment of TuDelft, The Netherlands. His research interest is specifically on innovative concepts to mitigate urban flood impacts, on flood-proofing building designs and technologies and on decision support tool development in urban planning with practical applications in urban flood management. He has a strong affinity with the ecological, socio-economic, institutional aspects of urban planning and water management. He has published/edited five books and more than 120 scientific publications in the field of environmental engineering and urban flood management.

Foreword – by Ruth Meinzen-Dick

This is an important book. In this era when development trends come and go, it is rare to have an assessment of an approach first applied 20 years earlier. It is even rarer to have critical reflections on an approach by people who have, in some cases, been involved in the process for two decades, and still rarer to have a comparative analysis of that approach across different countries and contexts. Yet, that is what we have for adaptive collaborative management (ACM). This matters, because many of the challenges that we face will not be solved overnight. We need evidence about what works over the long term.

ACM brings together the people who have interests in a forest not only to jointly plan, but also to learn from the implementation of those plans. While this may sound simple, it requires skilled facilitation to bring the different groups together, build communication and negotiation processes among them, as well as linking to higher and lower levels of organization (Colfer 2013). While there are broad principles for ACM, the *adaptive* part means that approaches are not only tailored to local contexts but also change over time. Instead of a mechanistic approach, this is organic; instead of 'social engineering' analogies, think of gardening – an ongoing process of bringing together elements so that they can grow in the particular environment.

I remember when I first heard of ACM, about 25 years ago. Devolution of natural resource management was happening in many natural resource management sectors, including forestry, fisheries, and irrigation. Recognition of the limited capacity of the state and technical experts to manage these resources led to transfers of responsibilities (and some rights) for management to communities and – it was hoped – a new relationship between state agencies and the people who depended on those resources for their livelihoods. My own work was on devolution of irrigation or what was referred to often as Participatory Irrigation Management, and I was involved in coordinating work across other natural resource management sectors, such as Fisheries Co-management or Joint Forest Management. The successful cases of devolution were often from pilot projects that had skilled facilitators to bring the resource users, government officials, and other technical experts together to build trust. But training people in the so-called 'soft skills' of facilitation is seen as less predictable than training in the so-called 'hard skills' of technical experts in engineering, water management, fisheries, or forestry.

Rather than investing in training facilitators and taking the time to develop locally adapted approaches, many of the devolution policies focused on structural approaches to 'get the incentives right' for officials and communities to carry out their roles, which were often predefined by outsiders. In many cases, this involved some form of mechanisms to make the government agencies formally accountable to resource users – a sound principle, but one that overlooked the differences in power and attitudes, both within and between agencies and communities.

ACM took a much more gradual, relational, learning approach to bring together the various actors and deal with the messy complexities involved, which made ACM more difficult to scale up. Then, as now, there was considerable demand from governments, donors, and international organizations for the so-called solutions that could be rapidly scaled up. But many of the structural programs that went for rapid scaling failed to deliver the expected improvements in resource conditions or people's livelihoods. Formal performance targets reduced incentives to accurately report, let alone learn from, what was happening in practice. And because they had not built buy-in from the government agencies or from the communities, the programs were abandoned. By contrast, this volume shows that ACM approaches have achieved substantial results, not only in terms of resources and livelihoods but also in terms of building the commitment, capacity and confidence of resource users in local communities as well as other actors. That, in turn, contributes to long-term resilience. While there are still pressures for 'scalable solutions', there is also appreciation for the need for social learning in natural resource governance (Pahl-Wostl and Patterson 2021). This volume provides lessons on how this can happen, in practice.

The problems that ACM seeks to address have never been more important – or more urgent. Climate change and massive biodiversity loss threaten not only local communities but also the world at large. We know that addressing these 'wicked problems' requires concerted action – but how is that to be achieved? Tropical forests are at the forefront of these issues, and externally imposed approaches relying on so-called 'expert' knowledge have proven inadequate. Yet, many development programs still have fixed approaches and performance indicators to show that they are being scaled in a consistent manner, which discourages honest reporting and assessment of what works and what does not and discourages learning and innovation. We urgently need to look beyond panaceas, to find ways to engage with complexity, draw on multiple sources of knowledge, knit together with effective facilitation and learning processes. This book provides guidance on how to do that.

The authors show how ACM is also not a panacea. Power differentials persist, even in collaborative management. Exclusions persist, even when the intent is to be inclusive. By reflecting on these issues, and even recognizing their own places in the power dynamics, the authors provide entry points and guidance on how to address these problems.

There are numerous specific, practical lessons provided, such as on how to develop the facilitation skills and capacities needed to bring people together to identify and address common problems. But beyond these is the change in ways of thinking about problems – and solutions. Rather than looking for nice, clean technologies or institutions that can be scaled across contexts, we should be recognizing the myriad combinations of technical and social innovations that people put together to adapt to local conditions through processes of 'bricolage' (Cleaver 2012). What is key is creativity, confidence, and collaboration for 'muddling through', accompanied by continuous reflection and learning about what works and what does not.

This volume is therefore an important resource for anyone interested in the forestry sector. But it has much broader applications as well. Other natural resource management sectors, such as water, fisheries, or rangeland management could benefit from the lessons on how to bring state agencies, communities, and other actors together. Agroecology seeks contextualized solutions based on co-creation of knowledge to enhance adaptive capacity of communities. Climate change mitigation and adaptation require collective action by different types of actors. These and many other types of development programs seek to develop multistakeholder forums (MSF), which risk becoming a new type of panacea, without looking at what kinds of facilitation and learning are required. All of these could benefit from the lessons presented in this volume. I hope it not only informs better policies and practices to address pressing challenges like climate change, biodiversity and habitat loss but also inspires more long-term studies of the implementation and effects of development approaches.

References

Cleaver, F. 2012. *Development Through Bricolage: Rethinking Institutions for Natural Resource Management*. London: Routledge.

Colfer, C.J.P. 2013. "The Ups and Downs of Institutional Learning: Reflections on the Emergence and Conduct of Adaptive Collaborative Management at the Center for International Forestry Research (CIFOR)." In *Adaptive Collaborative Approaches in Natural Resource Governance*, edited by H. Ojha, A. Hall and R.V. Sulaiman, 48–101. London: Routledge.

Pahl-Wostl, C. and J. Patterson. 2021. "Commentary: Transformative Change in Governance Systems: A Conceptual Framework for Analysing Adaptive Capacity and Multi-level Learning Processes in Resource Governance Regimes." *Global Environmental Change* 71: 102405. https://doi.org/10.1016/j.gloenvcha.2021.102405.

Acknowledgements

First and foremost, we thank the many communities and other collaborators with whom we have worked around the world these past two decades, particularly in Java, Sulawesi and Sumatra, in Indonesia; and Cameroon, Democratic Republic of Congo, Malawi, Uganda and Zimbabwe in Africa. We are grateful also for our institutions, CIFOR-ICRAF (in Bogor, Indonesia and Nairobi, Kenya); and also, in Colfer's case, Cornell University's Southeast Asia Program (in New York, USA). This book has benefited greatly from the constructive and insightful reviews of all chapters (though reviewers bear no responsibility for any remaining errors). We thank also Earthscan for its efficient conduct of the publication process. We particularly appreciate the careful eye and patience of Assunta Petrone, the copy editor. We are grateful as always to our families as well, who have accepted our extra work with equanimity and support.

Without the financial contributions from CIFOR-ICRAF, Program on Policies, Institutions and Markets (PIM), led by the International Food Policy Research Institute (IFPRI), the Cornell university's Open Access Publishing (COAP) fund (Ithaca, New York), and the East-West Center (Honolulu, Hawaii), along with Earthscan's willingness to reduce the price, we would not have been able to publish this book as an open access contribution. Given our sense of urgency regarding the environmental issues discussed herein, we have felt that broad access to the book was extremely important. We are grateful.

Abbreviations

ACM	adaptive collaborative management
AfDB	African Development Bank
ALL	*Africa Living Land*, a trademark
AMFN	African Model Forests Network
APL	*Area Penggunaan Lain*, Area for other uses (Indonesia)
ATR/BPN	Ministry of Agrarian Affairs and Spatial Planning (Indonesia)
B-Adapt	Business for adaptation to climate change, a project (Cameroon)
BCFS	Budongo Conservation Field Station – Uganda
BMC	Block Management Committee (Malawi)
C&I	Criteria and indicators
CAI	Critical Action Intellectual
CAPRI	Collective Action and Property Rights
CBFM	Community Based Forest Management (Malawi)
CBO	Community based organization
CCA	climate change adaptation
CFM	Collaborative Forest Management – Uganda
CIFOR	Center for International Forestry Research
CODECA	Community Conservation and Development Agency – Uganda
COP	Conference of the Parties (on Climate Change)
CSO	Civil Society Organization
DFO	District Forest Officer
DFS	District Forestry Service (Indonesia)
DoF	Department of Forestry (Malawi)
DRC	Democratic Republic of Congo
DRI	Displace, Replace, Intensify (DRC)
DRR	Disaster risk reduction
Ecotrust	Environment Conservation Trust Uganda
Exposure	The presence of people and assets that could be negatively affected by a hazard
FBE	Forest Based Enterprises

F4C	Facilitation for Change
FC	Forest Commission (Zimbabwe)
FELA	Freely Explored Landscape Agreements
FLEGT	Forest Law Enforcement, Governance and Trade
FMP	Forest Management Plans
FMU	Forest Management Unit
FOMOD	Dja and Mpomo Model Forest (Cameroon)
Forest paradox	The immense economic value and multifaceted wealth embedded in African natural environments and the extraordinary material poverty of the people living in those resource-rich places
FPIC	Free, Prior and Informed Consent
FPU	Forest Protection Unit
FR	Forest Reserve (Malawi)
FELA	Freely Explored Landscape Agreements (Cameroon)
FOMOD	Dja and Mpomo Model Forest (Cameroon)
FTLRP	Fast Track Land Reform Program (Zimbabwe)
GDP	Gross Domestic Product
GoI	Government of Indonesia
GoM	Government of Malawi
GVA	Gross Value Added
GWP	Global Water Partnership
HA	*Hutan Adat*, or customary law (Indonesian)
ha	hectares
Hazard	The frequency and intensity of natural disasters or long-term trends that may physically impact the welfare of a society
HTI	*Hutan Tanaman Industri*, or Industrial Forest Plantation (Indonesia)
ICDP	Integrated Conservation and Development Programme
ICRAF	World Agroforestry Center (now, CIFOR-ICRAF)
ICT	Information and Communication Technology
IFMSLP	Improved Forest Management for Sustainable Livelihood Programme (Malawi)
IPBES	Intergovernmental Science-Policy Platform on Biodiversity and Ecosystem Services
IUPHHK	*Izin Usaha Pemanfaatan Hasil Hutan Kayu*, Timber Forest Product Utilization Permit (Indonesia)
JCDS	Jakarta Coastal Defense Strategy
JGI	Jane Goodall Institute – Uganda
KICODA	Kapeka Integrated Conservation Development Agency – Uganda
KPH	*Kesatuan Pengelolaan Hutan*, Forest Management Unit (Indonesia)

KSNP	Kerinci Seblat National Park, in Sumatra (Indonesia)
LEF	Local Expert Facilitators
LULC	Land use/land cover
LULCC	Land Use Land Cover Change
LULUC	Land Use and Land Use Change
MCFW	Malawi College of Forestry and Wildlife
MDC	Movement for Democratic Party (Zimbabwe)
MF	Model Forests (Central Africa)
MK	Malawi Kwacha
MoEF	Ministry of Environment and Forests (Indonesia)
MOOCs	Massive Open Online Courses
MSF	Multistakeholder forum
MWLE	Ministry of Water, Land and Environment (Uganda)
NCICD	National Capital Integrated Coastal Development Masterplan (Indonesia)
NFA	National Forest Authority – Uganda
NFF	Nature's Future Framework
NFTPA	National Forestry and Tree Planting Act (Uganda)
NGO	Nongovernmental Organization
NTFP	Nontimber forest products
PAM	Perusahaan Air Minum, Drinking Water Company (Indonesia)
PAR	Participatory action research
PDO	Protected Designation of Origin
Perda	*Peraturan Daerah*, District Regulation (Indonesia)
Perda Khusus	*Peraturan Daerah Khusus*, Special District Regulation (Indonesia)
Perdes	*Peraturan desa*, village regulation (Indonesia)
PERFORM	Protecting Ecosystems and Restoring Forests in Malawi project
PFM	Participatory Forest Management (Malawi)
PLN	*Perusahan Listrik Negara*, National Electricity Company (Indonesia)
PME	Participatory Monitoring and Evaluation
PRAIS	Model Forest Practical Itinerant School (Cameroon)
PSHK-ODA	*Pusat Studi Hukum dan Otonomi Daerah*, Regional Autonomy and Law Research Center (Indonesia)
R&D	Research and Development
RMC	Resource Management Committee (Zimbabwe)
RSP	Resource sharing project
Sitting allowance	An amount of money paid to individuals for their participation in meetings, common practice in many developing countries (Africa)
SNUB	*Solidaritas Nasional Untuk Bulukumba*, National Solidarity for Bulukumba (Indonesia)

Stewardship Economy	An equitable system of exchange that rewards those managing land sustainably for the goods and services derived from those landscapes without disrupting the rights of people to food, nutrition, health, voice and decent livelihoods.
TfT	Training for Transformation (Zimbabwe)
Three Rs	Rights, roles and responsibilities (Uganda)
Transmigrant	The Transmigration Program was a Dutch colonial Government initiative, later continued by the Government of Indonesia; initially, it aimed to move landless people from densely populated Java to less populous areas of the country. More recently, it has been used for refugee resettlement.
UBI	Universal Basic Income, a program that delivers in cash, unconditionally, to everyone
UNFCCC	United Nations Framework Convention on Climate Change
VFA	Village Forest Area (Malawi)
Vlei	Seasonally flooded, marshy areas (Zimbabwe)
VNRMC	Village Natural Resource Management Committee (Malawi)
Vulnerability	The propensity to be negatively affected, including sensitivity to and the possible inability to adapt and respond to a hazard
WARSI	WARSI, a Jambi-based NGO (Indonesia)
Wicked	Problems Underlying values/causes of such problems may be ambiguous and contested. They are in essence unsolvable and addressing them is best done by improving problematic situations and learning from one's efforts.
WMO	World Meteorological Organization
YGB	*Yayasan Gita Buana*or Gita Buana Foundation (Indonesia)
ZANU-PF	Zimbabwe African National Union – Patriotic Front

1 A time to change direction

Carol J. Pierce Colfer and Ravi Prabhu

As Lao Tzu warned centuries ago, "If you do not change direction, you may end up where you are heading". Never before in history has the future intruded so violently into the present. Future shock is the term that Alvin Toffler created to describe some of this; he defined it as "…the shattering stress and disorientation that we induce in individuals by subjecting them to too much change in too short a time" (Toffler 1970). It isn't just individuals any longer though; whole systems – human and natural – are being stressed and disrupted by the anticipated footfalls of a chaotic future. As human beings, our instinctive response to what we perceive as chaos is to try and impose order through control. Often, this takes the form of technology or other forms of built capital. If fire is the risk, then we build fire towers, early warning systems, and fire breaks. If floods, then dykes, levees, bunds, and bridges. If drought, then wells, dams, and canals. If the changes are cultural, then the controls are social and often draconian.

Arguably, the future that we should dread most is the one portended by a rapidly, anthropogenically induced change in our climate. For good reason, if the data and the models are to be believed. Recently, at the United Nations' 26th Conference of the Parties (COP) on global climate change, these concerns were reiterated and several action streams identified. This book tries to respond to one of them in particular, the Glasgow Leaders' Declaration on Forests and Land Use (https://ukcop26.org/glasgow-leaders-declaration-on-forests-and-land-use/). The leaders explicitly recognize…

> that to meet our land use, climate, biodiversity and sustainable development goals, both globally and nationally, will require *transformative further action* in the interconnected areas of sustainable production and consumption; infrastructure development; trade; finance and investment; and *support for smallholders, Indigenous Peoples, and local communities, who depend on forests for their livelihoods and have a key role in their stewardship.*
>
> (2021, p. 1, our italics)

While the reiteration of the severity of the situation is welcome, we wonder what sorts of further transformative action will take place? Will these be efforts, as in the past, to impose order and exercise control? Or will they be more adaptive,

DOI: 10.4324/9781003325932-1

working more organically with nature and communities in these forested areas? Communities are already suffering from the shock of futures that have been imposed upon them and distrustful therefore of such efforts to control, order, and transform. Natural systems – systems like the forests of the Amazon or Borneo or Central Africa – are suffering losses to their resilience as a result of 'enrichments', 'improvements', and indiscriminate use and transformation, and are therefore struggling to survive, especially with the additional stress of climate change.

We recognize the utility of such pronouncements as statements of global intent, value, and cooperation. But we also note that the success of such proclamations at addressing the day-to-day problems of communities and the land use conflicts and confusions that exacerbate climate change everywhere has been underwhelming.

Our proposition, based on decades of experience in forests throughout the tropics and with forest peoples of varying ilk, is that such declarations and other top-down efforts cannot and will not achieve their expressed goals without the active involvement of local communities and other forest users, managers, and policymakers. These individuals and groups must themselves *want* to save, sustainably use, and restore forests. Efforts like criteria and indicators (C&I) for sustainable forest management, Forest Law Enforcement, Governance and Trade (FLEGT), or the NY Declaration on Forests, have either failed utterly or revealed serious shortcomings, where there have not also been a combination of more localized support (community, intermediate, and national levels) and a protection of local people's rights and claims to land. As we noted as far back as 2008:

> There is growing recognition that many efforts to address problems at local levels have in the past been unnecessarily passive, reactive, and/or purely technological. Effectively addressing climate change will require moving forward with more process-oriented approaches that look to the future, acknowledge local capabilities and opportunities, and build analytical and adaptive capacities at several levels.
>
> (Colfer 2008, p. 4)

Rather than imposing top-down order and exercising rigid control, efforts that enhance adaptation, collaboration, innovation, and work *with* natural and social systems are more likely to succeed because they deliver bespoke solutions that are contextually relevant. This has been the experience of the teams engaged in researching and developing adaptive collaborative management (ACM), over the past two decades. Processes, such as ACM, have been shown to spark initiative and generate original, collaborative solutions, resulting in more empowered communities, reducing the number of stressed and shattered individuals fearful of the uncertain futures that Toffler observed.

Indeed, the ACM experience of empowered and actively engaged communities is not unique. A whole host of authors have urged taking a systems perspective and implementing experiential and adaptive management (Gunderson and Holling 2002; Holling 1978; Lee 1993; Senge 1990; among others). Similarly, many have urged local level approaches based on structured, shared learning and collaborative action (Borrini-Feyerabend et al. 2004; Freire 1970; Guijt 2007; Leeuwis and Pyburn 2002; Uphoff 1996 and others). The unique contribution of this book is its longitudinal perspective on what has been learned over these last two decades.

In the pages to follow, we additionally ask how can an ACM approach contribute to making communities more resilient to climate change and other environmental problems? How can we reduce these impacts along with other unpredictable intrusions in these times of increasing uncertainty, while still continuing to aspire to a better future? What lessons can we draw from our experiences – often highly contextual and place- based – to apply to other, often larger, contexts? Is this even possible? Each chapter in this book will seek to do this, before we return in the last chapter to draw some insights from our experience to support the implementation of the Glasgow leaders' declaration. The chapters in this book represent a rare opportunity to look back at ACM-style efforts and see what has happened 5, 10, 20 years later. We look at what has endured and what has faltered. And we analyze the conditions that have contributed to and interfered with success as we consider how these might be scaled up or applied elsewhere, to reduce the shock of an uncertain future, without attempting to force order or control. Rather, we advocate collaborative nudging of people and systems into more adaptive and sustainable states.

We use climate change as an urgent example of a suite of 'wicked problems', environmental in nature and amenable to ACM-like approaches. We recognize that ACM takes time and resources – all in limited supply. It also requires skills – community analysis, facilitation, responsiveness to local interests and capabilities, incorporation of indigenous/local knowledge and more – that have not typically been activated in addressing land use issues. Another significant constraint, within development, conservation, and other land use management 'traditions' of relevance, is the recognition within ACM of our inherent inability to preplan exactly what will occur. ACM is an approach in which, *by definition*, the interests and capabilities of affected communities play key roles; those differ from community to community, from time to time, and from sub-group to sub-group. Significant differences also exist among communities in their *interest* in equity or environmental management – an issue addressed further in Chapter 11. An ACM approach is not compatible with rigid, preplanning structures like log-frames, for instance, so beloved by donors.[1] It is not, in short, a magic bullet.

But we have found that in the sites where ACM has been implemented, there are consistently valuable outcomes that include some of the actions needed to address climate change – whether restoration; more equitable access to resources; more sustainable management of forests and/or non-timber forest products; better maintenance of biodiversity; and/or clearer legal access to, or ownership of, land and forests. Importantly, communities engaged in ACM learn to exercise their agency more effectively, are more confident about their abilities to 'muddle through' with innovations – social and technological – that are bespoke and contextualized. Although the specific outcomes of ACM processes are impossible to predict, we have seen the approach regularly offering advantages not otherwise available:

- The opportunity to finetune responses and adaptations to the enormous variety and complexity that globally characterize local conditions (whether biophysical or sociocultural).
- Techniques for addressing local inequities and giving voice to otherwise silenced minorities (women, 'the poor', lower castes, marginalized ethnic groups, Indigenous Peoples, etc.).
- A mechanism for institutionalizing learning – and corrective action – as part of any development/conservation process designed to address climate change and other environmental problems.
- Responsiveness to local concerns, thereby activating people's natural motivations to protect their own habitats – for themselves and their descendants.
- Capacity building at both community level and among professional communities of practice, as all improve their abilities to collaborate within and across scale, analyze their own conditions, network to learn of external resources, manage conflicts, enhance equity, and build trust.

We argue that these pluses far outweigh the constraints we have identified and offer a pragmatic way forward – one that both complements the more common actions at higher scales (such as the Declaration with which we began this discussion) and provides concrete steps and procedures that can provide the varying advantages outlined above. We already know from decades of experience that depending on external policies alone – whether global, national, or even at an intermediate level – encounters serious difficulties in implementation. It is time to bite the bullet and focus on learning, collaboration, and good facilitation that genuinely empower the folks who either *de facto* manage the forests in which they live or are suffering their loss to the more powerful (plantations, logging concessions, mining, cattle ranching, etc., often in collusion with government actors).

In this collection, we focus on two world areas where development and climate change pressure are acute: four countries in Africa and three islands in Indonesia. These areas were selected due to serendipity, the availability of long-term ACM-style research, and the opportunities for our research collaborators to revisit them.

ACM and this book

The ACM research/action discussed in this book began with Ravi's and Carol's cogitations and conversations in the late 1990s. Ravi had brought Carol into the Center for International Forestry Research (CIFOR) for a project on C&I for sustainable forest management in the mid-1990s. She worked on the 'human wellbeing' aspects; and he, besides leading the project, focused on ecology and production (key findings available in, e.g., CIFOR 1999). As that project drew to a close, we began to consider what *really* needed to be done. We realized that perfecting sets of C&I, as we'd begun to do, was a never-ending process (and one that was not necessarily CIFOR's responsibility to undertake), since each forest, each community, would require some modification of any standardized set. At the same time, we wanted to **see** beneficial change in real forests and real communities, addressing the twin wicked problems of deforestation and disempowerment/disenfranchisement of local communities. ACM, as implemented by most of the authors represented in this book, grew out of this experience, this motivation, and these discussions.

In June of 2020, in the midst of the COVID pandemic, we approached onetime ACM team members, inviting them to contribute to an edited collection on their respective, post-ACM Program experience. This collection is the second of the two volumes that resulted, including a few authors not part of the original CIFOR work but using similar approaches. While the first of these two books spanned the globe (Colfer, Prabhu, and Larson 2022), this one narrows our focus to the two areas described above, areas that we, the editors, know well.

We developed these two books with the initial intention to assess what ACM researchers and practitioners had learned in the last 20 years. This book captures important elements of that learning. Predating the Glasgow Leaders' Declaration, our findings had convinced us to consider more deeply how the book's contents could help respond to the issues highlighted therein. In the late 1990s, our CIFOR teams began with a concern to apply what we had learned in the C&I project in communities and forest management units. We were as concerned about human well-being as we were about the ecological and economic implications of forest management; we sought sustainability of forest and human systems from the get-go; and we recognized the potential contributions local communities – in all their diversity – could make to better management. These local people were present; they had a stake in the future of their forests; they typically had experiential knowledge about their context and its history – surely, these were elements on which to build common cause, enhance human and forest well-being, and create a sustainable future through shared and systematic learning. This was our belief; and subsequent experience, as reflected in this volume and the last, have shown this to be largely true.

This is not to say that local communities always behave as stewards of their environments – as will be discussed in Chapter 11. Population growth can lead communities to over-use local lands; new economic opportunities, especially in cases where land tenure is insecure, can lead to unsustainable practices. Similarly, local customs can include extremely inequitable features that disadvantage certain groups and privilege others. These are some of the challenges that affect all efforts to improve human conditions and manage environments in a more benign manner; and they represent the 'art' that is an inherent part of our ACM approach (expressed most explicitly in Chapter 6).[2]

We now take, as a grand challenge, a grand speculation, the possibility of applying the lessons learned in this micro-scale context at a broader scale and in different contexts. Initially, we tried to address the complex, changing, and interconnected problems affecting forests and people. We considered, with local communities, how to make conservation and rural development more benign, management of forests more sustainable, gender norms more equitable, biodiversity more effectively maintained, etc. We found that progress *could* be made – though not in an entirely predictable manner – in those efforts (as shown throughout this book). And we found that much would endure over time but we also learned that the actions of actors at broader scales, while not undoing what had been accomplished, often rendered local efforts inconsequential in the broader scheme of things. Here, as a way of introducing the chapters to follow, we consider key elements of what our authors have learned, summarized below, and consider how these lessons can contribute to addressing climate change, one of the wickedest problems of all.

Common threads linking communities with global concerns

In 2008, Colfer reiterated four challenges already then noted by climate change researchers (2008):

- "Weak institutions and governance – particularly in need of greater downward accountability and transparency, which in turn call for new governance skills, forms and practice" (p. 1). The significance of this challenge has only increased and spurs us now to greater effort in this sphere.
- "Problems involving trade-offs among efficiency, effectiveness and fairness – requiring improved communication among stakeholders, the development of negotiation skills among the less powerful, and clearer definition of rights and responsibilities among forest users and abusers". These remain at issue, particularly 'fairness' but now, attention to trade-offs related to land use, authority/sovereignty, and tenure conflicts is just as urgent.
- "Lack of shared visions or harmonized plans among stakeholders – and the related need for consensus building strategies and joint measurement and monitoring mechanisms to assess progress". These are arenas in which ACM has shown its mettle time and again. Existing approaches now need to be used at a broader scale.
- "Population pressures – which can best be addressed by inclusive processes that involve community members (particularly women) directly in finding solutions". This topic remains somewhat taboo in the environmental world but has potential for reducing deforestation and degradation and opening up opportunities for women – both potentially contributing to addressing climate change (p. 2).

In this section, we highlight four additional themes with implications for addressing climate change. These have recurred in our subsequent ACM experience: (1) multiple scales and the exercise of power, (2) facilitation and learning, (3) collaboration in multistakeholder forums (MSFs), and (4) inclusivity and intersectionality. Discussion of these themes also serves as a means to introduce the chapters to come.

Multiple scales, the exercise of power

To address global concerns, the need to upscale, even multiscale, has become ever more obvious. We need to understand what's going on locally both to capitalize on the knowledge, experience, and enthusiasms of local communities and to avoid harm but we also need to work at higher levels. The examples highlighted below show why.

When we began to conceptualize CIFOR's version of ACM, we were aware, abstractly, of the relevance of scale. We anticipated that communities would need to deal with more powerful external actors. That recognition helped to propel us toward both involvement with outside actors and strengthening of institutions for community-level learning and collaboration, as mechanisms to deal more effectively with such actors.

But in the late 1990s and early 2000s, the forestry world was comparatively oblivious to community-level possibilities and dynamics. If communities were seen at all, it was as impediments to good management. Our emphasis in those early days involved focusing others' attention on the people living in forests, their desires, their capabilities, their interests, and their constraints. We tried to enhance learning and constructive collective action by – and give voice to – such local communities. Although we did not see communities 'through rose-colored glasses', we did seek to build on their potential constructive involvement in management and seek win-win options; and we recognized their internal diversity.

In the intervening 20-plus years though, forest villages in all countries have sprouted solid branches reaching to nearby towns and cities, to businesses and governmental agencies, even beyond, to countries across the sea. And these branches represent two-way communication channels; influence and power go both ways. Easy transport and the internet have widened almost everyone's networks and information access exponentially, and with this widening has come far more complexity. Whereas when we began, we could focus most of our attention at the village or forest management unit level (or thought we could), as time has gone by the power of external forces has hit home. And with that has come the recognition that our analyses and our actions – 'our' here referring both to ACM practitioners and to the participants with whom we work – needed to broaden. The collaboration and adaptive learning had to span scales.

The relevance of these multiple scales is particularly obvious in the two chapters that examine directly what has happened in our original ACM sites: Baru Pelepat in Jambi, Sumatra, Indonesia; and three communities in Gokwe, Zimbabwe. Yuliani et al. (Chapter 2) and Kozanayi et al. (Chapter 7)[3] conducted research in these villages over an intensive three years in the early 2000s.

Chapter 2, by Yuliani et al., shows the durability of many of the competencies gained during the original ACM project in Jambi.[4] They provide solid land use evidence of the sustainable management of the small protected area the community had struggled mightily and successfully to gain the legal right to manage; these rights were eventually recognized nationally. But, as Ridwan (2019) points out and Yuliani et al.'s maps show, despite the excellent successes the community achieved, the rest of the surrounding forest has been converted to oil palm, due to higher-level governmental policies.

Kozanayi et al. report in Chapter 7 the maintenance of several formal local Zimbabwean institutions that were initially strengthened during the ACM project but the authors also show how the chaos that has characterized Zimbabwe's national politics during these intervening years has wrought havoc on so much that was once accomplished. Early on, the team had high hopes for broader extension of the ACM approach as Forestry Commission officials expressed considerable enthusiasm (identifying additional pilot areas, helping produce and share a video explaining ACM, and more) – but the levels of political conflict just kept rising. Such Forestry Commission plans fell apart, in the midst of loss of foreign exchange, HIV-AIDS in epidemic proportions, political violence, and the loss of supportive bureaucrats who understood the approach (whether through death, out-migration, or simple transfers).[5] In the end, some village-level institutions and practices were sustained but the higher-level involvements faded away in the face of national chaos.

These two examples of effective ACM local action, seriously undermined over the long haul by decisions and conditions at higher levels, strengthen our conviction – an important finding of this book – that ACM-like processes need to be upscaled (*not* abandoned) – to *include* action at intermediate and higher levels.[6] Perhaps even better, they need to be carried out simultaneously at multiple, nested scales (as suggested in Chapter 5). This will be particularly important with a problem as global and urgent as climate change.

Three chapters also report examples of partial existing upscaling of ACM-like features. Diaw et al.[7] write of their experience meshing the ACM approach (as the 'software') with the regional level in their Model Forests leadership.[8] In Cameroon and the Democratic Republic of Congo, Diaw and his colleagues strove to upscale to a national, even supra-national level, a process they've had considerable success with.[9] But they argue that for the kind of success so sorely needed (economic success), we will need changes to the international economic system, which consistently hampers African economic advance. Besides noting the program's impressive economic successes, they, like Egunyu (Chapter 9), warn of the dangers of dependence on foreign donors.

Chapters 8 (by Kamoto, Missanjo, and Djenontin) and 9 (by Egunyu) consider national programs that strive to collaborate with local communities in forest management (Participatory Forest Management, PFM, in Malawi and Collaborative Forest Management, CFM, in Uganda). The Malawi case, coordinated by Kamoto, one of the original ACM researchers, shows how the national program has effectively institutionalized the learning element of ACM. Both the potential to institutionalize the approach and the adaptive quality of learning itself will be key to effective climate transformation – an observation also made by Ojha et al. (2022) who see transformation "both as an emergent (linked to structures), and deliberate process (linked to human agency)" (p. 622).

Chapter 9 describes the Ugandan CFM program, showing its considerable value to participants. Yet, it also demonstrates the shortcomings of an institutionalized process that relies on NGOs and unpredictable external funding to implement its plans.

In sum, these five cases make clear (1) the importance of upscaling or multiscaling ACM as we tackle complex environmental issues like climate change, forest restoration, and international structural inequities with environmental implications, (2) the durability of unpredictable effects of ACM efforts,[10] and (3) some examples of ways ACM has been upscaled, in both ad hoc and institutionalized ways.

Two strategies that can work, with a few caveats, as well at higher scales as at a village scale are good facilitation and MSFs.

The value of excellent facilitation and learning

The importance of good facilitation at the community level has been obvious for a long time. But attempts to institute adaptive and collaborative processes at broader scales will require even more skill and recognized legitimacy.[11] At the community level, facilitators are likely to have what might be called a societal 'prestige/power edge' in comparison to those facilitated. In our ACM experience, facilitators have been more educated, better connected, wealthier than most of the forest community people they have worked with.

At broader scales, this cannot be assumed. Facilitators working within bureaucracies, with industrial magnates or the COP, for instance, are likely to have to be even more skilled, prepared, and carefully selected. They will not be able to rely on prestige deriving from urbanity, education, or ethnicity alone; they are more likely to have to 'prove themselves' to participants in ways that were less necessary in villages or forests.

As we conducted the early research, we knew that we needed good facilitation from our own experience but also from advice and methodological guidance from others. The further into the ACM process we delved, the more central we saw excellent facilitation to be. The kind of skills we valued included the abilities to

- create an environment of safety and trust
- assume a position of relative neutrality among conflicting views and proponents (also stressed in Chapter 3, by Liswanti, Tamara, and Arwida)
- creatively address dilemmas and conflicts in a problem-solving manner
- manage domineering and reticent people so as to maximize diversity and equality of participant input
- serve as an information conduit from one group to another, often at different scales.

We also recognized the potential for 'facipulation' (the attempt to manipulate by skilled facilitation). Facilitation skills can be misused to mask power dynamics and/or to plant and nourish outsiders' goals. But a good facilitator can be something similar to what Ernstson et al. (2010) call a scale-crossing broker, though our own emphasis is on the linking among social categories as much as between physical sites.[12]

As Hagmann makes clear in Chapter 6, a good facilitator, by means of these capabilities and others, can contribute to maintaining interest in and commitment to a process; ensuring inclusivity; managing conflict, and preventing elite domination. Facilitation can also play a vital role in empowerment of participants, as they learn to analyze their own conditions (Freire 1970), envision a future they find desirable (Evans et al. 2006; Wollenberg, Edmunds, and Buck 2000), and then plan, implement, monitor, and revise as needed their chosen collaborative actions. These processes are likely to be just as relevant as we facilitate processes with climate change decisionmakers at any level. Power dynamics intrude even more obviously at higher scales.

McDougal and Ojha (2022) have provided a much-needed theoretical treatise on how ACM practitioners need to attend more self-consciously to power in their work. Briefly, they emphasize three concepts that help us understand the maintenance of power imbalances:

i privileged and marginalized social and gender identities (*unmarked categories*);
ii dominant beliefs (*doxa*) that are internalized and thus taken for granted and unnoticed; and
iii political representation (*delegation*) (p. 195).

They then "consider the persistence of these imbalances through the application of the more encompassing and interconnected concepts of *structure* and *agency*" (p. 195). Using this combination of social and feminist theory, they explain such persistence in community-based natural resource management and how an ACM approach may shift power-reproducing dynamics, partly via good facilitation. Such power imbalances are ubiquitous and account for much of the inequity we see contributing to unsustainable global practices. They will intrude even more, we suspect, as we move upward in scale.

Attention to learning and dependence on it are integral parts of ACM, by definition. Adaptation itself is built on the learning process: the iterative cycles of studying context, making plans, implementing them, monitoring what happens (all 'collaboratively'), and revising ('adapting') appropriately. Important in all the chapters, as a foundation of the approach, it is particularly highlighted in Chapters 4 (by Kusumanto et al.), 6 (by Hagmann et al.), and 8 (by Kamoto et al.). In the chapter most directly focused on addressing a climate change-related issue, flooding in Jakarta, Kusumanto et al. first set the stage by showing the complexity of the flooding problem. Then, building on their experience in Jambi, they emphasize the need to go beyond the technocratic 'solutions' that have dominated attempts to deal with flooding there so far. The authors follow this up with a 'thought experiment', imagining how one might begin to address the flooding issue, within an ACM-style approach that builds on social learning.

Hagmann integrates learning seamlessly, as he discusses the skills, attitudes, and approaches central to good facilitation. He looks at learning in a processual way and also emphasizes the importance of 'co-learning' – facilitators learn along with those involved in the processes they facilitate (as do ACM researcher/practitioners Cronkleton, Evans, and Larson 2022).

Kamoto, Missanjo, and Djenontin's chapter shows how Malawi's Department of Forestry has integrated routine studies to assess their progress on their goals ('Standards and Guidelines for PFM'), which they revise as new evidence comes in. It is an enlightening and encouraging example of an ACM-like process (termed Participatory Forest Management) that has successfully moved learning 'up-scale' – something that others have had difficulty doing (cf., Colfer 2013; Colfer and Pfund 2011).

Effective climate change mitigation and adaptation will require these facilitation, political, and learning abilities if local communities, districts, national actors, and others are to play active, equitable, and benign roles in planning, implementation, and ongoing monitoring of such processes.

Collaboration in multistakeholder forums (MSF)

MSFs have been the most consistently used mechanism for upscaling ACM-style approaches; and they are likely to be important in efforts to address climate change and other global natural resource-related problems. A key feature of MSFs is the capacity to bring together actors within nested scales: local people can come together with different sectorial actors at a particular level, and/or with actors from various levels of a particular institution. Such interaction is key for the development of trust, the relevance of which in collective action has been increasingly recognized of late (e.g., various chapters in Butler and Schultz 2019).

Liswanti, Tamara, and Arwida (in Chapter 3) build their examination of the development and endurance of trust on their study of a longstanding, multi-level, multi-sectoral MSF in the Jambi Province of Sumatra. They document the high levels of trust that were sustained among participants – among Jambi bureaucrats in varying departments, with international actors, and between them and Baru Pelepat community members – in a longstanding MSF that together addressed aspects of land use and land use change. Such trust is now widely recognized as fundamental for addressing climate change, stimulating forest restoration, community conservation, and other collaborative endeavors. Desirable effects of an MSF, first initiated in the early 2000s, in activities related to the Jambi ACM project described in Chapters 2 and 4, are shown to endure, some 20 years later. Collaboration within this district was further strengthened under the CAPRI project,[13] with ongoing encouragement from World Agroforestry researchers.

Fisher et al. (Chapter 4), who also focused much of their attention on this intermediate level (*Kecamatan*, County, and *Kabupaten*, District) in Sulawesi, made extensive use of an MSF. Using participatory action research (PAR) with bureaucrats, provincial and international researchers, NGOs, and community members in the Bulukumba District in Sulawesi in the mid-2010s, these authors facilitated the collaborative management of the Kajang Forest with relevant stakeholders. Chapter 5 describes a reflective study, clarifying the authors' own roles in facilitating this process initially, and showing how local actors continued to build on the cooperation and trust they had developed, after the facilitators had left the area. This MSF extended its action to two subsequent nationally emphasized topics (recognition of land rights and climate change adaptation, the latter still functioning as the authors' chapter was written). These MSFs, both in Sumatra and in Sulawesi, struggled to deal with the conflicting rights to forest lands that had been allocated to various parties (individuals, communities, oil palm companies) by higher-level government entities (see, e.g., Fisher et al. 2020).

Such MSFs have been seen as one of the few mechanisms available to deal with the common problem, in developing countries particularly, of lack of coordination among different segments of governmental bureaucracies (see Astutik, Pretzsch, and Kimengsi 2019, on medicinal plant management). Identifying common problems requires first mutual understanding; and addressing those problems requires a willingness to compromise and negotiate, engage in give and take. These discussions of MSFs show ACM's emphases on co-learning, collaboration, and monitoring and how it can work just as well at this intermediate level as at the village level (in which competition, factions, different goals, and understandings can also hinder good collaboration). All these processes are aided enormously by good facilitation.

One element of MSFs that has not received sufficient attention, in much of our own work as well as that of others, is its political dimension.[14] Ravikumar et al. (2018) acknowledge the widespread lack of coordination so commonly identified as a central problem in conservation and development spheres – most recently with regard to the SDGs. These were designed to intersect and mutually reinforce, but are often implemented separately within the usual governmental silos (Katila et al. 2019). That addressing climate change will require similar coordination seems uncontroversial. Ravikumar et al. argue persuasively, based on a study of 35 in-depth qualitative cases from Peru, Indonesia, and Mexico, that considerable coordination actually does exist. The problem is that the goals of some stakeholders are diametrically opposed to the goals of some others; and these authors argue that this must be recognized and taken onboard. The solution they've found (in these cases) has been the formation of alliances, coalitions, among those stakeholders with compatible goals. The authors conclude that

> To cast agricultural intensification advocates who extend irresponsible loans to smallholders, or even abusive oil palm companies that seek to grab land by any means available, as potential allies for environmental interests or smallholders is to misunderstand the political reality. In fact, such actors ought to be the targets of regulation, and their activities the object of organized political contestation.
>
> (p. 1452)

We accept this analysis but argue that MSFs can also perform such functions – first seeking win-win possibilities, and lacking such, they *can* provide a platform from which to strengthen constructive countervailing political alliances (as observed in the Sulawesi and Jambi cases). One can easily see such countervailing forces, political and economic, hampering global efforts to address climate change. Enhancing relatively neutral facilitation, providing advocacy for weaker groups to engage more fairly in processes, emphasizing co-learning, gaining listening skills, and other common elements of ACM can contribute at higher levels as well.

Another element of MSF success is related to participants' understanding of the central ideas of ACM, specifically the ideas of co-learning from the monitoring and adaptation of joint actions taken together. These ideas and practices can be very difficult to convey to participants who may be used to a much more hierarchical, directive, and top-down approach to development, conservation, climate change, and other forest-related endeavors. Mutimukuru-Maravanyika, Madzudzo, and Songe (2022) have analyzed the issues involved in ensuring that such understanding exists (and the dangers of not doing so) in Zambia; and Evans, Larson, and Flores (2020) and Cronkleton, Evans, and Larson (2022) each addressed the difficulties of overcoming the top-down expectations of field researchers new to ACM.

Inclusivity and intersectionality

At all scales, human beings differentiate themselves into more and less valued categories – as emphasized, for instance, by McDougall and Ojha (2022). The differentiations, while expanding cultural diversity (in itself of insurance value), are a recurrent inhibitor of mutual understanding and collaboration. This is as true in climate change negotiations and policymaking as it is within villages and families.

The authors of this book have, by and large, devoted large amounts of time and energy to the question of inclusivity. We have a strong understanding of its value – in the sense of

- gaining access to a holistic understanding of relevant contexts
- activating the creativity of a wide range of people with varying skills and knowledge
- responding appropriately to human and environmental needs
- gaining access to diverse networks and resources
- preventing harm arising from ignorance, and of course
- innately, as a human rights and ethical issue.

These values apply whether we are working in a village, a district, a country, or in the international arena. The difficulty of including marginalized groups, however, is likely to increase, the broader the scale (see, e.g., Colfer et al. 2015).

We have also come to realize more profoundly over the years the significance of looking at human difference from an intersectional perspective. We cannot, for instance, focus only on gender or race or nationality, in isolation. The way our world works, an older, educated, white, American woman has many advantages that a young, rural, black man (or woman) will not have. The various elements of a person's identity have differing implications for his or her power and life options. These vary also by context. Our emphasis on involving a variety of stakeholders and facilitating collaborative processes with care is ultimately inspired by our interest in countervailing such varying stereotypes and related structural disadvantages that we see in almost any human context. We have tried in ACM to address the many barriers to inclusivity in decisionmaking, in access to power and resources that we have encountered.

In moving the ACM processes 'upward' in scale, we encounter another difficulty: some see integrating individuals of varying identities and from multiple sectors as interfering with attempts to address more conventional inclusivity concerns (like gender or ethnicity per se). Within groups seeking broader ethnic inclusivity, for instance, women have been chastised for diluting this effort by their desire for gender equity (Asher 2016 and others). The variable barriers to women's involvement – such as norms against public speaking, lower access to education, lack of knowledge of national languages, etc. – are multiplied when the involvement of a woman from a marginalized ethnic group in a forest might be desirable.

Such impediments function at any scale. Many readers may be able to attest to the weighty disadvantage that non-native English speakers suffer in international arenas. Even the lack of *fluency* in English can be a serious impediment to shared decisionmaking. Input that would be articulate and germane in the proponent's native language is rendered inarticulate and halting – further reducing the proponent's courage to speak and the willingness of other participants to listen.

We have also tried from the beginning to 'walk the talk', which this book represents. We are proud to have assembled this group of mainly developing country researchers, many of whom were new graduates when CIFOR's ACM program began. Fresh out of college or graduate school, they tested their mettle in the humid forests of Indonesia and Central Africa or the dry forests of East Africa. They, like many of the individuals in the communities with whom we worked, have blossomed in the intervening years, seeking more education, meeting new challenges, and rising to positions of respect and power during these two decades. We have been humbled to recognize their continued commitments to understanding and improving the ideas of ACM, and to assessing their own work in as objective a way as possible. This book strives to portray both the successes and the failings of our efforts, so that we may all learn and improve as we address daunting challenges like climate change.

Ways forward

Although we further consider ways forward in Chapter 11, here we summarize our findings at three scales: individual, community, and among actors at broader scales.

At the individual level

Working with communities, the importance of individual action could never be discounted. Community actions themselves also require the commitment, intelligence, and perseverance of individuals; and many of the benefits that come from participation in an ACM process – such as increased self-confidence, analytical ability, conflict management skills, etc. – accrue to individuals, even as they may benefit the community at large. In every ACM site, these kinds of benefits accrued, in varying degrees, and many endured.

Facilitators, individuals as well, contribute mightily to any ACM process; and as we move forward and upward, we envision their roles as just as critical, as they make stronger links both between communities and various broader scale actors, and among broader scale actors themselves. Our original approach was to respond to local conditions, finding facilitators wherever we could – whether researchers, NGO personnel, or government officials; and such flexibility is likely to continue to be important, given the diversity of institutional contexts from country to country, even community to community, though we see clear advantages to institutions that can guarantee some longevity.

Besides the facilitation skills emphasized in Chapter 6, a broad perception of a facilitator's legitimacy will also be important. And that will be influenced by the institutions from which they come as well as their individual characteristics. Their success will depend on their abilities to 'infiltrate' (in a positive sense) the relevant institutions and convince those legally responsible for forest management and land use planning of the value of listening to communities, of explicitly and systematically *learning* from their actions, of being inclusive, and of doing so collaboratively with other sectors and other supportive individual actors. As Cronkleton, Evans, and Larson (2022) so poignantly demonstrate, ACM requires ongoing learning from all involved.

Such learning takes time – even on the part of the facilitator her/himself – and money. Both requirements imply attachment to some enduring institution, wherein individuals have time (and budget) first to learn how to do ACM and then to do it. Such institutions can be governmental or NGO, perhaps even some industrial entity (though we have not tried that).[15] Having 'champions' housed in relevant institutions can help enormously as well.

We should not discount the potential of action by people like ourselves either. Ojha[16] et al. (2022) wrote of Critical Action Intellectuals (CAI): "people who contribute to systemic change though their intellectual work and political engagement (praxis) in relation to fields of environmental governance" (p. 622).

> CAI are different from activists as they take knowledge creation and mobilisation as the primary vehicle for change. Unlike most academics, they engage directly with dominant actors and policy processes to tackle injustice and risks to sustainability in specific socio-environmental contexts.
>
> (p. 622)

All the authors of this book have been engaged in such work. Ojha et al. point out the critical roles such folks have played in making benign environmental and human improvements. As we move from the village level to incorporate actions at higher levels and try to address global concerns like climate change, the role of CAI is likely to become more important. Where else will the needed facilitators, researchers, managers, and policymakers – with sufficient 'legitimacy' – be found?

At the community level

Our work has focused on the community level, and we continue to consider it vital. Yet, we need to ensure that local action is linked more broadly than it has been in the past to the broader context. That can mean more 'look and learn' visits as were useful in Jambi; facilitation of shared learning between community members and political and economic actors at a broader scale, common in MSFs (e.g., in Sulawesi); training of community members in skills they may need in interaction with broader scale actors (e.g., self-confidence, speaking ability, literacy – these needs will differ from place to place, group to group) as was done in Zimbabwe; focused training of women (or other marginalized groups) in political action (as occurred in Uganda, Mukasa et al. 2022); or periodic assessment studies as implemented in Malawi. All have been tried, but expansion – and creation of new strategies – is needed, if we hope to succeed in mitigating and/or adapting to broader landscape scale change, including climate change. We should also remember that a community is no longer necessarily a place-based arrangement – as so many have learned within the context of COVID-19 and our shared dependence on Zoom, social media, and other electronic communication facilities.

At a broader scale

In our earlier work, we were not totally ignorant of the need to fashion or strengthen links between communities and broader actors (see Locatelli et al. 2008, e.g., pp. 13–19).[17] In some countries, we worked closely with forestry institutions, particularly Malawi, Nepal (McDougall et al., 2009, 2010), and Zimbabwe, from the start – with varying results.

We also convened national level committees in each country and an international steering committee, each composed of government officials, academics, and NGO personnel. We hoped that they could guide us and serve as a pathway through which our findings and understandings could filter throughout the institutions. The committees worked very effectively, while the projects (and their meetings) were funded, but dissipated afterward – though *individuals* had gained understanding which may have continued to influence thinking and action. We do not know.

ACM researchers in Cameroon took a different tack, initially focusing on six 'policy domains' – forest minorities, forest margins, community and communal forests, conservation and timber management (see also Diaw, Aseh, and Prabhu 2009). They tried from the start to bring ACM in at that scale, without sacrificing community-level involvement. This led to good interaction with government officials; and later to broader ACM acceptance as part of Canada's Model Forests Network, described in Chapter 10.

ACM researchers in Zimbabwe worked hard to integrate ACM in the Forestry Commission and had some significant success early on (with the government planning further pilot ACM projects, creating a video to disseminate the ACM approach, and including ACM methods in curriculum at the Forestry College).

Though these were encouraging for longer term institutionalization, Zimbabwe's political upheavals and the chaos that followed resulted in the dissipation of many of these efforts, as discussed in Chapter 7. Some similar successes and disappointments occurred in Nepal, as the Maoist uprising occurred just as findings were becoming available and useable by policymakers.

The example of Malawi, which seems to have taken on board at least the institutionalized learning aspects of ACM at a national scale, was a case of adding ACM to an existing forestry program committed to managing collaboratively. This does not seem to have been the case throughout Uganda (cf. the minimal progress reported in Chapter 8 of this volume to the more hopeful reports from Bomuhangi et al. 2022, also in Uganda).

All of these have shown some successes, at least in the short run, some in the longer term. In terms of future approaches, we may be able to use any or all of these approaches, depending on the context. But all must be strengthened, maintained, and funded over longer periods, with greater effort to create supportive alliances, and with greater attention to the systemic nature of interactions within and among levels.

A restoration program in the US (the 'Collaborative Forest Landscape Restoration Program') is having some success using a collaborative approach, due partly no doubt to its long-term funding (initially ten years, some of which have been renewed; Butler and Schultz 2019) – something none of our sites have had. The existence of considerable funding related to climate change may mean that similar allocations may be possible for ACM sites in the future, which would make a tremendous difference. Our two- to four-year[18] funding availability often meant that just as understandings and liaisons were ripening, the funding ended and many important undertakings lost the support they'd had (whether human or financial).

We began this chapter by saying that we needed to change direction. By that we referred to two issues: (1) moving away from purely technological solutions (as argued most persuasively in Chapter 4); and (2) expanding our adaptive and collaborative approaches at both local and broader societal scales. In this book, as we consider how best to build on these conclusions in addressing climate change and the other environmental problems that beset humanity and the Earth, we conclude that ACM is likely to be one suite of approaches of value. Specifically, the emphases on fair, respectful, and equitable collaboration, routine monitoring, and co-learning are crucial. The lessons we have learned from examining the past two decades of experience also identify the importance of up- or multi-scaling, excellent facilitation, the use/development of MSFs, and intersectional inclusivity as key features of any approach likely to be successful. As we implement ACM in particular contexts, we need to consider both how to institutionalize what's being learned and how to activate individual and collective agency. As we move upward and outward in our foci, dealing with ever-more powerful actors, the roles of researchers and other intellectuals are likely to expand in importance – especially if we can maintain the respectful, collaborative, learning attitudes so central to ACM.

Notes

1 Although from time to time, we have managed to phrase project goals in vague enough terms to allow for the needed flexibility in action.
2 As an anthropologist, Colfer has always seen her discipline as part art, part science, as it incorporates attention to values and other features not readily amenable to quantification or experimentation. Approaches that incorporate people and their sociocultural systems – such as ACM – are also likely to require attention to such features and thus require both art and science.
3 Yuliani, Kusumanto, Marzoni, Permatasari, Adnan, and Indriatmoko were all part of the original ACM team in Baru Pelepat, Jambi. Yuliani co-led from headquarters; Kusumanto co-led in the field. Kozanayi, Nyirenda, Mutimukuru, Matose, Ngwenya, and Sibanda were all part of the original ACM team in Gokwe, Zimbabwe; with Matose taking a lead. Colfer and Prabhu supervised the respective teams most consistently.
4 See, e.g., Diaw and Kusumanto (2005); Kusumanto et al. (2005); de Vries and Sutarti (2006), for more on this site.
5 See the collection by Mandondo, Prabhu, and Matose (2008); Vanclay, Prabhu, and Sinclair (2003, 2006), for key analyses from ACM's early days in Zimbabwe.
6 See Komarudin et al. (2012); and the Indonesian language collection by Adnan et al. (2008), which addresses the Bungo District (wherein Baru Pelepat is located) from a variety of perspectives; or more recently Sarmiento Barletti (2022) and Cronkleton, Evans, and Larson (2022) on the practicalities of such mid-level involvement, also a number of chapters in this volume.
7 Diaw led the ACM work in Cameroon and Ghana in the early 2000s. Nguièbouri was also a team member at that time.
8 The Central African Model Forest is part of the International Network of Model Forests, of Canadian fame.
9 See also Prabhu, Larson, and Colfer (2022) who describe two attempts to scale up many of ACM's elements in East Africa.
10 Although these are discussed in future chapters, examples include enduring features such as ongoing networking among levels (as in Chapter 2); maintenance of trust across scales and horizontally (in Chapters 3 and 5); continuation of strengthened institutions (as in Chapters 5 and 7); improved access to forests and their products (as in Chapter 8); see also examples in Colfer, Prabhu, and Larson (2022). Qualitative features like enhanced willingness of women to speak out in public and greater confidence of people from marginalized groups are harder to prove but are visible to those who have seen the changes arise and persist.
11 With regard to legitimacy, Louise Buck pointed out (in a review of this chapter) the following important issues: Who are these facilitators when ACM works well, who do they work for, in what type of institutions are they embedded? How are these positions funded? To what extent are they supportive of the dominant institutional actors who are legally responsible for forest management and land use planning? To what extent are they working for short-term projects only? These questions and more have important real-world effects and must be considered when choosing facilitators. See Chapter 6 for a more in-depth discussion of facilitation within the ACM context, including mechanisms for enhancing team members' abilities along these lines.
12 "We therefore define a scale-crossing broker as a social network position that links otherwise disconnected social actor groups which, through their social practices, interact with ecosystem processes at different ecological (and spatial) scales and at different physical sites" (Ernstson et al. 2010, no page numbers).
13 CAPRI is Collective Action and Property Rights, a multi-center research program within the Consultative Group of International Agricultural Research (CGIAR) centers.
14 Highlighted for ACM more broadly in Mutimukuru-Maravanyika's (2010) critical assessment of her own team's work in Gokwe, Zimbabwe. She emphasized the insufficient

attention within the team to power relations and impacts. Noting many of the same happenings recorded here, her observations strengthen our conclusion that multiple scales must be addressed.

15 See Rattanasorn, Fisher and Kugel (2012) for an example of collaborative restoration work with the military in Thailand.

16 Ojha himself was an early ACM partner and constructive critic.

17 We amended our definition of ACM to include,

> Working with a given group of people requires involving other people acting on other scales – usually at least one level down and one level up (e.g., user groups within a community and district officials above, as in Zimbabwe, Nepal, Indonesia, Philippines). Effective facilitation can act as a catalyst to empower communities to improve their own conditions, both human and environmental.
>
> (available in Colfer 2008, p. 2)

18 In a few cases, we had additional funding for a couple of years, but in each case, there was a gap in activities, in which much momentum was lost. Also in several cases, the next phase of activity was moved to a new location.

References

Adnan, Hasantoha, Djuhendi Tadjudin, Linda Yuliani, Heru Komarudin, Dicky Lopulalan, Yuliana L. Siagian, and Dani Wahyu Munggoro, eds. 2008. *Belajar Dari Bungo: Mengelola Sumberdaya Alam di Era Desentralisasi.* Bogor, Indonesia: CIFOR.

Asher, Kiran. 2016. "Tenure vs. Territory: Black Women's Struggles in the Pacific Lowlands of Colombia." In *Gender and Forests: Climate Change, Tenure, Value Chains and Emerging Issues*, edited by Carol J. Pierce Colfer, Bimbika Sijapati Basnett, and Marlene Elias, 206–218. London: Earthscan and CIFOR.

Astutik, Sri, Jürgen Pretzsch, and Jude Ndzifon Kimengsi. 2019. "Asian Medicinal Plants' Production and Utilization Potentials: A Review." *Sustainability* 11: 1–33 (5483). doi: 10.3390/su11195483.

Bomuhangi, Allan, Abwoli Yabezi Banana, Jimmy Bushoborozi, Concepta Mukasa, Alice Tibazalika, and Esther Mwangi. 2022. "Gender and Adaptive Collaborative Management in a Forested Ugandan Landscapes." In *Adaptive Collaborative Management in Forest Landscapes: Villagers, Bureaucrats and Civil Society*, edited by Carol J. Pierce Colfer, Ravi Prabhu, and Anne Larson, 87–104. London: Earthscan and Routledge.

Borrini-Feyerabend, Grazia, Michel Pimbert, M. Taghi Farvar, Ashish Kothari, and Yves Renard. 2004. *Sharing Power, Learning by Doing in Co-Management of Natural Resources throughout the World.* Teheran, Iran: IIED & IUCN, CEESP, CMWG, and Cenesta.

Butler, William H., and Courtney A. Schultz, eds. 2019. *A New Era for Collaborative Forest Management: Policy and Practice Insights from the Collaborative Forest Landscape Restoration Program.* London: Earthscan and Routledge.

CIFOR. 1999. *C&I Toolbox.* 9 vols. Bogor, Indonesia: Center for International Forestry Research.

Colfer, Carol J. Pierce. 2008. "Adaptive Collaborative Management Can Help Us Cope With Climate Change." *CIFOR InfoBrief* 13: 4.

Colfer, Carol J. Pierce. 2013. "The Ups and Downs of Institutional Learning: Reflections on the Emergence and Conduct of Adaptive Collaborative Management at the Center for International Forestry Research." In *Adaptive Collaborative Approaches in Natural Resource Governance: Rethinking Participation, Learning and Innovation*, edited by Hemant Ojha, Andy Hall, and Rasheed Sulaiman V., 48–102. London: Earthscan and Routledge.

Colfer, Carol J. Pierce, and Jean-Laurent Pfund, eds. 2011. *Collaborative Governance of Tropical Landscapes*. London: Earthscan and CIFOR.

Colfer, Carol J. Pierce, Ramadhani Achdiawan, Hasantoha Adnan, Moira Moeliono, Agus Mulyana, Elok Mulyoutami, James R. Roshetko, E. Linda Yuliani, with Balang and LepMil. 2015. "Preparing the Ground for Better Landscape Governance: Gendered Realities in Southern Sulawesi." *Forests, Trees and Livelihoods* 24 (1): 59–83. doi: 10.1080/14728028.2014.951002.

Colfer, Carol J. Pierce, Ravi Prabhu, and Anne Larson, eds. 2022. *Adaptive Collaborative Management in Forest Landscapes: Villagers, Bureaucrats and Civil Society*. London: Earthscan and Routledge.

Cronkleton, Peter, Kristen Evans, and Anne M. Larson. 2022. "Capacity Building for ACM: Lessons Learned from Training in Distinct Contexts." In *Adaptive Collaborative Management of Forest Landscapes: Villagers, Bureaucrats and Civil Society*, edited by Carol J. Pierce Colfer, Ravi Prabhu, and Anne M. Larson, 134–157. London: Earthscan and Routledge.

de Vries, Dede Wiliam, and Nurul Sutarti. 2006. "Gender Equity: Revealing the Reality for the Women of Jambi." *CIFOR Governance Briefs* 29: 6.

Diaw, Mariteuw Chimère, Tony Aseh, and Ravi Prabhu, eds. 2009. *In Search of Common Ground: Adaptive Collaborative Management of Forest in Cameroon*. Bogor, Indonesia: CIFOR.

Diaw, Mariteuw Chimère, and Trikurnianti Kusumanto. 2005. "Scientists in Social Encounters: The Case for an Engaged Practice of Science." In *The Equitable Forest: Diversity, Community and Resource Management*, edited by Carol J. Pierce Colfer, 72–112. Washington, DC: Resources for the Future and CIFOR.

Ernstson, Henrik, Stephan Barthel, Erik Andersson, and Sara T. Borgström. 2010. "Scale-Crossing Brokers and Network Governance of Urban Ecosystem Services: The Case of Stockholm." *Ecology and Society* 15 (4): 28.

Evans, Kristen, Anne M. Larson, and S. Flores. 2020. "Learning to Learn in Tropical Forests: Training Field Teams in Adaptive Collaborative Management, Monitoring and Gender." *International Forestry Review* 22 (2): 189–198.

Evans, Kristen, Sandra J. Velarde, Rocio P. Prieto, Sheila N. Rao, Sandra Sertzen, Karina Dávila, Peter Cronkleton, and Wil De Jong, eds. 2006. *Field Guide to the Future: Four Ways for Communities to Think Ahead*. Nairobi, Kenya: CIFOR, ASB, and World Agroforestry Centre.

Fisher, Micah R., Timothy Workman, Agus Mulyana, Balang Institute, Moira Moeliono, E. Linda Yuliani, Carol J. Pierce Colfer, and Urban El Fatih Bani Adam. 2020. "Striving for PAR Excellence in Land Use Planning: Multi-stakeholder Collaboration on Customary Forest Recognition in Bulukumba, South Sulawesi." *Land Use Policy* 99. December: 102997

Freire, Paulo. 1970. *Pedagogy of the Oppressed*. New York: Herder and Herder.

Guijt, Irene. 2007. *Negotiated Learning: Collaborative Monitoring for Forest Resource Management*. Washington, DC: RFF and CIFOR.

Gunderson, Lance H., and Crawford. S. Holling, eds. 2002. *Panarchy: Understanding Transformations in Human and Natural Systems*. Washington, DC: Island Press.

Holling, Crawford S. 1978. *Adaptive Environmental Assessment and Management*. New York: John Wiley and Sons.

Katila, Pia, Carol J. Pierce Colfer, Wil Dejong, Glenn Galloway, Pablo Pacheco, and Georg Winkel, eds. 2019. *Sustainable Development Goals: Their Impact on Forests and People*. Cambridge, UK: Cambridge University Press.

Komarudin, Heru, Yuliana L. Siagian, Carol J. Pierce Colfer, with Neldysavrino, Yentirizal, Syamsuddin, and Deddy Irawan. 2012. "The Role of Collective Action in Securing Property Rights for the Poor: A Case Study in Jambi Province, Indonesia." In *Collective Action and Property Rights for Poverty Reduction*, edited by Esther Mwangi, Helen Markelova, and Ruth Meinzen-Dick, 235–269. Philadelphia: University of Pennsylvania Press.

Kusumanto, Trikurnianti, Linda Yuliani, Phil Macoun, Yayan Indriatmoko, and Hasantoha Adnan. 2005. *Learning to Adapt: Managing Forests Together in Indonesia.* Bogor, Indonesia: CIFOR.

Lee, Kai N. 1993. *Compass and Gyroscope: Integrating Science and Politics for the Environment.* Washington, DC: Island Press.

Leeuwis, Cees, and Rhiannon Pyburn, eds. 2002. *Wheelbarrows Full of Frogs: Social Learning in Rural Resource Management.* Assen, The Netherlands: Koninklijke Van Gorcum.

Locatelli, Bruno, Markku Kanninen, Maria Brockhaus, Carol J. Pierce Colfer, Daniel Murdiyarso, and Heru Santoso. 2008. *Facing an Uncertain Future: How Forests and People can Adapt to Climate Change.* Bogor, Indonesia: CIFOR.

Mandondo, Alois, Ravi Prabhu, and Frank Matose, eds. 2008. *Coping Amidst Chaos: Studies on Adaptive Collaborative Management from Zimbabwe.* Bogor, Indonesia: CIFOR.

McDougall, Cynthia, Bishnu Hari Pandit, Mani Ram Banjade, Krishna Prasad Paudel, Hemant Ojha, Manik Maharjan, Sushila Rana, Tara Bhattarai, and Sushma Dangol. 2009. *Facilitating Forests of Learning: Enabling an Adaptive Collaborative Approach in Community Forest User Groups, a Guidebook.* Bogor, Indonesia: CIFOR.

McDougall, Cynthia, and Hemant Ojha. 2022. "How Adaptive Collaborative Management Can Leverage Changes in Power: Insights from Social Theory." In *Adaptive Collaborative Management in Forest Landscapes: Villagers, Bureaucrats and Civil Society*, edited by Carol J. Pierce Colfer, Ravi Prabhu, and Anne M. Larson, 186–216. London: Earthscan and Routledge.

McDougall, Cynthia, Hemant Ojha, Mani Ram Banjade, Bishnu Hari Pandit, Tara Bhattarai, Manik Maharjan, and Sushila Rana. 2010. *Forests of Learning: Experiences from Research on an Adaptive Collaborative Approach to Community Forestry in Nepal.* Bogor, Indonesia: CIFOR.

Mukasa, Concepta, Alice Tibazalika, Abwoli Yabezi Banana, Esther Mwangi, and Tendayi Mutimukuru-Maravanyika. 2022. "Strengthening Women's Tenure Rights and Participation in Community Forestry in Central Uganda." In *Adaptive Collaborative Management of Forest Landscapes: Villagers, Bureaucrats and Civil Society*, edited by Carol J. Pierce Colfer, Ravi Prabhu, and Anne M. Larson, 106–132. London: Earthscan and Routledge.

Mutimukuru-Maravanyika, Tendayi 2010. "Can We Learn Our Way to Sustainable Management? Adaptive Collaborative Management in Mafungautsi State Forest, Zimbabwe." PhD doctoral dissertation, Technology and Agrarian Development, Wageningen.

Mutimukuru-Maravanyika, Tendayi, Elias Madzudzo, and Mwansa M. Songe. 2022. "Researcher Collaboration Complexities in Participatory Action Research: Zambian Experiences." In *Adaptive Collaborative Management in Forest Landscapes: Villagers, Bureaucrats and Civil Society*, edited by Carol J. Pierce Colfer, Ravi Prabhu, and Anne M. Larson, 62–85. London: Earthscan and Routledge.

Ojha, Hemant, Andrea J. Nightingale, Noémi Gonda, Benard Oula Muok, Siri Eriksen, Dil Khatri, and Dinesh Paudel. 2022. "Transforming Environmental Governance: Critical Action Intellectuals and their Praxis in the Field." *Sustainability Science* 17: 621–635. doi: 10.1007/s11625-022-01108-z.

Prabhu, Ravi, Anne M. Larson, and Carol J. Pierce Colfer. 2022. "Circles and Spirals." In *Adaptive Collaborative Management of Forest Landscapes: Villagers, Bureaucrats and Civil Society*, edited by Carol J. Pierce Colfer, Ravi Prabhu, and Anne M. Larson, 238–259. London: Earthscan and Routledge.

Rattanasorn, Tawatchai, Bob Fisher, and Carolin Kugel. 2012. *Unusual Partnerships: Lessons for Landscapes and Livelihoods from the Doi Mae Salong Landscape, Thailand*. Gland: IUCN.

Ravikumar, Ashwin, Anne M. Larson, Rodd Myers, and Tim Trench. 2018. "Inter-Sectoral and Multilevel Coordination Alone do not Reduce Deforestation and Advance Environmental Justice: Why Bold Contestation Works when Collaboration Fails." *Environment and Planning C: Politics and Space* 36 (8): 1437–1457.

Ridwan, Isman. 2019. "Masyarakat Hukum Adat Datuk Sinaro Putih: Pasang Surut Kekuasaan Adat di Tengah Hegemoni Negara [Customary Community of Datuk Sinaro Putih: The Ebb and Flow of Customary Authority amidst National Hegemony']." *Jurnal Niara* 12 (1): 1–8.

Sarmiento Barletti, Juan Pablo. 2022. "Learning from Adaptive Collaborative Management: A Participatory Tool to Support Adaptive and Reflective Learning in Multi-Stakeholder Forums." In *Adaptive Collaborative Management of Forest Landscapes: Villagers, Bureaucrats and Civil Society*, edited by Carol J. Pierce Colfer, Ravi Prabhu, and Anne M. Larson, 159–184. London: Earthscan and Routledge.

Senge, Peter M. 1990. *The Fifth Discipline: The Art and Practice of the Learning Organization*. New York: Currency Doubleday.

Toffler, Alvin. 1970. *Future Shock*. New York: Random House.

Uphoff, Norman. 1996. *Learning from Gal Oya: Possibilities for Participatory Development and Post-Newtonian Social Science*. London: Intermediate Technology Publications.

Vanclay, Jerome K., Ravi Prabhu, and Fergus Sinclair. 2003. "Special Issue on Participatory Modelling of Community Forest Landscapes." *Small-Scale Forest Economics, Management and Policy* 2 (2): 117–326.

Vanclay, Jerry, Ravi Prabhu, and Fergus Sinclair. 2006. *Realizing Community Futures: A Practical Guide to Harnessing Natural Resources*. London: Earthscan.

Wollenberg, Eva, David Edmunds, and Louise Buck. 2000. *Anticipating Change: Scenarios as a Tool for Adaptive Forest Management (a Guide)*. Bogor, Indonesia: CIFOR.

Introduction to Chapter 2

Chapter 2, by Yuliani et al., makes a perfect lead-off to the contributions in this book. It and Chapter 7, by Kozanayi et al., in Zimbabwe are beautiful examples of what we hoped would emerge in our initial, open-ended inquiry to one-time ACM researchers: return visits and updates regarding what has happened in the locations we studied and among the peoples we collaborated with, in the early 2000s. As noted in Chapter 1, one of our initial hopes for ACM was that its intensive equitable interaction and facilitation, encouragement of learning processes and adaptation, and stronger links with outside actors might be the kind of process that could endure. Broadly, we sought sustainability; yet until recently, we have had no way to assess that.

The authors, Yuliani et al., are the researchers who initially formed the Jambi ACM team in the early 2000s. This experience has given them in-depth and longstanding knowledge of the context. They understand how things work there, they have friends and networks on which to draw, they have full-blown contextual knowledge of life in the village of Baru Pelepat; and they include both social and biophysical scientists, all of whom share at least one language with the community. Their assessment can be much deeper than could one by researchers coming in cold.

The flip side of this of course is that they bring to the assessment their own biases (as do we all). They may have unconscious inclinations to put the best light on the evidence they see, on what has transpired since their facilitation roles ceased. We are particularly fortunate in this case because we do indeed have the views, at least partially overlapping, of a team of external researchers, likely to be seen as more fully 'objective' – in Liswanti's team (Chapter 3), which worked more recently in the same area, though at a broader scale.[1]

Yuliani's team includes social scientists who assess what happened in interactions and sociocultural and political processes; and the biophysical scientists are able to use remotely sensed maps to show the dramatic, on-the-ground changes that have occurred, particularly in the community's surroundings, over time – so crucial in our efforts to assess likely climate change impacts.

Their analysis shows both the enduring aspects of capacities that were strengthened during the earlier ACM process but also the key roles of powerful outsiders

DOI: 10.4324/9781003325932-2

in controlling the trajectory of broader land use change. The Baru Pelepat community was able to withstand some assaults on their forests, through the collaborative skills they strengthened during and after the ACM project, but the ongoing onslaught spawned by more powerful governmental policies and related industrial practices was impossible for them to constrain – except within the small area they were able, through their own efforts with ACM facilitation, to protect (now an island in a sea of oil palm).

We remain strongly committed to the importance of empowering local communities via capacity building, learning processes, and building on local management. But this case makes crystal clear the importance also of gaining support at broader scales. If we are to contribute meaningfully and genuinely to addressing climate change and other 'wicked problems', we will have to initiate and/or strengthen such collaborative and adaptive processes at these broader scales.

Note

1 See also Bomuhangi et al. (2022) and Mukasa et al. (2022) for parallel assessments using quantitative and qualitative methods in Uganda.

References

Bomuhangi, Allan, Abwoli Yabezi Banana, Jimmy Bushoborozi, Concepta Mukasa, Alice Tibazalika, and Esther Mwangi. 2022. "Gender and Adaptive Collaborative Management in a Forested Ugandan Landscapes." In *Adaptive Collaborative Management in Forest Landscapes: Villagers, Bureaucrats and Civil Society*, edited by Carol J. Pierce Colfer, Ravi Prabhu, and Anne Larson, 87–104. London: Earthscan and Routledge.

Mukasa, Concepta, Alice Tibazalika, Abwoli Banana, Esther Mwangi, and Tendayi Mutimukuru-Maravanyika. 2022. "Strengthening Women's Tenure Rights and Participation in Community Forestry in Central Uganda." In *Adaptive Collaborative Management of Forest Landscapes: Villagers, Bureaucrats and Civil Society*, edited by Carol J. Pierce Colfer, Ravi Prabhu, and Anne Larson, 106–132. London: Earthscan and Routledge.

2 Revisiting Baru Pelepat

Life after ACM (Indonesia)

Elizabeth Linda Yuliani, Moira Moeliono, Trikurnianti (Yanti) Kusumanto, Marzoni, Effi Permatasari, Hasantoha Adnan, Yayan Indriatmoko, and Carol J. Pierce Colfer

Background

Of the more than 272 million people in Indonesia, 70 million are identified as 'customary people' (*masyarakat adat*; Andriarsi 2020), claiming more than 12 million ha of which some 2,403,485 are currently recognized (Badan Registrasi Wilayah Adat/Customary Territory Registration Agency/BRWA 2021). Many of these customary peoples depend on forest and natural resource products for their livelihoods, which are managed with local knowledge and customary rules. However, with Indonesian Law No. 5/1967, the state took full power to delineate and determine the use of forest areas and issue logging concessions. This law also stated that 'the use of forest resources by the indigenous peoples should not disturb the objectives of the law' (Awang 2006). Unclear tenure, unclear boundaries and overlapping claims have for decades – some would even say centuries – resulted in declines in the extent of the land people call their own. They have also exacerbated the degradation that contributes to climate change. This situation has continued under the current government. Declining land and resources, coupled with weakening of traditional values and customary institutions, have left customary peoples and the planet in a precarious condition.

In 1998, the New Order which had governed Indonesia since the mid-1960s was overthrown and a time of reform was anticipated. The Consultative Assembly issued a decree calling for agrarian reform. Yet, Law 41/1999 that replaced law 5/1967 in many ways reinforced state claims over forest (see, e.g., Wollenberg and Kartodihardjo 2002). In the same year, the new government issued the decentralization law providing a high degree of autonomy to district governments. For many local and customary people, this law was often understood as a return of their rights. Unfortunately, excesses in claiming and exploiting forests, supported by local governments, were used as an excuse to retract the law. Authority over forest governance was then allocated to provinces.

Subsequent years saw little progress in recognition of customary tenure in state forest land. Although in both the 1967 and 1999 laws, customary rights over customary forest or *Hutan Adat* were recognized in principle, in practice these customary forests remained part of state-controlled forests. *Hutan adat*, to local communities, means the area they have claimed as their own traditionally; yet to

DOI: 10.4324/9781003325932-3

the Indonesian government, it more commonly refers to a formal governmental program, which we abbreviate as HA. Recently, the Indonesian government has continued, even strengthened, its emphasis on economic growth. This has been most obvious in its promotion of oil palm and pulpwood plantations as well as coal mining (Adebayo et al. 2021). As a result, dependence on cash and pressure on land have only increased (Brockhaus et al. 2012).

These forces have also affected the village of Baru Pelepat, in Bungo District, Jambi Province, Sumatra. In 2000, the Adaptive Collaborative Management (ACM) team started their activities there. The team comprised CIFOR, Jambi University scientists (social scientists, ecologists, law and policy analysts) and local NGOs *Yayasan Gita Buana* (YGB/Gita Buana Foundation) and *Pusat Studi Hukum dan Otonomi Daerah* (PSHK-ODA/Regional Autonomy and Law Research Center). The ACM approach used Participatory Action Research (PAR) as a key method to facilitate learning, adaptation and collaboration among multiple stakeholders (Figure 2.1). Prior to the implementation of PAR, the team began with rapport building with the key stakeholders, and a series of context studies using a combination of participatory and conventional methods to gain a shared understanding of the local context, not only for the team but also among the key stakeholders themselves (Kusumanto et al. 2005).

Through the context studies and in the reflection stage of the PAR process, the main concerns of the local communities were identified: i.e., weak local institutions, particularly weak representation of and lack of opportunities for women, settlers and the nomadic Orang Rimba in decision-making, unclear village boundaries, the need to maintain the customary forest to preserve timber for descendants and to prevent landslides, and the inability to protect it against logging by both insiders and outsiders (concessionaires and other communities) in their customary territory. A concern of high relevance for the present study is the weak communication between community and government. Public decision-making and development planning tended to be minimally informed by the needs of those affected by government decisions (Kusumanto and Permatasari 2002). Another study found that at the time of our direct involvement, i.e., 2000–2006, the district administrations were incapable of intervening due to a weak institutional design for decentralization, particularly at the national level (Yasmi and Schanz 2010). Decentralization was not supported by sufficient capacity building for effective monitoring at district levels; nor was there sufficient funding for implementation and monitoring at these levels. Consequently, district administrations were unable to intervene and the conflict between Baru Pelepat and the timber

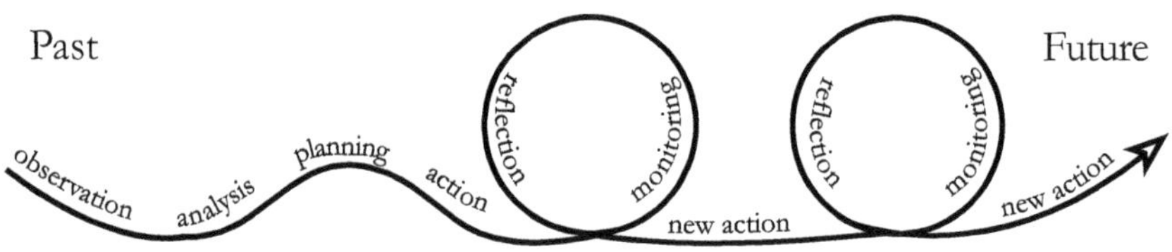

Figure 2.1 PAR cycle

corporations was allowed to escalate. Diaw and Kusumanto (2005), Kusumanto (2007a, 2007b) and Kusumanto et al. (2005) all document how the ACM team responded to each issue and the outcomes of their facilitation. Box 2.1 reflects some of the reasons the community aspired to obtain formal recognition for their customary forest (i.e., delineation as an HA).

BOX 2.1 Baru Pelepat community rationales for obtaining legal recognition for their HA

In a community meeting, customary leaders said,

> How can we apply our customary rules if the boundary of our territory and rights are unclear?

Meanwhile, in in-depth interviews, community members said,

> How can the community respect customary leaders if leaders' relatives break the rules without being sanctioned?

Subsequent community meetings then identified one of the solutions with the greatest potential, i.e.,

> Getting legal recognition as a HA.

An intense period followed wherein the ACM team facilitated community representatives in approaching the Bungo District Government for obtaining recognition for their customary forest as a HA (see the next section for summary of processes, including ACM involvement). For detailed processes, see Adnan et al. (2008), Dobesto (2008), Kusumanto et al. (2005), Marzoni (2008) and Pariyanto (2008).

In October 2006, the Government of Bungo District signed a District Regulation (*Peraturan Daerah/Perda*) to legally recognize the customary people of Baru Pelepat and nearby Batu Kerbau, known as the *Datuk Sinaro Putih* people. This was long before the 2012 Constitutional Court's decision which declared all customary forests in principle outside state forest; to take true ownership though, the HA had to be legally recognized. The signing of this regulation in Bungo District was an impressive and unusual achievement. The *Perda* gave rights to the people to manage the customary territory, including the *Rimbo Adat* Datuk Rangkayo Mulyo (*Rimbo Adat* is Minangkabau for customary forest).

In 2006, funding for the ACM project ended. Prior to completion of the project, the team, together with community representatives, partners, the donor, external evaluator and the ACM Steering Committee members, had developed some elements of an exit strategy: building the capacity of potential community members, including women, to become pioneers and good leaders; continuing to facilitate collaboration among intra-community groups; and reaching their goals, including maintaining their natural resources, in particular timber, NTFPs and clean water from the HA, and fish from *lubuk larangan* (a protected river area) sustainably.[1]

Have the strategies worked? What happened after the project ended? Moreover, although the project ended, the ACM approach continues to be used under different names. A recent iteration is the landscape approach of which the following four principles are shared with ACM: the need to identify a common concern, facilitate multistakeholder processes and encourage continuous (social) learning and implementation of the resulting adaptive management. In this chapter, we analyze how the ACM strategies (including the proposed capacity building) have worked – or have not – and what legacy they left in one site (see also Liswanti et al., this volume). We aim to understand whether ACM remains a relevant approach and how it has evolved.

The site

Baru Pelepat is located in Bungo District, 64 km from the district capital Muara Bungo, Jambi Province (Figure 2.2). The village is situated by the Pelepat River, which was used as the main transportation route until the late 1990s. In early 2000, reforms and decentralization processes at the national level affected village lives all over Indonesia. Baru Pelepat became a village in a new district, and issues

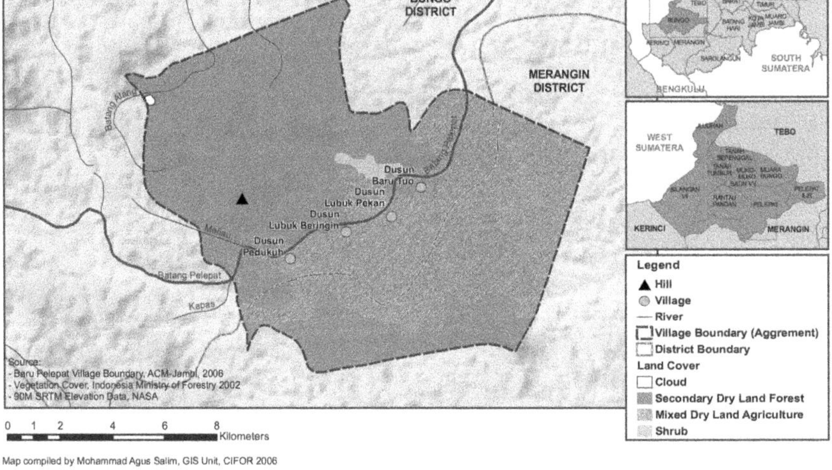

Figure 2.2 Map of Baru Pelepat Village

of land tenure, governance, gender and natural resource management were negotiated at the different levels.

Like the neighboring villages, Sungai Beringin in the north, Batu Kerbau to the west, Rantel to the east and Merangin District to the south, parts of Baru Pelepat were located within the State-designated forest area. Large-scale logging concessions operated in the area between 1975 and 1999 and continued with 'illegal' logging[2] until early 2000. The main population of the village is descended from a group of Minangkabau from West Sumatra led by *Datuk Sinaro Putih*. About one third of the people living in the Baru Pelepat territory are transmigrants, mainly from Java. Some came through the government's transmigration program in 1997 (Kusumanto and Indriatmoko 2001); others moved in spontaneously over time. At least four nomadic Orang Rimba groups, each consisting of approximately 25 members totaling around 100 members, call the area part of their territory (Indriatmoko 2001).

Summary of the ACM process of getting legal recognition[3]

In the early 2000s, the first few years of Indonesia's decentralization, there was a big movement to ensure that local and customary rights to land and forest resources were recognized. Indeed, the work in Malinau (then East Kalimantan Province, now North Kalimantan Province) revealed the perception that decentralization included the devolution of land and forest resources to customary and local communities (Moeliono and Limberg 2012). Gaining formal recognition within the designated state forest area was not easy. In 2002, the only legal basis was the IX/2001 Decree of the People's Consultative Assembly (*Majelis Permusyawaratan Rakyat*/MPR) on agrarian and tenure reform and Law No. 41/1999 on forestry.

Yet, it was already clear that to gain recognition of rights – in this case clear rights over the customary forest or designation as a HA – the government would have to formally recognize the customary community through a regional regulation. To work with government to formulate and issue such a regulation, Baru Pelepat asked the ACM team to facilitate the process.

The process started in 2002 through a series of multistakeholder meetings. After more than a year of long multistakeholder discussions, in August 2003, all community members and a neighboring village agreed to reinforce customary rules in managing their forest through legal recognition. However, this was a period when formal, village-level processes throughout the country were very much determined at district level; governmental processes were very top-down. At the same time, a District Decree (*Surat Keputusan Bupati*) was considered legally weak.

To find solutions, the team collaborated with other institutions having similar initiatives in the District and Province, e.g., WARSI, a Jambi-based NGO, and ICRAF (the World Agroforestry Center), as well as holding discussions among government officials at multiple levels (i.e., the District Forestry Service, District Legal Bureau, Directorate General of Social Forestry and Ministry of Forestry). Participants in discussions concluded that a special District Regulation (*Perda*

Khusus) on the confirmation of local people's claims could be the solution. Meanwhile, as development of the *Perda Khusus* would take time, Baru Pelepat people conducted participatory village planning, incorporating HA management and rules into such planning. The HA management planning was then legalized through Village Regulation No. 2/2005.

The *Perda Khusus* was not the only process initiated through ACM. Managing the customary territory also required attention to the strengthening of local institutions. This was done with the community taking leadership in boundary negotiations, mapping and building a long-term village management plan.[4] The community also provided information and data on the customary territory for the HA assessment and verification, and participated in preparing a *naskah akademik* (literally an 'academic script', a study used as a basis for drafting a regulation) and a draft formulation of the regulations for the *Perda Khusus* (see details in Section 'Local Institutions: Decision-Making, Inclusivity and Conflict Management'). After a long and winding road, the District Regulation was issued in October 2006.

Meanwhile, higher legislation was needed to strengthen the *Perda*. Eleven years after the *Perda*, such legislation was issued at the national level. The Constitutional Court's Decision No. 35/2012, which declared this HA (and all others) to be outside the national forest estate, became the basis for the Bungo District Forestry Service to register Datuk Rangkayo Mulyo Customary Forest (HA) by Ministerial Decree. It should be noted that the lack of clarity of tenure also applies to state forest as the state never finished the demarcation and gazettement processes required by law.

It still took five years and the start of the national social forestry program to obtain actual legal recognition from the State. On 25 October 2017, a Ministry of Environment and Forestry (MoEF) decree recognizing the Baru Pelepat forest as a HA was handed to the customary leaders directly by the President of Indonesia, Joko Widodo, in his palace (Figure 2.3), together with eight other HAs from all over Indonesia. Baru Pelepat's forest was one of the first to be granted such a decree. The total extent of Baru Pelepat's HA after verification is 821 ha, which originally consisted of 245 ha of Production Forest and 576 ha of other uses.

Yet, a lot of questions remain. Have the people's concerns been addressed with the issuance of the *Perda* and Ministerial Decree? Has the regulation led to better protection of their forest? How has it affected their livelihoods? Has the legal recognition and issuance of the decree empowered local people? In what ways and why? This chapter describes what has happened with the HA and the people after the decree was granted, including important lessons, successes and challenges.

Methods

To collect information on what has happened on the ground, we rely primarily on the observations of two of the authors, members also of the original ACM team (Marzoni and Permatasari). In this, we apply a reflexive method whereby we take into consideration the position and experience of the researchers. Marzoni has been a resident of the village since 2006 and was a member of the ACM team

Figure 2.3 Samsuri, head of Baru Pelepat, receiving the Ministry's Decree from the President

before that. We document his observations using the following guiding questions: What are the outcomes of ACM activities that are still present or performed now? What are the examples/evidence? What ACM outcomes have disappeared or been discontinued? What community goals were not achieved? How has the customary forest management institution engaged women in decision-making? Have they been represented in this institution and been involved in communication and negotiations with government and companies? Is there any evidence to that effect?

Permatasari facilitated Baru Pelepat women in interactions among the women themselves as a heterogeneous group and with other community stakeholders during the original ACM program. Her current engagement as an agricultural extension official of Jambi province has brought her in contact with individuals in the Bungo Government who have been involved in ACM in one way or another. We document her observations on gender issues, using the following questions: Referring to ACM's achievement to include the diverse Baru Pelepat women in the public domain, has this social change continued to prevail, with what impact at wider scales? If so, based on what evidence? How has the (sub) district government experienced this development? What has been their response (e.g., any attitudinal or policy changes)? Is there any confirming evidence?

As the ACM project aimed to facilitate sustainable and equitable forest management through social learning, adaptation and equitable collaboration, we specifically document the following elements: forest cover, performance of local institutions, particularly in relation to stakeholder representation and inclusion, the institutions' capacities to foster collaborative learning and adaptation; and conflict management, social capital and livelihood assets. To triangulate their observations, we compare and complement them from the scholarly literature, e.g., Komarudin, Siagian, and Colfer (2012), Larson and Sarmiento Barletti (2019), Syamsuddin, Komarudin, and Siagian (2007), Wiliam-de Vries and Sutarti (2006) and Wiliam-deVries (2006).

As part of the triangulation, we also performed land use/land cover change analysis of the village from 2006, 2010, 2013 and 2020. We chose these dates in accordance with the timeline of major changes explained by Marzoni. We used Landsat Thematic Mapper (TM) 5 images, path 126/row 61 (acquisition dates 30 May 2006 and 25 July 2009) and Landsat TM 8, path 126/row 61 (acquisition dates 15 April 2013 and 10 November 2019). Other sources included boundaries of the HA resulting from participatory mapping in 2005, the boundaries of Production Forest from the MoEF, road and river networks from the Geospatial Information Agency, and the stable boundaries of the village resulting from mapping of the natural objects known by the community as the boundaries in 2005. In accordance with the objectives of this analysis, i.e., to triangulate with the qualitative information, we focused on the following land use categories: forest (covering natural and secondary forests), traditional cultivation areas (covering swidden dry rice fields, young fallow, old fallow and small-scale rubber and mixed gardens), new land/soil exposure, settlement and water. The images were manually interpreted and validated using authors' knowledge of field conditions, combined with high-resolution satellite imagery from Bing maps aerial view (dated 11 January 2016) and Google Satellite provided in QGIS at zoom level 17–18, which were equal to spatial resolution 1.2–0.6 m (https://docs.microsoft.com/en-us/bingmaps/articles/bing-maps-tile-system).[5]

Results

Land use/land cover change

Results of Baru Pelepat Land Use and Land Cover Change analysis between 2006 and 2020 show some major changes, in particular the loss of forest, in total 6,668 ha, and the loss of traditional cultivation area, 1,750 ha – clearly reflecting conditions exacerbating climate change. These declines occurred along with the increasing extent of plantations, i.e., from 0 ha in 2006 to 4,649 ha of large-scale and 2,674 ha of small-scale oil palm plantations, and 1,116 ha of industrial plantation forest (*Hutan Tanaman Industri*/HTI). These changes are shown in Figure 2.4.

Figure 2.5 shows where and how the major changes have occurred.[6] The land converted into oil palm was mainly forest on land designated for other uses (*Area Penggunaan Lain*/APL), a category which is under the authority of the district government, and Production Forest, a category under the authority of the Ministry of Forestry. In 2008, an oil palm company, PT CSH, obtained a 'location permit'

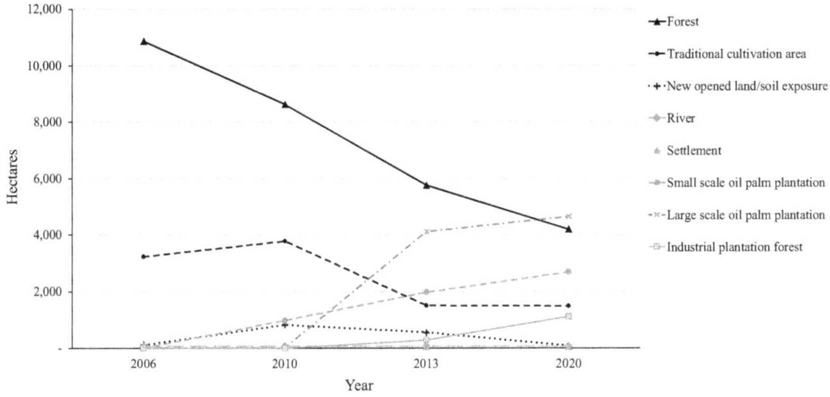

Figure 2.4 Land use/land cover changes in Baru Pelepat between 2006 and 2020

(*Ijin Lokasi*), an early step in the process of gaining a cultivation permit, in Baru Pelepat. According to Law No. 18/2004 on Plantations, the company had to get approval from the customary people and pay the agreed compensation. The company therefore recruited several village members (all married Minangkabau men) to determine and settle any disputes over land[7] within the company's designated area.

Yet, recognition of territory alone is not enough, requiring follow through in terms of efforts to deal with poverty, equity and representation (Larson, Barry, and Dahal 2010). Despite the established concerns about conservation in the community, driven by socio-economic needs, the community also sought financial gain from this development. They cleared and cultivated some parts of the APL and applied for a license, the Sporadik (*Surat Pernyataan Penguasaan Fisik Bidang Tanah*/Letter in Evidence of Physical Authority over Land) from the village administration. As in so many areas, insecurity of tenure encouraged short-term thinking. People said quite realistically, 'It will be cleared anyway by the company. But if we clear it first, we can get some money, and we can also become the company's partner'. By doing this, they were compensated by PT CSH for the land they released at IDR 3 million per ha (USD 310/ha at the 2008 average exchange rate) and the registered owners automatically had the rights to become the company's partner (*mitra perusahaan*) and to receive a share of the benefits. The community's heartfelt need for financial gain and its opportunistic response to future tenure uncertainties illustrate the rationality of local people, but also that ACM was weaker in fostering learning at scales beyond the community and in facilitating learning across boundaries (between sectors, institutions, disciplines and knowledges). See also Kozanayi et al., this volume.

The 2020 map shows that despite forest loss in the surrounding area, most of the HA has been maintained, except in the northern and southeastern boundaries. When the oil palm company cleared land in 2010, they encroached beyond the boundary of their concession and cleared around two ha of HA, including

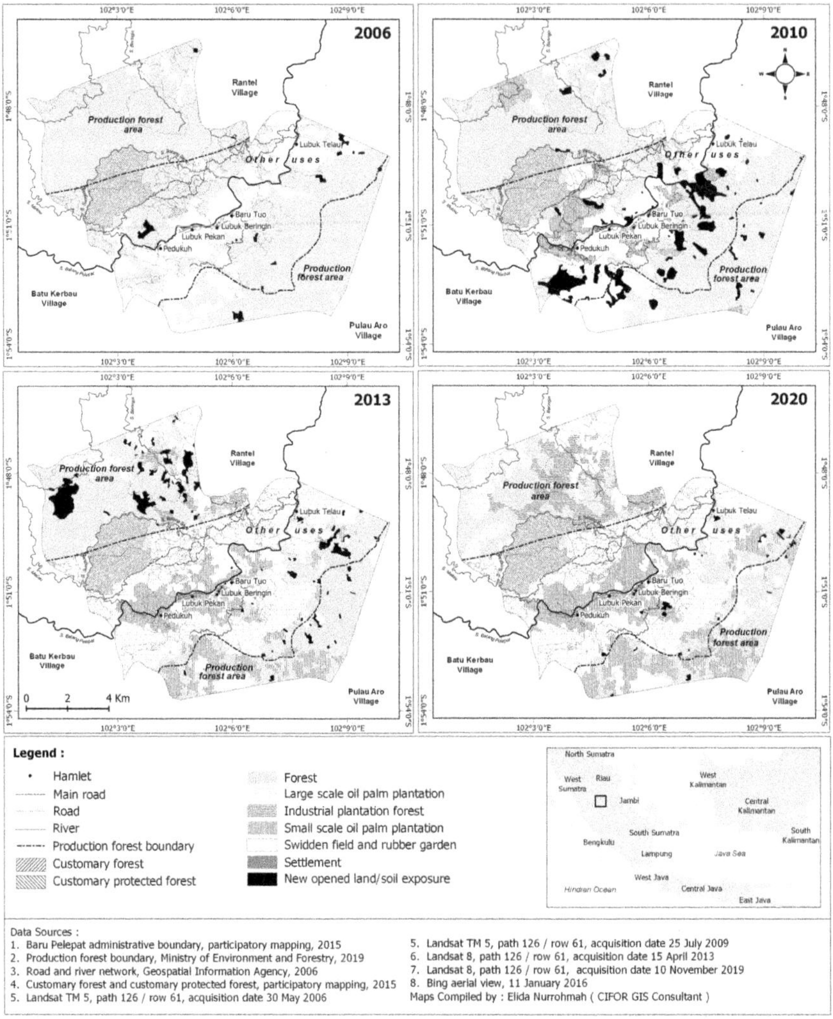

Figure 2.5 Maps of land use/land cover changes in Baru Pelepat between 2006 and 2020

cutting big trees. Abu Nazar, the head of the HA managing group at that time, together with four community members, including Marzoni (resident co-author), went to the location and asked the company to stop clearing the land. However, the company referred to the map attached to their permit from the district government, which showed that part of the HA was included in the area allocated for their use.

The HA managing group then reported the company to the District Forestry and Plantation Service. Together, they inspected the location and found that the company did trespass on the HA boundary, and that the map used by the company contained incorrect coordinates. After a village meeting involving the company and village institutions, mediated by the District Forestry and Plantation Service, all parties agreed that the company had to pay customary fines in the form of one goat, twenty *gantang* of rice (62.5 kilos), four bolts of fabric and a variety of side dishes, whose value in total was equal to IDR 4.6 million (USD 510 at the 2010 average exchange rate).

This provides one example of the HA managing group actively maintaining the forest and enforcing the rules. The managing group's sanctioning of a community member illegally logging in the forest represents another good example of their continuing active management. A community member who logged commercially in the HA insisted that the status of the forest was Production Forest. The person was put on customary trial by the HA managing group, customary leaders and village administration. After explanations from the customary leaders and the District Forestry Service about the status of the forest, the person was fined in accordance with HA regulations.

The seriousness of the HA managing group in preserving their HA has been acknowledged by the district government. As a reward for their good management, the village received a clean water supply installation. The national government provided piping of clean water from its source in the HA to the village. However, the pipes only reach the hamlet closest to the source, and the village administration and HA managing group are still looking for additional funding to add the pipes to reach all houses in the four hamlets. Other rewards were funding for patrolling the forest and renewing boundary markers. The District Forestry Service also provided a replanting program. Unfortunately, these programs were an all too familiar process of only providing materials without follow-up. As one village member said:

> Support for planting only was not enough. There should be a program [funding] for maintenance, e.g., to replace dying seedlings, otherwise the newly planted seedlings lose out in competition with other plants [*Bantuan tidak cukup hanya untuk penanaman. Harusnya ada untuk perawatan juga, misalnya untuk mengganti yang mati. Kalau tidak, lama-lama bibit itu kalah, tertutup oleh semak dan pepohonan di sekelilingnya*].[8]

Local institutions: decision-making, inclusivity and conflict management

Our early ACM facilitation of social learning was designed according to the needs and problems that the stakeholders considered important (Diaw and Kusumanto 2005). The people expressed the need to learn about their own diversity. They identified issues through which they learned that there are dissimilar experiences, histories and knowledges within the community. Through the learning process, stakeholders became more aware of the needs and interests of other stakeholders (Indriatmoko et al. 2006). The facilitation and monitoring of democratic decision-making encouraged inclusion of all groups without exception (Moeliono

2006). Moreover, by using monitoring questions instead of indicators, the monitoring of decision making processes proved more locally relevant and operational. Hence, a question-based ACM approach can be of particular importance for gender-sensitive and inclusive monitoring in development and research initiatives, many of which also deploy indicators (see also Sarmiento Barletti's (2022) use of questions in a similar fashion).

These processes of institutional strengthening have continued until now, some 15 years after closing-off of the ACM team's facilitation. One of the main outcomes of the ACM project, as acknowledged by Baru Pelepat people and Bungo District Government in 2006, was the strengthened capacity of local institutions, i.e., village administration, customary institutions, including the HA managing group, the village legislative assembly (*Badan Perwakilan Desa*/BPD) and the women's group, to perform participatory, transparent and democratic decision-making at the village level. These mechanisms differ from those in neighboring villages, where decision-making and conflict management are largely still the responsibility of village elite men with very little (if any) engagement of other community members.[9]

Over time, these above-mentioned local institutions have remained inclusive of the diverse community groups, including those which, at the start of ACM's initial project in the 2000s, were largely excluded from village and stakeholder decision-making. Particularly women, the settler or migrant group and the nomadic Orang Rimba were among the marginalized. Moreover, female 'newcomers' were at the time often confronted with double exclusion on village matters; they had the least voice, due to their lower socio-cultural status compared to original Minangkabau women[10] and, being female, compared to male villagers.

With regard to women's inclusion in public decision-making, it is noteworthy that in the 2015 election for village head, a woman was among the nominated candidates. Nonetheless, local institutions have been less successful in shaping inclusivity for the nomadic Orang Rimba. In theory, ACM would promote the inclusion of this group also in the multistakeholder negotiations. The group's engagement in village affairs has remained a pertinent institutional challenge, most fundamentally due to the group's geographical mobility, but also to cultural difference and negative stereotypes. Orang Rimba were, and probably still are considered primitive by the Malays and as a result have distanced themselves (Persoon and Wardani 2017) – creating a vicious cycle of exclusion and avoidance. This necessitates focused policy support from district and higher-level government and changing attitudes.

Local institutions have also developed a conflict management mechanism that is based on community-wide learning and knowledge development. During earlier research, the ACM team, village leaders and negotiators learned that in conflicts involving external parties – at the time (2004), logging companies and the District Forestry Agency – it was essential that all community stakeholders be well informed about the negotiations so that they were aware of any conflicts, especially if these should escalate (Yasmi and Schanz 2010). Unfortunately, at that time, the community lacked negotiation experience, let alone training in explicit conflict management. Yasmi and Schanz (2010) describe six months of conflict in 2004 that evolved from the level of 'anxiety' to 'intimidation' – a level of extreme conflict.

However, when disagreement arose in 2010 between the community and oil palm companies on an overlapping HA and plantation area as described above, the local institutions were better prepared. Using prior experience, they were able to organize the necessary knowledge needed for the negotiations. An example is the fieldwork that Baru Pelepat villagers carried out to collect data, delineate their HA and develop maps for use in the negotiations with the oil palm companies, a process mediated by the District Forestry and District Plantation Agencies. The disagreement between the community and oil palm companies did not escalate and was eventually settled. In the entire process, all community stakeholders were kept informed on the fieldwork and the subsequent negotiations through their representatives in the village legislative assembly. See Li and Semedi (2021) for a more typical exploration of conflictual community – plantation relations.

Our past and more recent observations reveal that the social organization of such conflict management is key whereby community-wide communication and learning in formal meetings and informal gatherings played complementary roles in managing specific 'parts' of conflicts. The regional policy and institutional conditions were supportive for successful conflict handling. District Forestry and District Plantation Agencies effectively mediated the conflict at their own discretion to conform to decentralized regional development policies. In 2019, however, district forestry offices no longer existed, having been dissolved in 2014. On the ground, forestry staff presence was expected to be supplied through the forest management units, FMUs (*Kesatuan Pengelolaan Hutan/KPH*) administered from the provincial level. In this case, the FMU in charge is KPH Bungo.

Land use decisions: evidence of learning from past experience

Land use changes into the large-scale oil palm plantation described above provide an interesting opportunity to investigate how the Indonesian Law on Plantations No. 18/2004 (Government of Indonesia 2004)[11] was implemented on the ground. The law stated that a plantation planning to operate on customary land would require approval from the customary right holders, as well as agreement on the conditions of the land transfer and the amount and nature of compensation.

We investigated whether or not the compensation and conditions of land transfer had been transparently discussed and negotiated between the people and the company, as well as between the community representatives and the community they represent. We considered what the capacity of the people was in doing so, especially without further facilitation by the ACM team in the location. The results are described below.

When PT CSH came in 2008, they offered a partnership scheme with the initial offer being 80% of net income for the company, 20% for the landowner, deducting plantation operational costs from the landowner's share. The people negotiated a 70–30 split, and the company agreed. All such decisions were written in an agreement, signed by both parties before a notary. Soon, however, people found that the 30% share was disappointing. The amount varied across time, ranging between IDR 50,000 and 100,000 per month depending on operational costs, price of the oil palm fruit, taxes, etc. When another company, PT PML came in 2010, people had learnt from past experience with PT CSH, and negotiated a benefit sharing of 60–40 with PT PML, based on the selling price of the fruit, with no deduction of operational costs (in a different area). The company agreed.

Successful negotiation between village members with powerful actors without NGO facilitation is unusual in Indonesia and rarely reported in the literature. Many studies from other locations in Indonesia reported that during land transfer discussions, companies did not clearly describe compensation, conditions and risks, and even in some cases the land transfer process involved trickery and bribery (see, for example, Anderson 2013; Colchester, Jiwan, and Chao 2013; Colfer 2021; Li and Semedi 2021; Yuliani et al. 2020). Free, Prior and Informed Consent (FPIC) was seldom, if ever, implemented.

Baru Pelepat people, however, negotiated based on knowledge and collective action from past experience: for example, when an oil palm company almost tricked the people in a neighboring village (see Endah 2008). Baru Pelepat residents learned to unite and make collective decisions transparently in order not to be so easily tricked by powerful actors. The important roles of collective action in securing the rights of the poor in the region were also reported by Adnan et al. (2008), Komarudin, Siagian, and Colfer (2012) and Syamsuddin, Komarudin, and Siagian (2007). Further, the people also learned benefit sharing calculations from their own experience with the earlier oil palm company that came in 2004 (Adnan et al. 2008).

ACM, *development and conservation: gender, economics, biodiversity and collective action*

Two issues particularly highlighted in this section hold special relevance for global efforts to address climate change. Concerns about gender equity, addressed first, are important in our attempts to activate and make use of the energies, skills and knowledge of the feminine half of the human population (besides potentially strengthening efforts to stabilize population sizes, Colfer, Dudley, and Gardner, 2008), in seeking locally acceptable solutions.

Similarly, the value of collective action, addressed last, should be clear from all the studies discussed in this book. Climate scientists and activists agree that to address climate change in a meaningful fashion, huge swathes of the human population will have to transform their ways of life and change their behavior. Forest communities, such as Baru Pelepat and others, demonstrate less consumptive ways of life that are less damaging to the planet; and they have shown their potential for conservation-oriented collective action in maintaining part of their own forest (recognizing that they have also cooperated with outside groups when insecurity of tenure encouraged short-term thinking/action for private and immediate benefit) – even in the face of the wholesale attack on surrounding forests.

The challenge now is to extend this kind of collective action, up the ladder also to those who have the power and authority to make decisions to care rather than to exploit.

Gender

As noted above, ACM facilitation in the 2000s challenged the gender social constructs at community and local stakeholder levels. PAR learning processes were successful in breaking down institutional barriers and gender-biased mindsets that confronted women in partaking in decision-making and in benefitting from development (cf. Mukasa et al. 2022). This success has nonetheless been largely limited to the local community. In our past work, intense *local* facilitation did not allow us to reach out more widely, either horizontally (building linkages around gender issues with other communities) or vertically (linking up gender issues with supra-village government and stakeholders).

Some anecdotal information – including the recent woman candidate mentioned above – indicates changing views on gender of, predominantly male, government officials and villagers in the region. For example, some district government staff have found that Baru Pelepat women do not shy away from speaking in formal meetings and more freely convey their views and feelings. Additionally, a number of district level officials were exposed to more equitable gender-related ideas in their participation in a second wave of ACM.[12] A multistakeholder forum routinely took place in the city of Jambi, facilitated by CIFOR, ICRAF and local NGO partners, connected with CIFOR's CAPRI project (2004–2006). Nonetheless, the degree to which ACM has meaningfully contributed to gender-sensitive regional development and forest institute decision-making remains incompletely answered.[13]

Past analyses on gendered forest decision-making in the region found the need to (a) practice gender sensitivity in decision-making circles at the district, subdistrict, village and customary community institution (*adat*) levels and (b) facilitate the implementation of gender mainstreaming at the local level (Siagian et al. 2005; Wiliam-de Vries and Sutarti 2006).[14] Nevertheless, the implementation of gender mainstreaming – which has been a national government policy since 2000 – whether carried out by government or civil society may be challenged in the long run because it has been largely externally driven.

Economics

Local people might want to protect their forests for many reasons, e.g., for cultural reasons or as a reserve for later use. National and regional policies, however, are prioritizing economic development over other objectives (Maryani et al. 2021). The Social Forestry program, for example, clearly says that it provides access to forests to improve livelihoods (Government of Indonesia 2016). Indeed, despite FPIC, people are not questioning the way government programs are implemented, especially as this program does allow the earning of much needed cash income.

When deciding to protect the HA, Baru Pelepat residents were hoping that the HA could become a source of alternative income from ecotourism. However, this has not been achieved. According to the HA managing group, ecotourism to the HA was considered too dangerous, as there were signs and sightings of the Sumatran tiger. Therefore, the people and the Government of Bungo District decided not to develop further plans to promote ecotourism. For some community members, the failure to achieve the economic objectives was disappointing, as has been commonly reported, e.g., by the people of Guguk village (Sari 2013) and the Iban of Sungai Utik in Kapuas Hulu (Heri Valentinus, pers. comm. in 2020).

However, recognition of the HA for economic purposes is not always consistent with customary rules. For example, in Bulukumba, Sulawesi,[15] the recognition of the Kajang HA was targeted by the District Cultural and Tourism Service as a tourist destination, thus encouraging mass tourism. This opened up areas that the community had considered taboo.

These cases provide important lessons, applicable at regional and national levels, that a project's objectives and motivations have to be clear from the beginning and aligned with the meaning of the HA for the people. If the meaning is more cultural services (such as local identity) and indirect or non-cash benefits (such as timber for village facilities and clean water, prevention of floods and landslides), economic plans to manage the forest should be aligned with these customary rules and meanings.

Biodiversity

The Baru Pelepat Village Regulation on the HA only regulates the use of timber. Protection of biodiversity, particularly endangered or protected species, unfortunately was not regulated. The HA is inhabited by many endangered species such as the Sumatran tiger (*Panthera tigris sumatrae*), listed as a critically endangered species in the IUCN Red List (Linkie et al. 2008), and *kuau*, the Great Argus pheasant (*Argusianus argus*), listed as a vulnerable species (BirdLife International 2020) – both are protected under Presidential Regulation No. 7/1999. As the surrounding forest areas have disappeared, the HA has become one of the last refuges for wildlife, making the animals even easier to hunt. In the last few years, people from the village and outside hunted the *kuau* for sale. Tens of birds were killed in one hunt, but, as has often been observed when wildlife regulations are disobeyed, no legal action has been taken by law enforcement officials. So far, no tiger hunting has been reported in the HA.

When located on APL, the drive for economic opportunities such as oil palm has left the HA an island of forest surrounded by this commodity. Such isolated small patches of forest protected by traditional belief systems contribute to upholding some ecosystem services such as crop pollination and seed dispersal. However, such places are likely to be very vulnerable (Bodin et al. 2006), insufficient to support certain wildlife species (Deere et al. 2020), pose a danger to the people, and increase risks of human-wildlife conflict (Nyhus and Tilson 2004). Large mammals require large home ranges; for instance, the Sumatran tigers' home range for adult females ranges from 40 to 70 km^2 and adult males around 180 km^2 (Franklin et al. 1999), and Sumatran orangutans' home ranges for adult females are ca. 850 ha, whereas subadult and adult males' home ranges are at least 2,500 ha (Singleton and Schaik 2001).

This provides an important lesson that biodiversity protection should receive better attention in HA management and rules, e.g., by linking with traditional beliefs and norms about certain wildlife species (Asba, Rahman, and Evita 2019; Yuliani et al. 2018), and in broader landscapes such as industrial plantations. But this will not suffice unless the district and provincial government allow for a landscape approach to biodiversity conservation and include protection of corridors to allow connectivity between protected areas (see Bennett 2003) or land-sparing strategies between plantations (Yue et al. 2015). One attempt has been the requirement for oil palm companies to protect high conservation value (HCV) areas but these HCV areas need to be legally secured (Colchester et al. 2009), connected and not become isolated islands (Purwanto 2019).

Collective action

Over time, Baru Pelepat's local institutions have been able to maintain and facilitate learning and collective action in a context wherein, driven by the commercial oil palm sector, the forest landscape has been rapidly changing. ACM thus offers a pathway for a kind of regional development grounded in the facilitation of learning for adaptation and collective action among stakeholders at different levels and scales (see below). It necessarily, if not urgently, embraces diverse, complementary social organizations (including formal and informal, regulated and unregulated, structured and less structured, public and civic forest and development organizations) and schemes in a region-wide effort to more democratically and fairly involve forest stakeholders. By and large, government forest policies and programs tend to emphasize formal institutions for the involvement and consultation of local communities, such as *musrenbang*[16] (Komarudin, Siagian, and Colfer 2012). ACM can potentially play an important, complementary role to the formal institutions and programs in paying more attention to processes of learning, communication and action among various social domains, levels and scales. An approach like ACM also allows for attention to a key variable, ethnicity, something so far impossible within the Indonesian policy context.

Discussion and conclusions

The early work promoting ACM clearly left a legacy, most noticeably in the people's increased self-confidence in negotiating with outsiders and pride in maintaining their HA. The community appears to have internalized the plan-act-evaluate mode (Figure 2.1) blended with social learning as a basis for its decision-making and, externally, for communicating and negotiating with oil palm companies (see Chapter 4 on social learning in a broader context). After more than a decade, without external facilitation, their HA has been well maintained, as demonstrated through the official 'verification and validation' required for Ministerial recognition of local rights to manage the HA. In addition, the community has made use of opportunities that emerged with changes in policy. While the current agrarian reform is not making much progress, it did revive the Social Forestry program and started to implement Constitutional Court Decision No. 35/2012, that decreed HA were the property of customary communities and outside the state forest estate. When in 2017, the HA was finally recognized by the state, the communities gained full authority over its use and management. However, as the case in Kajang has shown (Fisher et al. 2020), the state also has had no more obligation to protect the forest. When a community is not strong enough to exclude outsiders, it can still lose its HA.

The fact that the HA was known and registered by MoEF shows the usefulness of keeping the story alive through good documentation, combined with proper acknowledgment to the involved/contributing parties. By getting legal status, the local communities' concerns and aspirations were addressed, including the needs to preserve timber for descendants and to prevent landslides.

When the HA boundary was violated by the oil palm company, the customary managing group reported to the District Forestry Service, which responded and took action to solve the problem. Local people who know where and to whom to report, and district officials who respond and support the communities are rather uncommon. In other studied locations, e.g., Kapuas Hulu, West Kalimantan, local people did not know how to address concerns regarding large private companies, while the district officials relied only on written complaints, rarely visiting communities (Moeliono et al. 2021; Yuliani et al. 2020). Colfer found a similar inability, without higher-level support, to manage powerful external actors in several East Kalimantan communities in 2019. The multistakeholder process of the *Perda Khusus* facilitated by the ACM project in Jambi has helped break the powerful communication barriers between villagers and the government. District officials learned how to respond constructively to perspectives that conflicted with their own. They felt encouraged to seek alternative ways (Yuliani 2004) of building understanding and improving program performance (Kusumanto 2006). District officials are aware and respect the legitimacy of the HA. Larson and Sarmiento Barletti (2019) also reported the usefulness of multistakeholder processes facilitated by the ACM team in Baru Pelepat from the people's perspectives – specifically building their capacities, encouraging their participation, along with simultaneous strong and consistent advocacy with the government to recognize local rights.

HA legal recognition can become a powerful tool to strengthen local rights and empower local customary people, though there are risks. Some fear that individual rights to and use of their resources may encourage individuals to seek economic compensation in return for their property (ownership or use rights), seen in Dharmasraya District, West Sumatra (Andiki, Sukirno and Prabandari 2019).

Although this 'customary community' has full rights to manage and use this forest, the requirement to have formal governmental recognition of the community as a *Masyarakat Hukum Adat* ('customary community') has been problematic. Though forest areas are managed by the MoEF, local and customary communities are administered under the Ministry of Home Affairs. Most, unlike the Kajang of Sulawesi or the Baduy of Java,[17] have been visually integrated into the larger society, while keeping many of their customary rules. Customary rules are by definition informal and only apply to members of the community. To secure and protect property rights, a formal decree that also binds outsiders is necessary. But this could result in a community losing the unique cultural basis of their rules.

Moreover, although the government is unlikely to allocate timber or oil palm concessions, neither are these forests considered seriously in spatial and development plans. While village forests (*hutan desa*, another government program) – officially part of state forests – are expected to be managed as a source of income through cooperatives or village enterprises (Government of Indonesia 2016), HAs are more often considered sacred forests to be protected through local wisdom. These various governmental social forestry schemes are part of the larger landscape and their good management, even their existence, will depend on the way the surrounding land is used.

Thus, while the ACM intervention was very successful at the village level to set up management rules and at the district level to obtain the *Perda Khusus*, the same multistakeholder processes need to be widened to landscape scale[18] if we expect to contribute substantively to addressing climate change; and they will have to include more consideration of environmental constraints. The mapping exercise carried out by the people of Baru Pelepat, including negotiating village boundaries with neighboring communities, has potentially laid the groundwork for landscape-related activities in collaboration with adjacent villages.[19] The early focus was strongly on mapping village administrative boundaries, and discussing rules for internal village management of resources, with minimal attempts to engender inter-village cooperation in managing forests. Making such inter-village collaboration a reality will require more efforts at subdistrict and district levels to encourage a wider, transboundary landscape perspective where activities in one village might impact the neighboring village (see, e.g., Feintrenie and Martini 2011).

Kusumanto et al., this volume, discuss the question of whether social learning as it unfolded in Jambi could evoke wider learning and thereby address the complexities inherent in sustainability processes at broader scales. Scholars note that the probability that sustainability transitions occur depends on whether an intervention can encourage learning applicable beyond the boundaries of its initial niche (van Mierlo and Beers 2020). The observation that actors at wider scales often co-opt villagers, resulting in conflict, suggests that relational and institutional changes would be locally welcome. ACM, an approach that effectively induces organizational and administrative changes, may be able to 'uplift' social learning to wider scales (Vinke-de Kruijf, Pahl-Wostl, and Knieper 2020). As discussed above, ACM has evidently paved the way for wider social-ecological changes.

Our use of ACM in this case also shows the 'adaptive' part of the methods. Decentralization processes in 2001 opened space for districts to formulate regulations. This opportunity was perceived by team participants who then pushed for the *Perda Khusus* to include not only the recognition of the customary communities but linked it directly to people's rights. Such adaptive adjustments to opportunities make ACM such a powerful tool.

Researchers' double and triple loop learning: what have we learnt?

The importance of researchers' learning is stressed by Evans, Larson and Flores (2020). A key feature of ACM, through its PAR framework, is the various loop learning of the researchers – also reflected in CIFOR's ACM series (Diaw, Prabhu, and Aseh 2009; Hartanto et al. 2003; Kusumanto et al. 2005; Mandondo, Prabhu, and Matose 2008; McDougall et al. 2010; see also Cronkleton, Evans, and Larson (2022) and Kusumanto et al., this volume). ACM activities provided many important lessons over the past two decades. Through double loop learning, our team found that multistakeholder processes are not simply meetings or forums involving multiple stakeholders; instead, they must be carefully designed to promote learning, information exchange, mutual understanding, break communication and power barriers, and strengthen joint decision-making and 'buy-in' from key stakeholders. Previous views and assumptions have made way for new perspectives and knowledges (see Hagmann et al., this volume).

The ACM approach can contribute to conflict management, facilitating divergent views that can morph into less complex and more shared understandings. Shared social learning can transform unstructured situations to more structured and manageable ones (Kusumanto 2007b). The anticipatory character of such conflict management lies in dealing with potential conflicts before they emerge and with existing ones before they become intractable.

Faced with highly diverse perspectives and interests, facilitating multistakeholder processes using the ACM framework clearly provided us with rich experiences. We learned (a) the need to recognize and anticipate different perspectives or signs of conflict; (b) how to respond to stakeholders' unrealistic expectations of the project or the project team; and (c) how to facilitate more meaningful participation to build participants' self-reliance, self-motivation and self-efficacy to achieve their goals (e.g., via Appreciative Inquiry; see Yuliani et al. 2015).

Furthermore, in our case, ACM proved to be an effective approach for researchers to engage in triple loop and double loop learning. We learned that the ways our research team and local stakeholders interacted and communicated were conditional on data and information sharing and interpretation. As such, we obtained insight on how to do research in the first place (triple loop): what research to do, what questions to ask and how to find answers. Additionally, the ACM framework encouraged us to revise and adapt research objectives, questions, methods, all requiring us to revisit our presumptions on local phenomena (double loop). Conditions were thus shaped for any research endeavor to become locally relevant and social-ecologically inclusive – as we also strove to understand the broader natural resource political economy and its trajectory.

These experiences led us to conclude that, besides the decrees or 'new policies', sometimes prioritized as key 'outputs or outcomes', the process itself is equally important; it too needs time and commitment. Developing democratic local institutions and building people's capacity to partake in public decision-making should become key priorities in obtaining policy and political support, engaging relevant government officials/policy makers in the process – all crucial if we are to effectively combat climate change in an equitable and sustainable manner.

Our ACM experiences in Jambi and those of others elsewhere have led to the recognition that facilitating upward to include broader scale actors in the grounded community efforts where we all began is essential. ACM's Jambi work has also provided tools and lessons learnt to move us forward to implementing broader scale action, e.g., in

- facilitating collective action and gender equity in Tanjung Jabung Barat District of Jambi Province (Komarudin, Siagian, and Colfer 2012; Wiliam-deVries 2006), in Uganda (Bomuhangi et al. 2022; Mukasa et al. 2022), and in Zimbabwe (Kozanayi et al., this volume),
- turning conflict into collaborative management of Nipa-Nipa Grand Forest Park in Southeast Sulawesi (Moeliono et al. 2015),
- getting legal recognition for the Ammatoa Kajang customary forest in South Sulawesi (Fisher et al. 2018, and this volume) and
- operationalizing the Landscape Approach in Ghana, Indonesia and Zambia (Reed, Ros-Tonen, and Sunderland 2020).

We remain impressed with the longevity of many of the efforts instituted some 20 years ago and committed to the importance of starting at the local level, understanding local realities and institutionalizing co-learning processes as we attempt to bring our own ACM learnings to broader scales and changed political-economic situations. If we are serious about a democratic world in which people's voices are heard, more progress is needed, however, on rendering institutions and individuals operating at broader scales more responsive to local voices, desires and capabilities, and more adaptive to changing social and economic circumstances.

Acknowledgment

The Jambi ACM activities in 2000–2003 were supported by ADB (grant no. RETA 5812), and in 2003–2006 by the UK Department for International Development/DFID-Multistakeholders Forestry Program/MFP. The writing of this chapter was supported by the International Climate Initiative of The German Federal Environment Ministry (IKI, grant No. 18_IV_084). The authors also wish to thank the people of Baru Pelepat, Government of Bungo District, local partners (Gita Buana Foundation and PSHK-ODA), WARSI, ACM Indonesia Steering Committee members Dani Munggoro, Prof. Mustofa Agung Sarjono, Sih Yuniati, Dr. Tetra Yanuariadi, Dr. Yunita T. Winarto, and our home institutions (CIFOR-ICRAF, TYK research & action consulting The Netherlands, Kemitraan/Partnership for Governance Reform, Indonesia Infrastructure Finance, Jambi Provincial Service on Food Crops Horticulture and Livestock, State Vocational High School 13 of Bungo, and Cornell University's Southeast Asia Program), Dr. Agus Justianto, Tri Nugroho, Iman Budi Setiawan, Dr. Amy Ickowitz, Wiwit Siswarini and Elida Nurrohmah. We also thank the two anonymous reviewers for their valuable, constructive input.

Notes

1 These local conservation goals were ascertained through a community visioning process, also seen in a video that the community and the ACM team prepared (Munggoro, Yuliani, and Indriatmoko 2006).
2 Such logging was considered illegal by the state, as a result of Law 41/1999 which solidified the classification of customary land as part of state forest. Conflict about ownership/tenure between the national government and local traditions is the source of an important global discourse (see Marfo et al. 2010, for example), and is still one of the major problems of forest governance in Indonesia. The 2012 Constitutional Court's Decree No. 35/2012 declared that customary territory should be taken out of the state forest.
3 For details, see Adnan et al. (2008), Kusumanto et al. (2005), Dobesto (2008), Marzoni (2008) and Pariyanto (2008).
4 ACM's participatory mapping overlapped with that of the World Bank financed Integrated Conservation and Development Project (ICDP). Baru Pelepat was one of the 134 villages adjacent to the Kerinci Seblat National Park (KSNP) which were involved in the ICDP. ICDP's mapping served as a tool to plan for biodiversity conservation and village development in KSNP's buffer zone. Differently, ACM saw participatory mapping more as a vehicle for stakeholder dialogue and democratic forest management through the customary and local institutions. Initial mapping activities were collaboratively carried out between the ICDP team (WWF, WARSI, KSNP and Bungo District Government) and the ACM team. Implementation of the ICDP in Baru Pelepat was from 2001 to 2002. At the time of ICDP's conclusion, the ACM team had just begun advocacy for district legal recognition for the HA.
5 Bing source imagery varies from 7 cm per pixel to a few meters per pixel depending on the global location. Detailed explanation of ground resolution, map scale, etc., can be found here: https://docs.microsoft.com/en-us/bingmaps/articles/bing-maps-tile-system. For our analysis, we had zoom level 17–18 (both Bing and Google Satellite), scale 1:3,000–1:1,500. From the link above, zoom level 17 is equal to spatial resolution 1.2 meters, and zoom level 18 is equal to 0.6 meters. We used these satellite images to see the most recent land cover, and as an additional reference to compare a spectral pattern in the latest Landsat image against higher resolution images. By doing this, we could identify that certain spectral patterns are analogous to particular land covers.
6 Ridwan (2018, 2019) notes the significant indigenous knowledge and management of natural resources in Baru Pelepat, while at the same time pointing out the adverse impacts of externally determined policies.

> The nation, in the name of increasing regional income, has given permission to plantation companies to manage part of the Datuk Sinaro Putih traditional forest land as an oil palm area. In the end, [Baru Peletpat's] river has become polluted, every rainy season a flood comes
>
> (*...mereka terus terancam karena kekuasaan mereka atas hutan adatnya tidak lagi penuh. Negara dengan mengatasnamakan untuk meningkatkan pendapatan daerah terus memberikan izin pada perusahaan perkebunan untuk mengolah sebagian lahan hutan adat Datuk Sinaro Putih menjadi kawasan perkebunan sawit. Akhirnya sungai mereka menjadi tercemar, setiap musim hujan tiba terjadi banjir.*).

While the HA has been maintained, by and large, the surrounding area, outside local control, has suffered mightily.

7 According to customary law, cultivated land and fallow, plus a 300 m belt around the land called 'kepalo umo' (Lit. 'ricefield head'), is considered private or family land. If these lands fall under the APL category (district government's land), the people could claim private or family ownership using *Sporadik* or SKT (*Surat Keterangan Tanah*/ Letter of Land Clarification) to get compensation. The Minangkabau are a matrilineal group. In this case, lands are typically inherited along the female line, though much

decision-making power rests with women's brothers (rather than their husbands). There are also variations (and resulting confusions) in this region related to the more patrilineal Jambi group and the bilateral Javanese.

8 Researchers focused on restoration routinely and globally complain about this common problem with such programs (see Katila et al., In Press; or Gramling 2021).

9 Sari (2013) who worked in two nearby villages summarizes gender dynamics thusly: "[W]omen in the village interacted closely with other women, and strong social norms and religious background were to a certain extent limiting the interaction between women and men" (p. 36); and

> Women were mostly not present in meetings about forest management even though invited due to heavy workloads at home. Participation exclusion of women prevail[ing] in Guguk village has hindered meaningful contribution of women to make inputs and influenc[e] the decision making process for forest management. (p. 65)

10 It is worth noting that the majority of 'newcomers' were of Javanese descent and part of a government transmigration project. Others spontaneously settled there and were of Sumatran ethnicities other than Minangkabau. On a national scale, due to public support to transmigrants, Javanese women tended to have higher socio-economic status than local women and spontaneous settlers. Within the Jambi context, however, the status of in-migrants suffered locally.

11 In 2014, Law No. 18/2004 was replaced by Law No. 39/2014. As the process of land and forest conversion into the oil palm plantation occurred in 2008–2010, it was regulated under the 2004 law.

12 This work is described in Komarudin et al. (2012); and Irawan et al. (2008), for instance, discuss one of the policy issues they addressed.

13 See also Liswanti et al., this volume, which discusses recent and broader views of the multistakeholder participants in the ACM process.

14 The CAPRI project, coordinated by Komarudin (Komarudin, Siagian, and Colfer 2012), attempted to involve women in multi-level decision-making and governance in the nearby village of Batu Kerbau, with some success in the mid-2000s; we have no evidence on whether these changes were sustained.

15 See Fisher et al. (2018, 2020); or Fisher et al., this volume, for a discussion of the multistakeholder processes undertaken there.

16 *Musrenbang* stands for *musyawarah rencana pembangunan* or community discussion on development planning. Musrenbang is an annual process during which residents meet to discuss the issues facing their communities and decide upon priorities for short-term improvements.

17 These communities are 'visible' because they wear particular clothing and insist on particular customary rules more vocally than do most Indonesian villagers.

18 The CAPRI project did just that, but when the project ended, so did much of the collaboration among district agencies – though the findings of Liswanti et al., this volume, suggest that some of the collaboration and cooperation at that level continued. In any event, such collaboration needs to be institutionalized.

19 Colfer and Pfund (2011) provide a number of cases, including in Jambi, showing the complexity of actually doing this.

References

Adebayo, Tomiwa Sunday, Gbenga Daniel Akinsola, Dervis Kirikkaleli, Festus Victor Bekun, Sukru Umarbeyli, and Oseyenbhin Sunday Osemeahon. 2021. "Economic Performance of Indonesia Amidst CO_2 Emissions and Agriculture: A Time Series Analysis." *Environmental Science and Pollution Research* 28 (35): 1–15.

Adnan, Hasantoha, Djuhendi Tadjudin, Elizabeth Linda Yuliani, Heru Komarudin, Dicky Lopulalan, Yuliana L. Siagian, and Dani Wahyu Munggoro. 2008. *Belajar dari Bungo: Mengelola Sumber Daya Alam di Era Desentralisasi (Learning from Bungo: Managing Natural Resources in the Decentralization Era)*. Bogor, Indonesia: CIFOR.

Anderson, Patrick. 2013. "Free, Prior and Informed Consent? Indigenous Peoples and the Palm Oil Boom in Indonesia." In *The Palm Oil Controversy in Southeast Asia. A Transnational Perspective*, edited by Oliver Pye and Jayati Bhattacharya, 244–258. Singapore: Institute of Southeast Asian Studies.

Andiki, Febri, Sukirno, and Adya Paramita Prabandari. 2019. "Peralihan Hak Tanah Ulayat di Kabupaten Dharmasraya (Transfer of Customary Land Rights in Dharmasraya District)." *Notarius* 12 (2): 856–865.

Andriarsi, Melati Kristina. 2020. *Sebaran Masyarakat Adat* (*Distribution of Customary Communities*), edited by Padjar Iswara. Jakarta, Indonesia: Aliansi Masyarakat Adat Nusantara/The Alliance of the Indigenous Peoples of the Archipelago. https://katadata.co.id/padjar/infografik/5f8030631f92a/sebaran-masyarakat-adat.

Asba, A.R., F. Rahman, and A.L. Evita. 2019. "Save the Forest and Biodiversity: A Cultural Anthropology Perspective on Kalumpang Customary to Preserve their Ecological Living." *IOP Conference Series: Earth and Environmental Science* 270: 012005.

Awang, SanAfri. 2006. *Sosiologi Pengetahuan Deforestasi: Konstruksi Sosial dan Perlawanan* (*The Sociology of Deforestation: Social Construction and Resistance*). Yogyakarta, Indonesia: Debut Press.

Bennett, Andrew F. 2003. *Linkages in the Landscape: The Role of Corridors and Connectivity in Wildlife Conservation* (2nd ed.). Gland, Switzerland and Cambridge, UK: IUCN.

BirdLife International. 2020. *Argusianus Argus*. The IUCN Red List of Threatened Species 2020: e.T22725006A183255774.

Bodin, Örjan, Maria Tengö, Anna Norman, and Thomas Elmqvist. 2006. "The Value of Small Size: Loss of Forest Patches and Ecological Thresholds in Southern Madagascar." *Ecological Applications* 16 (2): 440–451. doi: 10.1890/1051–0761(2006)016[0440:TVOSSL]2.0.CO;2.

Bomuhangi, Allan, Abwoli Yabezi Banana, Jimmy Bushoborozi, Concepta Mukasa, Alice Tibazalika, and Esther Mwangi. 2022. "Gender and Adaptive Collaborative Management in Forested Ugandan Landscapes." In *Adaptive Collaborative Management of Forest Landscapes: Villagers, Bureaucrats, and Civil Society*, edited by Carol J. Pierce Colfer, Ravi Prabhu, and Anne Larson, 87–104. London: Earthscan and Routledge.

Brockhaus, Maria, Krystof Obidzinski, Ahmad Dermawan, Yves Laumonier, and Cecilia Luttrell. 2012. "An Overview of Forest and Land Allocation Policies in Indonesia: Is the Current Framework Sufficient to Meet the Needs of REDD+?" *Forest Policy and Economics* 18: 30–37.

BRWA (Badan Registrasi Wilayah Adat/Customary Territory Registration Agency). 2021. *Infografis Status Pengakuan Wilayah Adat di Indonesia* (*Infographic of Status of Recognition of Customary Territories in Indonesia*). Jakarta, Indonesia: Badan Registrasi Wilayah Adat/Customary Territory Registration Agency.

Colchester, Marcus, Norman Jiwan, and Sophie Chao. 2013. "Conflict or Consent? Oil Palm Expansion and Community Rights." Presented in The Annual World Bank Conference on Land and Poverty. Washington, DC: World Bank.

Colchester, Marcus, Patrick Anderson, Norman Jiwan, Andiko, and Su Mei Toh. 2009. "Report of an Independent Investigation into the Effectiveness of the Application of High Conservation Value Zoning in Palm Oil Development in Indonesia." Public discussion document. Jakarta, Indonesia: Forest Peoples Programme, Sawit Watch, Wild Asia and HuMa.

Colfer, Carol J. Pierce. 2021. *Masculinities in Forests: Representations of Diversity*. London: Earthscan and Routledge.

Colfer, Carol J. Pierce, and Jean-Laurent Pfund, eds. 2011. *Collaborative Governance of Tropical Landscapes*. London: Earthscan and CIFOR.

Colfer, Carol J. Pierce, Richard G. Dudley, and Robert Gardner. 2008. "Forest Women, Health and Childbearing." In *Human Health and Forests: A Global Overview of Issues, Practice and Policy*, edited by Carol J. Pierce Colfer, 113–133. London: Earthscan.

Cronkleton, Peter, Kristen Evans, and Anne M. Larson. 2022. "Capacity Building for ACM: Lessons Learned from Training in Distinct Contexts." In *Adaptive Collaborative Management Revisited*, edited by Carol J. Pierce Colfer, Ravi Prabhu, and Anne Larson, 134–157. London: Earthscan and Routledge.

Deere, Nicolas J., Gurutzeta Guillera-Arroita, Philip J. Platts, Simon L. Mitchell, Esther L. Baking, Henry Bernard, Jessica K. Haysom, G. Reynolds, D.J. Seaman, Z.G. Davies, and M.J. Struebig. 2020. "Implications of Zero-Deforestation Commitments: Forest Quality and Hunting Pressure Limit Mammal Persistence in Fragmented Tropical Landscapes." *Conservation Letters* 13 (3): e12701.

Diaw, M. Chimere, Ravi Prabhu, and Tony Aseh, eds. 2009. *In Search for Common Ground: Adaptive Collaborative Management of Forest in Cameroon*. Bogor, Indonesia: CIFOR.

Diaw, Mariteuw Chimère, and Trikurnianti Kusumanto. 2005. "Scientists in Social Encounters: The Case for an Engaged Practice of Science." In *The Equitable Forest: Diversity, Community and Resource Management*, edited by Carol J. Pierce Colfer, 72–109. Washington, DC: Resources for the Future.

Dobesto, Ismal. 2008. "Jalan Panjang Partisipasi: Proses dan Pembelajaran dalam Penyusunan Peraturan Desa secara Partisipatif (The Long Road of Participation: Processes and Learning in Participatory Development of Village Regulations)." In *Belajar Dari Bungo: Mengelola Sumberdaya Alam di Era Desentralisasi (Learning from Bungo: Managing Natural Resources in the Decentralization Era)*, edited by Hasantoha Adnan, Djuhendi Tadjudin, Elizabeth Linda Yuliani, Heru Komarudin, Dicky Lopulalan, Yuliana L. Siagian, and Dani Wahyu Munggoro, 419–443. Bogor, Indonesia: CIFOR.

Endah, Riya Dharma Datuk Rangkayo. 2008. "Hutan Adat Batu Kerbau: Sisa-sisa Kearifan Lokal." In *Belajar dari Bungo: Mengelola Sumber Daya Alam di Era Desentralisasi*, edited by Hasantoha Adnan, Djuhendi Tadjudin, Elizabeth Linda Yuliani, Heru Komarudin, Dicky Lopulalan, Yuliana L. Siagian, and Dani Wahyu Munggoro, 65–82. Bogor, Indonesia: CIFOR.

Evans, Kristen, Anne M. Larson, and Selmira Flores. 2020. "Learning to Learn in Tropical Forests: Training Field Teams in Adaptive Collaborative Management, Monitoring and Gender." *International Forestry Review* 22 (2): 189–198.

Feintrenie, Laurène, and Endri Martini. 2011. "Role of the District Government in Directing Landscape Dynamics and People's Futures: Lessons Learnt from Bungo District, in Jambi Province." In *Collaborative Governance of Tropical Landscapes*, edited by Carol J. Pierce Colfer and Jean-Laurent Pfund, 55–78. London: Earthscan.

Fisher, Micah R., Moira Moeliono, Agus Mulyana, E. Linda Yuliani, Andi Adriadi, Kamaluddin, Junaid Judda, and M. Aalif K. Sahide. 2018. "Assessing the New Social Forestry Project in Indonesia: Recognition, Livelihood and Conservation?" *International Forestry Review* 20 (3): 346–361.

Fisher, Micah R., Timothy Workman, Agus Mulyana, Balang Institute, Moira Moeliono, E. Linda Yuliani, Carol J. Pierce Colfer, and Urban El Fatih Bani Adam. 2020. "Striving for PAR Excellence in Land Use Planning: Multi-stakeholder Collaboration on Customary Forest Recognition in Bulukumba, South Sulawesi." *Land Use Policy* 99: 102997.

Franklin, Neil, Bastoni, Sriyanto, D. Siswomartono, J.D. Manansang, and R. Tilson. 1999. "Last of the Indonesian Tigers: A Cause of Optimism." In *Riding the Tiger: Tiger Conservation in Human-dominated Landscape*, edited by J. Seidensticker, S. Christie, and P. Jackson, 113–125. Cambridge, UK: Cambridge University Press.

Government of Indonesia. 2004. *Undang-Undang Republik Indonesia No. 18/2004 tentang Perkebunan (The Indonesian Act No. 18/2004 on Plantations)*. Jakarta, Indonesia: House of Representative of the Republic of Indonesia.

Government of Indonesia. 2016. *Regulation of the Minister of Environment and Forestry No. 83 on Social Forestry*. Jakarta, Indonesia: Ministry of Environment and Forestry.

Gramling, Carolyn. 2021. "Why Planting Tons of Trees Isn't Enough to Solve Climate Change." *ScienceNews*. https://www.sciencenews.org/article/planting-trees-climate-change-carbon-capture-deforestation.

Hartanto, Herlina, Ma. Cristina Lorenzo, Cecil Valmores, Lani Arda-Minas, Linda Burton, and Ravi Prabhu. 2003. *Learning Together: Responding to Change and Complexity to Improve Community Forests in the Philippines*. Bogor, Indonesia: CIFOR.

Indriatmoko, Yayan. 2001. Planning of Community Action Research. Internal Report. Unpublished. Bogor, Indonesia: CIFOR.

Indriatmoko, Yayan, Elizabeth Linda Yuliani, Hasantoha Adnan, Eddy Harfia Surma, and Novasyurahati. 2006. Reassessment of Criteria and Indicator of Sustainable and Equitable Forest Management in Baru Pelepat Village. Internal Report. Unpublished. Bogor, Indonesia: CIFOR.

Irawan, Deddy, Umar Hasan and Heru Komarudin. 2008. "Penataan Ruang untuk Memperkuat Hak Properti Masyarakat (Spatial Planning to Strengthen Community Property Rights)." In *Belajar Dari Bungo: Mengelola Sumberdaya Alam di Era Desentralisasi (Learning from Bungo: Managing Natural Resources in the Decentralization Era)*, edited by Hasantoha Adnan, Djuhendi Tadjudin, Elizabeth Linda Yuliani, Heru Komarudin, Dicky Lopulalan, Yuliana L. Siagian, and Dani Wahyu Munggoro, 388–403. Bogor: CIFOR.

Katila, Pia, Carol J. Pierce Colfer, Wil Dejong, Glenn Galloway, Pablo Pacheco, and Georg Winkel (eds). In press. *Restoring Forests and Trees for Sustainable Development: Policies, Practices, Impacts and Ways Forward*. England: Oxford University Press.

Komarudin, Heru, Yulia L. Siagian, and Carol J. Pierce Colfer. 2012. "The Role of Collective Action in Securing Property Rights for the Poor: A Case Study in Jambi Province, Indonesia" (Chapter 8). In *Collective Action and Property Rights for Poverty Reduction*, edited by H. Markelova, E. Mwangi, and R. Meinzen-Dick, 235–269. Philadelphia: University of Pennsylvania Press.

Kusumanto, Trikurnianti. 2006. Social Learning, Policy Making and Customary Forest: Lessons from Jambi (Central Sumatra). Research Report. Unpublished. Bogor, Indonesia: Yayasan Gita Buana, PSHK-ODA, MFP-UK, and CIFOR.

Kusumanto, Trikurnianti. 2007a. "Learning to Monitor Political Processes for Fairness in Jambi, Indonesia." In *Negotiated Learning: Collaborative Monitoring in Forest Resource Management*, edited by Irene Guijt, 124–134. Washington, DC: Resources for the Future.

Kusumanto, Trikurnianti. 2007b. "Shaping Opportunities for Improving Forest Quality and Community Livelihoods in Central Sumatra and East Kalimantan, Indonesia." *Adaptive Collaborative Management of Community Forests in Asia: Experiences from Nepal, Indonesia and the Philippines*, edited by Robert Fisher, Ravi Prabhu, and Cynthia McDougall, 93–133. Bogor, Indonesia: CIFOR.

Kusumanto, Trikurnianti, and Effi Permatasari. 2002. Up-and-Down the Ladder: A Jambi Case Study of Policy-Related Information Flows. Unpublished Report. Bogor: CIFOR.

Kusumanto, Trikurnianti, Elizabeth Linda Yuliani, Phil Macoun, Yayan Indriatmoko, and Hasantoha Adnan. 2005. *Learning to Adapt: Managing Forests Together in Indonesia.* Bogor, Indonesia: CIFOR, YGB, PSHK-ODA.

Kusumanto, Trikurnianti, and Yayan Indriatmoko. 2001. Stakeholder Analysis. In *Context Study Report: Adaptive Collaborative Management (ACM) Research in Jambi, Indonesia.* Internal Report. Unpublished. Bogor, Indonesia: CIFOR.

Larson, Anne, and Juan Pablo Sarmiento Barletti. 2019. The Role of Multi-stakeholder Forums in Subnational Jurisdictions: The Adaptive Collaborative Management Multi-Stakeholder Forum, Jambi, Indonesia. Bogor, Indonesia: CIFOR. https://www.cifor.org/publications/pdf_files/flyer/7453-Flyer.pdf.

Larson, Anne M., Deborah Barry, and Ganga Ram Dahal. 2010. "Conclusions and Reflections for the Future of Forest Tenure Reform." In *Forests for People: Community Rights and Forest Tenure Reform*, edited by Anne M. Larson, Deborah Barry, Ganga Ram Dahal, and Carol J. Pierce Colfer, 210–223. London and Washington, DC: Earthscan.

Linkie, M., H.T. Wibisono, D.J. Martyr, and S. Sunarto. 2008. *Panthera tigris* ssp. sumatrae. The IUCN Red List of Threatened Species 2008: e.T15966A5334836.

Li, Tania M., and Pujo Semedi. 2021. *Plantation Life: Corporate Occupation in Indonesia's Oil Palm Zone.* Durham, NC: Duke University Press.

Mandondo, Alois, Ravi Prabhu, and Frank Matose. 2008. *Coping Amidst Chaos: Studies on Adaptive Collaborative Management from Zimbabwe.* Bogor, Indonesia: CIFOR.

Marfo, Emmanuel, Carol J. Pierce Colfer, Bocar Kante, and Silvel Elias. 2010. "From Discourse to Policy: The Practical Interface of Statutory and Customary Land and Forest Rights." In *Forest for People: Community Rights and Forest Tenure Reform*, edited by Anne Larson, Deborah Barry, Ganga Ram Dahal, and Carol J. Pierce Colfer, 69–89. London, UK: Earthscan and CIFOR.

Maryani, Retno, Deden Djaenudin, Mega Lugina, Indartik, Moira Moeliono, and E. Linda Yuliani. 2021. "Perspectives on Landscape Approaches in Indonesian Policies." In *P3SEKPI Infobrief.* Bogor, Indonesia: Center for Research and Development for Social, Economy, Policy and Climate Change of the Forestry Research Development and Innovation, Ministry Environment and Forestry (P3SEKPI) and CIFOR.

Marzoni. 2008. "Menata Ruang Desa: Jalan menuju Komunikasi dan Kolaborasi." In *Belajar dari Bungo: Mengelola Sumberdaya Alam di Era Desentralisasi (Learning from Bungo: Managing Natural Resources in the Decentralization Era)*, edited by Hasantoha Adnan, Djuhendi Tadjudin, Elizabeth Linda Yuliani, Heru Komarudin, Dicky Lopulalan, Yuliana L. Siagian, and Dani Wahyu Munggoro, 444–459. Bogor, Indonesia: CIFOR.

McDougall, Cynthia, Hemant Ojha, Maniram Banjade, Bishnu Hari Pandit, Tara Bhattarai, Manik Maharjan, and Sushila Rana. 2010. *Forests of Learning: Experiences from Research on an Adaptive Collaborative Approach to Community Forestry in Nepal.* Bogor, Indonesia: CIFOR.

Moeliono, Ilya. 2006. *Laporan Pengkajian Program ACM di Kabupaten Bungo, Jambi.* Assessment Report of ACM Program in Bungo District, Jambi. Jakarta, Indonesia: DfID-MFP.

Moeliono, Moira, Elizabeth Linda Yuliani, Valentinus Heri, and Nehemia Ngilah. 2021. Report on Analysis of Stakeholders, Multi-Stakeholder Platforms, and Capacity Building Needs. COLANDS Baseline Studies Report. Unpublished. Bogor: CIFOR.

Moeliono, Moira, and Godwin Limberg. 2012. "Tenure and Adat in Malinau." In *The Decentralization of Forest Governance: Politics, Economics and the Fight for Control of Forests in Indonesian Borneo*, edited by Moira Moeliono, Eva Wollenberg, and Godwin Limberg, 221–239. London: Earthscan.

Moeliono, Moira, Hasantoha Adnan, Agus Mulyana, and E. Linda Yuliani. 2015. "Beneath a Leaking (Legal) Umbrella: An Experiment in Collaborative Management of the TAHURA (Grand Forest Park) Nipa-Nipa." *Brief* 54. Bogor, Indonesia: World Agroforestry Centre (ICRAF) Southeast Asia Regional Program. http://apps.worldagroforestry.org/sea/Publications/files/policybrief/PB0105-15.pdf.

Mukasa, Concepta, Alice Tibazalika, Abwoli Banana, Esther Mwangi, and Tendayi Mutimukuru-Maravanyika. 2022. "Strengthening Women's Tenure Rights and Participation in Community Forestry in Central Uganda." In *Adaptive Collaborative Management of Forest Landscapes: Villagers, Bureaucrats, and Civil Society*, edited by Carol J. Pierce Colfer, Ravi Prabhu, and Anne Larson, 106–132. London: Earthscan.

Munggoro, Dani W., Elizabeth Linda Yuliani, and Yayan Indriatmoko. 2006. *The People of Baru Pelepat: Learning to Change* (video CD). Bogor, Indonesia: CIFOR.

Nyhus, Philip J., and Ronald Tilson. 2004. "Characterizing Human-Tiger Conflict in Sumatra, Indonesia: Implications for Conservation." *Oryx* 38 (1). doi: 10.1017/s0030605304000110.

Pariyanto. 2008. "Jalan Menikung: Proses Multipihak dalam Pengakuan Hutan Adat (A Windy Road: Multistakeholder Processes in Formal Recognition of Customary Forest)." In *Belajar dari Bungo: Mengelola Sumber Daya Alam di Era Desentralisasi*, edited by Hasantoha Adnan, Djuhendi Tadjudin, Elizabeth Linda Yuliani, Heru Komarudin, Dicky Lopulalan, Yuliana L. Siagian, and Dani Wahyu Munggoro, 404–418. Bogor, Indonesia: CIFOR.

Persoon, Gerard A., and Ekoningtyas Margu Wardani. 2017. "Projected Futures of the Orang Rimba in Sumatra (Indonesia). Heritage and Rights of Indigenous Peoples." *Archaeological Studies Leiden University* 39: 61–77.

Purwanto, Edy. 2019. *HCV Mainstreaming in Indonesia: Tropenbos Indonesia's Experience (2004–2017)*. Bogor, Indonesia: Tropenbos Indonesia.

Reed, James, M.A.F. Ros-Tonen, and T.C.H. Sunderland. 2020. *Operationalizing Integrated Landscape Approaches in the Tropics*. Bogor, Indonesia: CIFOR.

Ridwan. 2018. "Dualisme Kepemimpinan Adat Di Desa: Dilema Kekuasaan Adat Lokal Pada Masyarakat Hukum Adat Datuk Sinaro Putih, Dusun Batu Kerbau, Kecamatan Pelepat, Kabupaten Bungo (The Dualism of Customary Village Leadership: A Dilemma for Local Customary Authority in the Customary Community Datuk Sinaro Putih, Batu Kerbau Hamlet, Pelepat County, Bungo District)." *Jurnal Administrasi Sosial dan Humaniora* 3 (1): 90–98. doi: 10.5281/zenodo.2447097.

Ridwan, Isman. 2019. "Masyarakat Hukum Adat Datuk Sinaro Putih: Pasang Surut Kekuasaan Adat di Tengah Hegemoni Negara (Customary Community of Datuk Sinaro Putih: The Ebb and Flow of Customary Authority amidst National Hegemony)." *Niara* 12 (1): 1–8.

Sari, Intan Maya. 2013. "Community Forests at a Crossroads: Lessons Learned from Lubuk Beringin Village Forest and Guguk Customary Forest in Jambi Province – Sumatra, Indonesia." PhD diss., University of Oslo.

Sarmiento Barletti, Juan Pablo. 2022. "Learning from Adaptive Collaborative Management: A Participatory Tool to Support Adaptive and Reflective Learning in Multi-Stakeholder Forums." In *Adaptive Collaborative Management of Forest Landscapes: Villagers, Bureaucrats, and Civil Society*, edited by Carol J. Pierce Colfer, Ravi Prabhu, and Anne Larson, 159–184. London: Earthscan.

Siagian, Yuliana L., Brihannala Morgan, Yentirizal, and Neldy savrino. 2005. "Strengthening Women's Participation through Collective Action for Inclusive Decision Making Processes: Lessons Learned from Two Villages in Jambi Province, Sumatra, Indonesia."

International Research Workshop on "Gender and Collective Action", 17–21 October 2005, Chiang Mai, Thailand.

Singleton, I., and C.P. van Schaik. 2001. "Orangutan Home Range Size and Its Determinants in a Sumatran Swamp Forest." *International Journal of Primatology* 22: 877–911. doi: 10.1023/A:1012033919441.

Syamsuddin, Neldysavrino, Heru Komarudin, and Yulia L. Siagian. 2007. "Are Community Aspirations Being Accommodated in Development Plans? A Lesson from Collective Action in Jambi." *CIFOR Governance Briefs* 34 (34): 12.

van Mierlo, Barbara, and Pieter J. Beers. 2020. "Understanding and Governing Learning in Sustainability Transitions: A Review." *Environmental Innovation and Societal Transitions* 34: 255–269. doi: 10.1016/j.eist.2018.08.002.

Vinke-de Kruijf, J., C. Pahl-Wostl, and C. Knieper. 2020. "Wider Learning Outcomes of European Climate Change Adaptation Projects: A Qualitative Comparative Analysis." *Environmental Innovation and Societal Transitions* 34: 270–297. doi: 10.1016/j.eist.2018.09.004.

Wiliam-de Vries, Dede. 2006. *Gender Bukan Tabu: Catatan Perjalanan Fasilitasi Kelompok Perempuan di Jambi.* Bogor: CIFOR.

Wiliam-de Vries, D., and N. Sutarti. 2006. "Gender Equity: Revealing the Reality for the Women of Jambi." *Governance Brief* 29. Bogor, Indonesia: CIFOR. https://www.cifor.org/publications/pdf_files/GovBrief/GovBrief0624E.pdf.

Wollenberg, Eva, and Hariadi Kartodihardjo. 2002. "Devolution and Indonesia's New Forestry Law." In *Which Way Forward? People, Forests and Policymaking in Indonesia*, edited by Carol J. Pierce Colfer and Ida Aju Pradnja Resosudarmo, 81–109. Washington, DC: Resources for the Future.

Yasmi, Yurdi, and Heiner Schanz. 2010. "Managing Conflict Escalation in Forestry: Logging Versus Local Community Interests in Baru Pelepat Village, Sumatra, Indonesia." *International Journal of Biodiversity Science, Ecosystem Services and Management* 6 (1–2): 43–51.

Yue, Sam, Jedediah F. Brodie, Elise F. Zipkin, and Henry Bernard. 2015. "Oil Palm Plantations Fail to Support Mammal Diversity." *Ecological Applications* 25 (8): 2285–2292.

Yuliani, Elizabeth Linda. 2004. Decentralization, Deconcentration and Devolution: What Do They Mean? [Lists of Definitions of Terms Used in Papers Presented at the Interlaken Workshop on Decentralization, 27–30 April 2004, Interlaken, Switzerland]. Bogor, Indonesia: CIFOR.

Yuliani, Elizabeth Linda, Hasantoha Adnan, Carol J. Pierce Colfer, and Yayan Indriatmoko. 2015. "Problem-solving Versus Appreciative Inquiry Approaches in Community-based Conservation." *Forests, Trees and Livelihoods* 24 (2): 97–111.

Yuliani, Elizabeth Linda, Hasantoha Adnan, Ramadhani Achdiawan, Denny Bakara, Valentinus Heri, Jim Sammy, M. Agus Salim, and Terry Sunderland. 2018. "The Roles of Traditional Knowledge Systems in Orangutan (*Pongo* spp.) and Forest Conservation: A Case Study of Danau Sentarum, West Kalimantan, Indonesia." *Oryx* 52 (1): 156–165. doi: 10.1017/s0030605316000636.

Yuliani, Elizabeth Linda, Wouter T. De Groot, Luuk Knippenberg, and D.O. Bakara. 2020. "Forest or Oil Palm Plantation? Interpretation of Local Responses to the Oil Palm Promises in Kalimantan, Indonesia." *Land Use Policy* 96: 104616.

Introduction to Chapter 3

This chapter continues analyzing what has happened over the years in the Jambi site, broadly speaking. Where Chapter 2 focused on what has happened in the community of Baru Pelepat, a focus of CIFOR's early ACM activity; here, Liswanti, Tamara, and Arwida shift gears, examining what happened at the *Kabupaten* (district) level over these same years, following multi-stakeholder efforts to coordinate and share information related more to the district. The multi-stakeholder group included active participation and agency on the part of NGOs, universities, government officials, and community members, as well as the World Agroforestry Center (ICRAF). The CIFOR ACM team was part of this group, playing an important role, along with a number of other people and institutions.

This chapter represents a significant shift in the researchers' 'gaze'. Whereas Chapter 2 was analyzed and written by participants in the original research, focused specifically on the community, the research discussed in Chapter 3 was conducted, analyzed, and written by researchers who had not been involved initially – providing a valued external view. Chapter 3 also focused on a broader scale (the district as much as the village), one which had supported the community of Baru Pelepat in formalizing their village regulation (*Perdes*) and district regulation (*Perda*). One interesting disjuncture comes partially at least from this difference in experience and perspective: Whereas those researchers actively involved in the initial research considered the decision to try to maintain the forest to be one considered and taken by the community itself, the more recent researchers seem to consider that that decision did not come from the community. They emphasized the local community's focus on economic considerations (e.g., logging); and the more proactive and directive encouragement from the organizers and other stakeholders through awareness raising about maintaining the forest in a sustainable way. It may well be that local community members considered that they were making a more independent determination, whereas district-level stakeholders emphasized their support and encouragement of this decision. The bottom-up orientation of the original research has morphed, through time, into a more top-down process – one in which community members are consulted and participate but where the work is defined and often led by the external actors. Both assessments agreed, however, that both ACM organizers and MSF participants built strong communication that included routine dialogue with the community on the impact of logging and other future forest activities.

DOI: 10.4324/9781003325932-4

The degree to which local decisions are truly local is something of a sticking point in almost any participatory approach. How neutral can a facilitator actually be? In the CIFOR cases, team members were always cognizant of the fact that CIFOR was an institution concerned (broadly) with forests. We did not feel free to address absolutely anything that a community might choose to address (like Matuk et al. 2020), despite our efforts to be truly as non-directive as possible, given our affiliation. The constraints imposed by donors were often still more restrictive (see Colfer et al. 2011 or Ojha, Hall, and Sulaiman 2013). In the eyes of many villagers and other inhabitants of remote, forested areas, the power and prestige assigned to external (educated, urban, sometimes foreign) facilitators also lend credence – willy nilly – to whatever they may express. Villagers are likely to be attuned to subtle, even inadvertent indicators of the preferences of (often beloved) facilitators, whom they can also see as sources of funds and other benefits (see Johnson and Pokorny 2022). This adds both nuance and uncertainty to the determination of who is actually making decisions. Such uncertainties are likely to be just as evident when climate change is the focus of adaptive and collaborative effort. The passion facilitators may feel, the urgency institutions may reiterate, will encourage actors at various levels to *insist*, where a sustainable process requires instead *listening* first and then *adapting creatively and collaboratively*.

References

Colfer, Carol J. Pierce, Etienne Andriamampandry, Stella Asaha, Emmanuel Lyimo, Endri Martini, Jean Laurent Pfund, and John Watts. 2011. "Participatory Action Research for Catalyzing Adaptive Management: Analysis of a 'Fits and Starts' Process." *Journal of Environmental Science and Engineering* 5, no. 1: 28–43.

Johnson, James, and Benno Pokorny. 2022. "Local People's Perspectives on Action Learning: Impressions from the Amazon." In *Adaptive Collaborative Management of Forest Landscapes: Villagers, Bureaucrats and Civil Society*, edited by Carol J. Pierce Colfer, Ravi Prabhu, and Anne Larson, 31–60. London: Earthscan and Routledge.

Matuk, Fernanda Ayaviri, Esther Turnhout, Luuk Fleskens, Eufran Ferreira do Amaral, Moacir Haverroth, and Jelle Hendrik Behagel. 2020. "Allying Knowledge Integration and Co-Production for Knowledge Legitimacy and Usability: The Amazonian SISA Policy and the Kaxinawá Indigenous People Case." *Environmental Science and Policy* 112: 1–9.

Ojha, Hemant, Andy Hall, and Rasheed V. Sulaiman, eds. 2013. *Adaptive Collaborative Approaches in Natural Resource Governance: Rethinking Participation, Learning and Innovation*, edited by Hemant Ojha, Andy Hall, and Rasheed V. Sulaiman, 48–102. London: Earthscan and Routledge.

3 Trust building in a multi-stakeholder forum in Jambi, Indonesia

Nining Liswanti, Ade Tamara and Shintia Arwida

Introduction

A participatory approach has been understood as one of the tools to address complex issues such as land use and land use change (LULUC) (Bourgoin et al. 2012; Kearney et al. 2007; Sayer et al. 2013). There is an increasing interest in using participatory approaches to address climate change. Increased public participation instead of a top-down approach in policy making or management decisions has contributed to more effective problem solving by involving those likely to be affected by the decisions (Bourgoin et al. 2012). Hence, engaging multiple stakeholders has also been perceived to lead to more sustainable and equitable results (Berkes 2010; Davies and White 2012). Likewise, support toward participatory decision making from donors and practitioners persists, leading to the continued application of such approaches, including multi-stakeholder forums (MSFs) (Sarmiento Barletti and Larson 2019a).

In Adaptive Collaborative Management (ACM), MSFs are commonly used to support participatory decision making and function as a mechanism for accomplishing some of ACM's goals. They hold much promise for addressing climate change as well. The MSFs are institutional coordination mechanisms that enable discussions, negotiations, and joint planning between stakeholders from various sectors in a given landscape to seek actions to address common problems or to achieve common goals for their benefit (Sarmiento Barletti et al. 2020; see also Fisher et al., this volume). More importantly, MSFs allow for negotiation between historically underrepresented actors (e.g., indigenous peoples and local communities) and powerful actors (e.g., corporations, NGOs, and government agencies; Edmunds and Wollenberg 2002). The benefits range from upholding human rights and participatory democracy – which defend the key roles that local people play in the sustainability of policies and projects – to improving coordination among different sectors (Bäckstrand 2006; Reed 2008; Reed et al. 2009). Findings from the literature show how MSFs can promote equal opportunities for speaking, listening, negotiating, and planning together among participants, as well as allowing knowledge transfer and paving the way to solving common challenges (Rondinelli and London 2003; Selsky and Parker 2005).

Both ACM and MSFs have been driven by similar optimistic goals, including democratic principles and ideas like justice, equity, and local communities'

DOI: 10.4324/9781003325932-5

empowerment (Sarmiento Barletti 2022). "Inclusivity and shared decision making" are central for both approaches (Sarmiento Barletti 2022). McDougall and Ojha (2022) even argued the need to reframe ACM into adaptive collaborative governance due to its focus on improving decision making and equity rather than "technical management". Collaborative strategies, for instance, capacity building, were thus necessary especially in engaging those who were not familiar with the concept or had never been involved in such a process, be it an ACM process or even a discussion with various actors present (Cronkleton, Evans, and Larson 2022). These strategies are needed to overcome challenges resulting from power imbalances (McDougall and Ojha 2022).

For MSFs specifically, critiques have arisen following their implementation, such as their failure to challenge inequality, even legitimizing unequal practices (Edmunds and Wollenberg 2002; Sarmiento Barletti et al. 2020). The notion of "bringing everyone to the same table" is not enough without acknowledging power differences and developing strategies to address them (Edmunds and Wollenberg 2002). Furthermore, MSFs regularly adopt soft mechanisms for the implementation of agreements with no legal obligations or sanctions, which can lead to increased conflict among participants (Mena and Palazzo 2012). Therefore, a nurturing environment for collaboration is needed to face such challenges.

Here, we focus on building the trust that has been widely recognized as an element that can positively influence collaboration in any participatory process (Alfantoukh, Ruan, and Durresi 2018; Ansell and Gash 2008; Berkes 2009; Edelenbos and Klijn 2007; Emerson, Nabatchi, and Balogh 2011; Fisher and Brown 1989; Stern and Coleman 2015). Sarmiento Barletti et al.'s (2020) study on models of participation in MSFs found power relations and trust as conditions affecting quality of participation, specifically for forums that focus on granting local communities more control over natural resources – as is likely to be important if we hope to involve them meaningfully in climate change mitigation and adaptation. Armitage et al. (2009) have also mentioned trust building, among others, as factors that influence deliberation and learning processes. In addition, trust has been understood as a precondition for transparency and shared understanding, allowing for development of knowledge of local contexts and willingness to work on solving common problems (Komarudin et al. 2011; Kusters et al. 2018).

Given how scholars have emphasized the importance of trust in collaboration, this chapter aims to explore more deeply how trust contributes to the development of a nurturing environment for collaboration in MSFs. The MSF studied focused on supporting local communities in their decision to protect customary forest and natural resources (see Yuliani et al., this volume). We identify conditions and activities that contribute to the development of trust and the implications of trust for MSF processes and achievement of outcomes. This study was part of a Center for International Forestry Research (CIFOR) global comparative study of 13 MSFs in Brazil, Ethiopia, Indonesia, and Peru. For this chapter specifically, we focused on an MSF that was part of CIFOR's ACM project (2000–2006) in Jambi Province, Indonesia. The ACM project employed an MSF approach, among others, in its project activities, especially in gathering local communities and other stakeholders, such as NGOs and government, to discuss and negotiate on issues related to customary forests.

Trust in the MSF

Trust has been studied as a factor that can sustain collaboration and strengthen stakeholders' commitment toward a process (Ansell and Gash 2008; Emerson, Nabatchi, and Balogh 2011). In addition, trust can also foster innovation and problem solving as a perception grows that others have the same desire to work and find common solutions (Edelenbos and Klijn 2007). The role trust and distrust play in influencing relationships among stakeholders involved in a system may thus "serve as a potentially important source of institutional resilience" (Stern and Baird 2015).

As a dynamic process, trust or distrust appears as a product of interaction in the past and present. Together, this experience will shape people's expectations about the future; and often becomes a basis – considered a cost-benefit evaluation by Hamm (2017) – of whether one decides to cooperate with others or not (Gray, Shwom, and Jordan 2012; Hamm 2017; Vangen and Huxham 2003). Trust is thus defined in this study as an individual's positive expectations of others' behavior, thoughts, and decisions regardless of the uncertainty of those expectations (de Vries 2014; Stern and Coleman 2015).

To understand how trust can emerge, four types of trust identified by Stern and Baird (2015) were used here to identify factors that can facilitate trust building in the ACM process. The four bases of trust are (1) dispositional, where trust is developed based on "personal histories, general hearsay and innate tendencies", (2) rational, where trust is formed based on one's expectation of outcomes resulting from a trustee's actions, (3) affinitive, where trust is based on feelings of shared values or "social connectedness", and (4) system-based, developed due to fair and transparent procedures. These different types of trust will not be treated separately but rather are used to shed light on how different sources of trust and distrust can develop and enhance each other.

Method and context

Study area

Bungo District, with the capital city of Muara Bungo, has a total area of 4,659 km^2, equal to 9.8% of Jambi Province. The poverty rate in this district has decreased from 5.82% in 2017, to 5.78% and 5.6% in 2018 and 2019, respectively. These changes are partially due to various efforts of the local government in supporting communities in the agriculture, forestry, and fisheries sectors, such as providing them with fish spawn, rubber, oil palm, and rice. Oil palm (60,628 ha) and rubber plantations (98,000 ha) are still the main economic activities, followed by the manufacturing industry, trade/hotels/restaurants, and financial/real estate/business services providing the main incomes for the people of Bungo

District (Badan Pusat Statistik Kabupaten Bungo 2020). Previously, the mining sector had a stronger role (coal and gold are spread throughout most of Bungo's territory).

Baru Pelepat village, Bungo District, Jambi – the location of the ACM MSF (see Figure 3.1) – is inhabited by Minangkabau people along with Orang Rimba, a group of previously hunters and gatherers. Bungo District is in a transition zone from the matrilineal land inheritance tradition of West Sumatra to the more patrilateral and bilateral patterns seen elsewhere in Sumatra: paddy rice fields tend to be inherited through the female line, while rubber gardens are inherited through either male or female lines (Suyanto and Otsuka 2001; Suyanto, Tomich, and Otsuka 2001).

In Baru Pelepat, as elsewhere in Indonesia, unclear forest boundaries have caused numerous problems. In 1997, the Transmigration[1] Program from the central government was implemented in the village of Baru Pelepat, allocating certain land rights to the transmigrants. This resulted in disparate visions/perceptions among the forest stakeholders (transmigrants and local people) about how they should manage the land (Adnan et al. 2008). While customary communities knew two types of land ownership, individual and customary, migrants focused on accumulating individual lands. Often, irresponsible actors within the community sold customary common lands to migrants, exacerbating conflict between the two parties (Adnan et al. 2008).

Private sector actors such as timber and oil palm companies, who have received legal rights to manage forest lands, have often blamed local problems

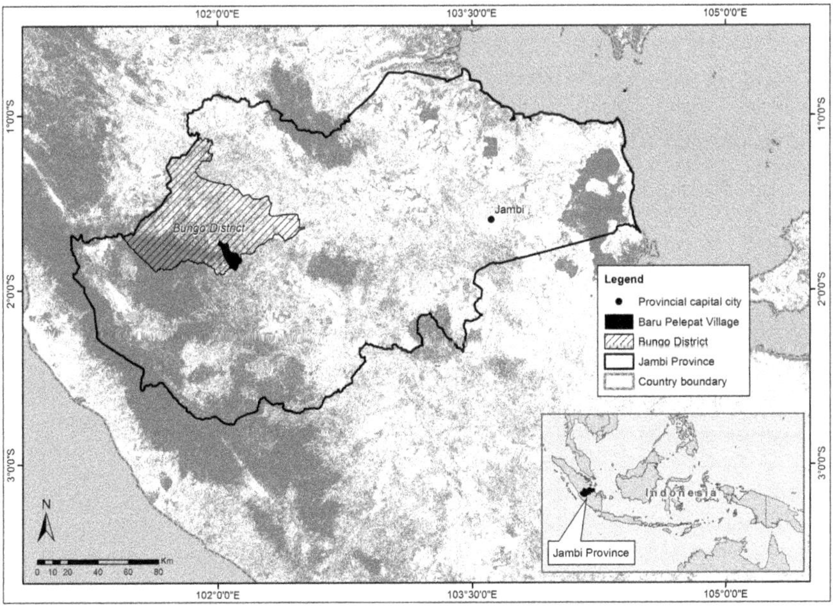

Figure 3.1 Map of Bungo District

on community conflicts over forestland. Forests have been claimed by both the community, which has inhabited the area for generations (*de facto*), and the logging companies. Logging operations – some legal, some not – were often carried out without the knowledge of the wider community (Smith et al. 2003). Although in the 2000s, social forestry schemes offered management rights to local communities, according to Fisher et al. (2018), there was little political will on the government's side and no mechanisms to hand over forestland management to local communities. Without legal rights, neither the communities' role in managing the forest nor their ownership of it was legally recognized. It was not until 2012 that customary communities were able to gain ownership rights through enactment of constitutional court ruling No.35/PUU-X/2012, which recognized the role of customary communities (*masyarakat adat*) in managing forests and natural resources.

The customary forest of Baru Pelepat, known locally as Datuk Rangkayo Mulio customary forest, was by 2021, surrounded and directly adjacent to oil palm plantations, an IUPHHK area (*Izin Usaha Pemanfaatan Hasil Hutan Kayu*/Timber Forest Product Utilization Permit), and mining sector expansion.

The MSF

Figure 3.2 shows that between 1970 and 2000, prior to the ACM forum's establishment, LULUC – so central in attempts to address climate change – had become crucial issues in the Bungo District forests. This included oil palm development planning that began in 1990 and eventually occurred near Baru Pelepat in 2007. Some villagers accepted this for economic reasons, with the oil palm plantation located outside their customary forest.

The ACM project worked directly in Baru Pelepat for six years (2000–2006), in collaboration with subnational stakeholders such as a university and NGO. In addressing issues of LULUC, the ACM team conducted action research to support

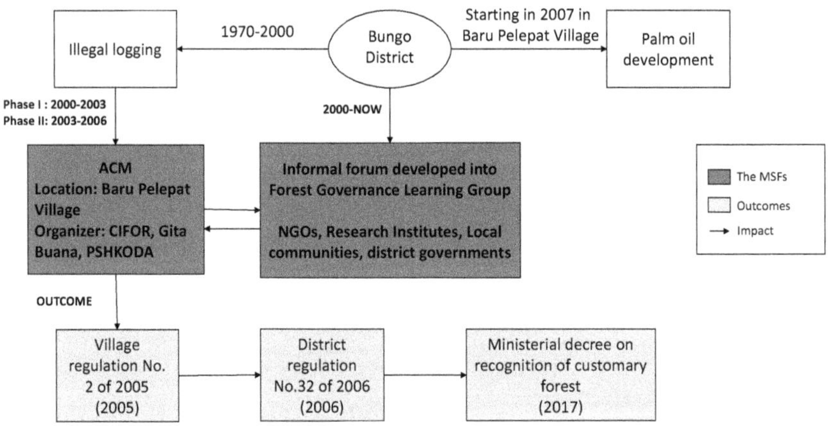

Figure 3.2 Timeline of the ACM project and MSF

development of strategies to achieve sustainable forest management. Several respondents told us that in the early 2000s, an informal MSF began functioning in this region, a knowledge sharing platform among several actors, including local NGOs, the university, subnational (district) government, and research centers. Local communities were included in discussion forums as part of a learning process about the importance of preserving forests for themselves. This MSF gathered frequently to discuss various issues related to their work in Bungo District. These actors saw the potential of collaboration in working toward similar objectives as well as creating more synergy and effectiveness. In addition, they also contributed to the discussion about ACM. This MSF, here called ACM MSF, worked alongside the ACM project.

Data collection and analysis

The original ACM MSF was organized at a subnational level and was selected as part of our MSF study due to its work that dealt with LULUC problems at the local level. Structured interviews were carried out in 2018 with four different respondent groups: key context respondents (to gain understanding about the context in which the MSF was set up), MSF organizer(s) and participants and those relevant stakeholders who did not participate in the MSF for various reasons (see Sarmiento Barletti 2022; or Sarmiento Barletti and Larson 2019b for complete methods and a tool for MSF research). Using stakeholder mapping and snowball sampling, a total of 49 respondents from different sectors were identified and interviewed: 5 key context respondents, 6 organizers, 28 participants of the ACM MSF and 10 MSF non-participants.

The Q-methodology was also carried out with participants and MSF organizers to understand their perspectives about the MSF as a participatory process. Participants and MSF organizers were asked to place 42 statement cards on a grid (Figure 3.3) based on the strength of agreement or disagreement. The 42 statements describe different aspects of MSFs, covering topics ranging from design of MSFs to opportunities and benefits. These statements were developed based on a literature review and scoping research on MSFs in Brazil, Peru, Ethiopia, and Indonesia. As their perspectives might be based on opinion rather than the real conditions of the MSF, follow-up interviews were conducted to understand their sorting and to carry out a more in-depth analysis of their perceptions. A detailed explanation about this method can be found in Sarmiento-Barletti and Larson (2019b).

The analysis follows an inductive content analysis where the collected data was systematized and coded. A diagram was developed to show how data analysis was done (Figure 3.4).

The questions we asked didn't directly address the trust issue. But some respondents mentioned trust explicitly when they were asked about the MSF success. In addition, responses from questions about "MSF effectiveness" and "MSF equity" were also coded to find patterns of activity in the forum. Answers that portrayed positive perceptions of the MSF process were categorized as having trust-related implications. This reflected our assumption that positive perceptions resulted

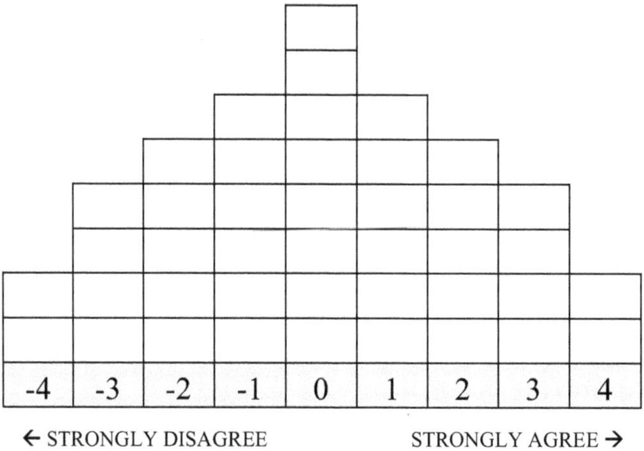

Figure 3.3 Q-methodology grid

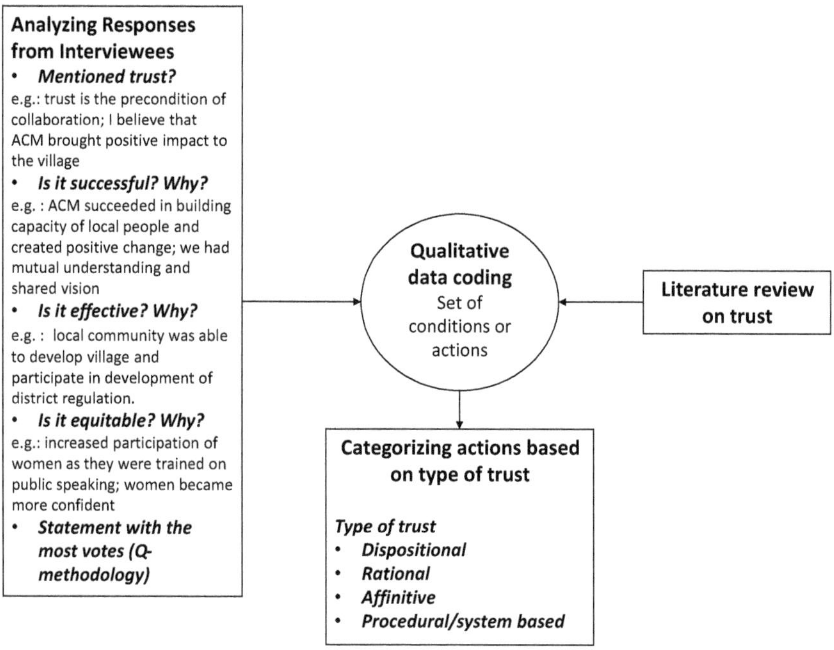

Figure 3.4 Data analysis

from alignment of participants' personal views and the MSF process which could contribute to the emergence of trust. Their continued participation within the project and MSF process that was in place from 2000 to 2006 also portrayed their optimism in achieving positive outcomes, despite the uncertainty of the future. Our coding system was also checked with relevant literature on trust.

We did not conclude that trust is the absolute reason for the success of the ACM project and MSF, but rather we acknowledge the dynamic presence of this concept that implicitly played a significant role in the collaborative process. To complement the quantitative results, relevant quotes from the interviews were added for clarification.

What did we find?

The ACM MSF: success and challenge

The ACM project started with participatory action research that focused on engaging local communities in community-determined activities.[2] Stakeholder mapping was done in early days to identify key participants for the project and MSF. Besides paying close attention to power and interests, organizers sought to improve social capital by engaging with actors who could introduce the project approach and MSF to wider relevant stakeholders.

In the beginning of the project, 90% of the local communities in Baru Pelepat depended on unsustainable logging (locally called *bebalok*) as a complement to their livelihoods which according to one participant was an easier way to obtain money than in agriculture, their usual source of subsistence. Many locals also considered this illegal activity "a collective one": half of local men went into the forest and extracted timber, thus attracting others to do the same. Although they considered themselves a customary community, prior to ACM, the customary rules that regulated the use of forest and resources were not being practiced.

The ACM's facilitator as an organizer and some community members in the beginning talked about the implications of deforestation for the future. They also discussed the possibility of forest loss if deforestation continued, leaving no trace to future generations and whether or not this was the preferred condition. The discussions were not a one-time activity but took place both in regular conversation and in formal focus group discussions (more about this in the next section), attracting more locals who previously refused to join. Several discussions with other stakeholders from government and NGO sectors were also held in parallel. Through the long series of discussion forums, the MSF achieved several outcomes, including enactment of village regulation No. 2 of 2005[3] and district regulation No. 32 of 2006.[4]

To finally arrive at the decision to protect the customary forest was not instantaneous. It followed a long and challenging process especially in bringing together the divergent opinions within the community. The efforts of the ACM MSF to protect the customary forest and its surroundings – initially proposed by members of the Baru Pelepat community – created a dilemma for other Baru Pelepat community members whose main livelihood came from their rubber gardens and was proving inadequate due to the fall of rubber prices in the market. Additionally, there was a surprising increase in the price of oil palm. This condition contributed to rapid land conversion. Fourteen percent of present-day MSF participants expressed their concern about the lack of economic benefits from the customary forest and saw it as a challenge for more sustainable outcomes. Despite this, the local community is still active in protecting their customary forest. In addition, they continue to use knowledge gained from their ACM participation even though the ACM project ended long ago; local champions trained during ACM continue to facilitate village discussions and negotiations between communities and other actors such as district government and private sector. A decade and a half later, Baru Pelepat received legal recognition for its customary forest, Datuk Rangkayo Mulio, from the Government of Indonesia in 2017. This recognition gives full rights to the local communities for managing their customary forest, changing the previously *de jure* status of customary forest from state-owned to customary ownership.

The next section focuses on activities that contribute to the emergence of trust. Seven interviewed participants explicitly saw trust as having shaped their cooperation and willingness to support and participate in the project and MSF. The facilitators' approach in engaging local communities played an important role in shaping communities' trust toward the people and the process (see the next sections). Two local community participants expressed how the inclusion of local communities, whom they themselves considered uneducated compared to the other groups of participants, and informal day-to-day interaction had convinced them that the ACM team brought a positive message to the people in the village. Moreover, other participants from different sectors described the close relationships that had developed among participants: local community, government staff, and NGOs, which allowed them to collaborate and help each other. Further, they were confident that collaboration through MSF and other related work would lead to common solutions.

Finally, five activities have been identified that potentially contribute to the emergence of different trusts: capacity building activities, women's empowerment, emergent informal relationships, past positive collaboration, and a development-oriented shared vision (Figure 3.5).

Strategies that influenced trust building in the MSF process

Capacity building

Capacity building is one strategy to encourage more active participation of local communities with an idea that learning and knowledge sharing can motivate

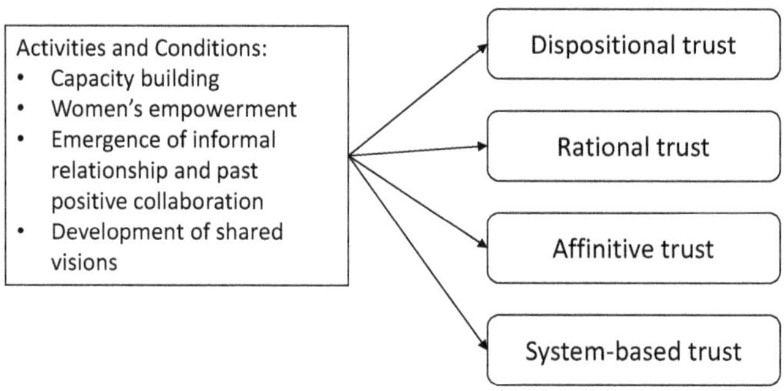

Figure 3.5 Activities and conditions that contributed to trust development

change within communities (see Hagmann et al., this volume). The ACM process initiated a shift from dependency on deforestation as an income-generating activity to a more sustainable one.

> The objective of the forum was to discuss and share information and the point was that they [the local community] were more ready to make a well-informed decision.
>
> (ACM Jambi participant, Man, Academia, 2018)

Capacity building covers different activities such as improving literacy and practicing public speaking. Local communities were also equipped with technical knowledge about sustainable forest management and development of village regulations. In addition, the ACM team organized field visits to several other villages, allowing Baru Pelepat residents to reflect on their own conditions and gain other perspectives. These activities allowed local communities to gain insight and contributed to their decision to protect their customary forest for the next generation, improve skills related to non-timber forest product use, reduce logging activity, and restore benign traditional practices. In addition, they also decided to document a set of rules in customary forest management.

Those activities contributed to meaningful participation of local communities in the forum wherein 94.7% of participants interviewed perceived the MSF process as equitable. In addition, there are several changes perceived by most of the interviewed participants, from an increased number of local community members who were more confident in speaking in the forum compared to before the ACM project, to an increased number of women who joined the discussion (women's participation is discussed in the next section) and took part more actively in village institutions. Facilitators also played a big part in improving forum dynamics by supporting communities' preparation prior to discussions and encouraging communities to speak in the forums.

> The forum provides an equal arena for everyone involved. Through capacity development, local communities have confidence to speak, express objections or give inputs in the forum. Despite a rough beginning where there was domination from higher figures in the village and from the government, this tendency changed gradually.
>
> (ACM Jambi Participant, Man, NGO, 2018)

Women's empowerment

Women's empowerment became one focus of the ACM project. The women, who were previously marginalized, had been empowered and facilitated so that they participated meaningfully in the forum and decision-making processes. Prior to the ACM project, women were always excluded from village meetings or decision-making processes. There was a general assumption, from men and women themselves, that women would always accept any decision as they were seen to have inadequate knowledge to have a say or oppose or support any proposition.

> Prior to ACM, women only played a role in the domestic arena. We were never invited to any events. So in ACM women were empowered and trained, especially to be able to speak our minds. Slowly, we became more confident in giving inputs in any meetings, to voice objections. I was so happy with the ACM program and team and I cried when the program was over.
>
> (ACM Jambi Participant, Woman, Local community, 2018)

Engaging women required different strategies, including the presence of a facilitator with whom local women felt most comfortable. In this case, this meant having a woman facilitator instead of a man, from a similar religious background. There were also separate groups and activities for local men and women. In the women's group, women would take turns voicing their opinions, which slowly helped with developing women's confidence in public speaking.

Forty-six percent of interviewed participants mentioned improved women's participation as one important achievement from the ACM project and the MSF. Not only did women become more confident in joining discussions with other outside actors, men's perception of women's ability and position in society also changed.

> ACM invested time, human and financial resources to improve the capacity of women villagers in giving active and meaningful participation. For the case of women villagers, we had a clear number of seats on the forum with delegated powers to make decisions.
>
> (ACM Jambi organizer, Woman, Academia, 2018)

Informal relationships and past collaboration

The original ACM team began its work with broad interests in more sustainable management of natural resources (forests) and improved human well-being. Within those broad goals, decisions about what to address were left to the communities. Considering the significant role timber and other natural resource extraction played in local community livelihoods in Baru Pelepat, taking the decisions they did meant real economic losses for certain members of the community and required considerable negotiation but they decided to protect a number of hectares of customary forest where commercial logging would be prohibited.[5] The concern with equity (as part of human well-being) meant introducing the foreign concept of gender equality, which was not part of their culture. Care was taken to avoid rejection from local groups in the village. The ACM team began by getting to know the local communities, their practices and trying to build trusted relationships with them. Some of the team went together into the forest, following locals' activity in *bebalok* (logging) and mining. They also held casual visits to people's homes in the village. A woman facilitator joined weekly religious gatherings with local women. These strategies of building informal relationships through fishing, hunting, or a mere conversation over a cup of coffee and cigarettes supported knowledge exchange and played a part in getting community interest and acceptance of the program's collaborative and learning approach. More importantly, this condition helped in creating a more comfortable atmosphere and mutual respect among them during discussion in the forum.

> I was an illegal logger before ACM came into our village. In the beginning, I rejected these [ACM team] people, maybe because I didn't know them well yet. But slowly with more interaction I developed a close relationship and became willing to collaborate.
>
> (ACM Jambi participant, Man, Local community, 2018)

Past collaboration among several stakeholder groups had also influenced the ACM MSF process. An informal multi-stakeholder platform was formed prior to the ACM MSF, which aimed to bring together all stakeholders working in Bungo District (NGOs, research institutes, government staff) to talk about various issues, from gender to land use change issues. This multi-stakeholder platform was perceived as a good start for forming a lasting and good relationship among its participants, which, in turn, influenced the ACM project.

Past collaboration and informal relationships outside of the project and MSFs helped in ensuring a more effective process of the MSF. A participant from a research institution admitted that they were not part of the ACM project but willing to collaborate in the ACM discussion and capacity building program. Past positive collaboration between this organization and those involved in the ACM program and the MSF supported the development of social capital that mutually benefitted both sides.

Shared vision

A shared vision becomes another factor that played a role in strengthening collaboration in the ACM project and MSF. A participant from an NGO explicitly mentioned how the decision to collaborate was based on the understanding that everyone shared the same vision to help government achieve sustainable forest management and to improve local communities' agency in the long run. With this goal in mind, each could synergize the different resources and programs that strengthened the MSF process and outcomes. For instance, an NGO participant mentioned how synergy through the MSF process could help the ACM team to build its network at the district level, which was needed for their efforts to gain support for the development of village and district regulations.

> … Nobody claims the MSF succeeded because of only certain parties. (…) We built a strong foundation as we realized that we had common interests and that's what made the MSF work (…). Besides, stakeholder mapping was

> done to identify potential stakeholders to be invited to the MSF, for instance those who had strategic positions in the government.
>
> (ACM Jambi Organizer, Man, NGO, 2018)

Through an MSF, work could be divided such that each sector had its own role while complementing the others. In addition, participants were aware of how they needed each other to carry out their different programs in Bungo District, leading to improved cooperation among participants where each would help based on their own capabilities.

> I love the idea of collaborating as we did in Bungo, also through informal interactions prior to the formal forum; it enabled us to connect more easily with the government. We understood that we needed each other to do our job so that we could help the government to put together a program that was pro-conservation and environment and, at the same time, supported the marginalized community by providing them with useful capacities and information.
>
> (ACM Jambi Participant, Woman, NGO, 2018)

Overview of the quantitative results

Using the results from the Q-method, we explored from 34 respondents (participants and key context respondents) the four statements with the most votes, from both highly positive to highly negative views of the MSF in relation to trust (Table 3.1).

Table 3.1 Summary of the four Q-method statements receiving the most extreme evaluations (N = 34)

Four statements the most respondents *disagreed strongly* with (−4)	A (%)	B (%)
It doesn't matter what the MSF decides because it will never be implemented.	88	50
MSFs are just a way to create the appearance that participants are equals, which makes things worse for the less powerful.	79	35
Because MSFs only address immediate problems, rather than their underlying causes, their outcomes will never change the status quo.	85	41
Making laws simpler to comply with is a better solution than an MSF.	65	26
Four statements the most respondents *agreed strongly* with (+4)		
Successful MSFs make decisions based on the common good.	94	65
Successful MSFs have an unbiased facilitator.	85	38
MSFs improve information sharing and transparency.	82	44
MSFs can empower indigenous people/local communities (IP/LCs) and/or previously marginalized groups (by, e.g., gender, race, caste)	79	41

A: total percentage of respondents agreeing with this statement (−1 to −4)
B: percentage of respondents strongly agreeing with this statement (−3 to −4)

The four statements most strongly disavowed (voted −4 and −3) showed participants' confidence in the MSF process. Participants tended to disagree with the statement that MSF decisions would never be implemented. This was perhaps assessed in this way because it was not in line with their experience during the MSF where their common agreement to protect the customary forest has remained operational until now. One of the local community participants acknowledged how the MSF had played a role in facilitating the local community to protect their customary forest from encroachment and other illegal activities. The decision to protect the customary forest was taken by the local community, who decided to revive and impose customary rules and sanctions. Indeed, statements about MSFs as a platform that only creates an appearance of equality received strong disagreement from the ACM MSF participants. The MSF has provided opportunities for the community, especially women, to actively participate in forums by facilitating their involvement with training, workshops, and field visits and to represent themselves in the forum. Likewise, the statement about the role of MSFs in empowering local communities, indigenous people, or other previously marginalized groups was valued positively by most of the study participants. In addition to capacity development, participants also agreed on the role of a relatively unbiased facilitator in the MSF's success. This perhaps was based on the experience in the MSF process, in which according to the interviews, the facilitator tried to remain neutral and minimally intervene in the process. An unbiased facilitator was seen in a discussion session as helping the local community to better adjust to the presence of different (and higher status) stakeholder groups. Facilitators would encourage local community members to speak and keep the discussion from being dominated by certain stakeholders.

Participants argued that the MSF informal setting allowed the development of collaboration and cooperation among participants. This informal and flexible setting also helped in decreasing domination from powerful actors and increasing participation for the less powerful ones. Results from the interviews showed how participants expressed the benefit of this informal setting of MSF, from its contribution to "break the wall of doubt among participants so that they start collaborating with each other" to how the local community "felt less pressure in participating in the discussion". In addition to that, stakeholders' interaction prior to the forum had positively influenced the MSF process, which eased the collaboration process (also reported in several contributions in Butler and Schultz (2019)).

As a result, statements about transparency and information sharing are perceived to have been improved through the MSFs by most of the participants. One participant from the NGO sector mentioned that despite the possibility of data leakage that could benefit one party or another, this did not prove to be a challenge. Information was shared and transparency could be pursued as trust had been built among the participants involved.

The development of cooperation and further collaboration among participants influenced the MSF process and contributed to their achievements, which was in line with participants' disagreement over the statement, "Because MSFs only address immediate problems, rather than their underlying causes, their outcomes will never change the status quo". Although one participant mentioned the role of MSF only as a platform for knowledge sharing and discussion, others argued that participants in the MSF also identified the underlying problem of deforestation that occurred in the village at that time and found solutions agreed by all stakeholders. Participants' realization of the solutions formed through the MSF was also reflected from the most common statement considered extremely positive, that is, "successful MSFs make decisions based on the common good".

> With this collaboration, one memorable outcome was the district regulation about the community forest. Even though this forum had no designated budget/funding. Also, with the connections built from this forum, we could benefit each other in many more things.
>
> (ACM Jambi Participant, Woman, Government Agency, 2018)

In addition to that, the statement on "making laws simpler to comply with is a better solution than an MSF" received strong disagreement from participants, showing how the MSF was still perceived to serve an important function in addressing problems faced by the regions.

> I would say that one factor that played a role in the success of the MSF process is the "emotional closeness" built among participants and I think this kind of MSF is preferable to any other approach.
>
> (ACM Jambi participant, Man, Government Agency, 2018)

Discussion

We describe here the importance of trust in the ACM MSF as a factor that contributes to a collaborative environment. Our findings showed some factors influencing trust as stressed by scholars and some lessons for developing trust building during the MSF process. These factors have affected the MSF's achievements and their impact a decade and a half after the project ended.

On trust building in the MSF

The following discussion further elaborates the trust building aspect as related to the four categories of trust developed by Stern and Baird (2015).

Dispositional trust

Emborg, Daniels, and Walker (2020) argued that dispositional trust is based on a "person's personality or outlook". Dispositional trust derives from the trustor's assessment toward the trustee, which can be based on position, institution, or context. In addition, context also plays a role in the emergence of dispositional trust; for instance, supportive conditions can lead to the development of trust more readily than competitive environments (Fulmer and Gelfand 2012).

Findings from this study did not clearly show the presence of dispositional trust, which fits with Coleman and Stern's (2018) observations on the absence of dispositional trust in natural resource governance. The local community did not directly accept the presence of the ACM team. Instead, it needed almost two years for their trusted relationship to form. An early invitation sent to the local community was accepted as it was quite common for local people to attend an event or meeting when they received an invitation. Several participants admitted that they were curious about the ACM program and thus agreed to come to their first meeting. As for the other participant groups, they had previous collaboration prior and outside of the ACM project and MSF, thus setting a basis for further collaboration within the ACM efforts.

Rational trust

Rational trust is developed when trustors expect positive outcomes derived from efforts put forth by the trustees (Stern and Baird 2015). There are several conditions where rational trust could emerge in the ACM MSF. First is the increased number of women participating in the ACM MSF, as expressed by several participants and pointed out as one of the achievements of ACM in general. This perception of improvement compared to the situation before the ACM program and MSF took place contributed to increased participants' trust that the activities brought a positive change in the community. Second is the enactment of the district regulations, following the formulation of village regulations. These legal products were seen as achievements of the ACM project and MSF as they formed the basis for state legal recognition of the customary forest of Datuk Rangkayo Mulio. These outcomes, both the regulations and empowerment of the local community and women, created changes on the ground where deforestation and inequality had been major problems. These outcomes helped shape participants' trust in the ACM process.

In addition, trust building in the ACM MSF was influenced by their past collaboration and relationship outside of the MSF. This can be confirmed through findings from our interviews where the participants acknowledged the role of positive collaboration outside of the MSF in improving cooperation and collaboration within the MSF processes (see Vangen and Huxham 2003 and Gray, Shwom, and Jordan 2012). Prior relationships outside of the MSF between actors provided a chance for participants to judge the trustworthiness of others. Participants' positive judgment of others thus became a starting point for them to at least sit at the same table in the discussion.

Affinitive trust

Affinitive trust is based on feelings of shared values or "social connectedness". The objectives of the ACM MSF included instilling in communities greater awareness of the importance of managing customary forests in a sustainable way and developing strategies to achieve that. The process of engaging local communities was not easy due to their dependence on illegal timber extraction. Early engagement was dominated by communities' lack of trust toward the ACM organizers. Nevertheless, several facilitators were assigned to approach local communities, including women, especially in the initial process, to build personal bonds with the communities, and to understand their system holistically. This was done through village or religious events and other informal interactions, where eventually local community members grew accustomed to the presence of facilitators and became more open with new information. Local community members' perceptions of the organizer and other participants' goodwill contributed to the development of affinitive trust. In addition, the informal atmosphere of the MSF with activities such as field visits (as part of capacity building) was acknowledged to enhance the formation of emotional closeness among participants. This is in line with findings from Emborg, Daniels, and Walker (2020) and Stern and Baird (2015) who also mentioned the role of field trips in the development of "personal bonds" in further building trust in MSFs. This informal relationship also contributed to a feeling that everyone in the room shared the same values. This was confirmed through interviews with several participants who considered the forums not to be led by any specific organization or individual but rather developed based on understanding of common needs. Additionally, study participants also agreed that outcomes produced by the MSF were based on the common good.

Stern and Coleman (2015) mentioned how affinitive trust is stronger compared to the other types, influencing people's actions more than rational trust. Our case demonstrates this notion. Despite being the most implicit as it is formed based on "feelings", it has contributed to cooperation and willingness to join hands in the ACM process. Through continuous discussion, whether in ACM MSF or through informal conversation, participants could understand the vision and motivation of their counterparts, further learn to appreciate and respect them (see Emerson, Nabatchi, and Balogh 2011). The discussion process also allows for different perspectives to be identified and solutions to be formulated (see Juerges et al. 2017). Willingness to respect and listen attentively, especially to those less powerful actors, like the local community, influenced participants' perceptions of others' personality and integrity.

System-based trust

System-based trust derives from perceptions of fair and transparent procedures. The objectives and activities of the ACM MSF in promoting conservation and sustainable forest management through community empowerment and deliberation have also influenced the trust building process in the MSF. Power differences among actors were addressed by organizing capacity development activities to help the local community members participate in the MSF more effectively. This helped to level the playing field with the other more powerful actors. The presence of facilitators perceived as unbiased, coupled with a flexible dialogue setting also encouraged active participation of the local communities who were not used to being at the same table with government or other stakeholder groups. Likewise, involvement of less powerful actors, in this case the local community, went beyond just sitting at the same table with the other actors (see Riley et al. 2018). As a result, the MSF successfully fostered an environment where participants felt as equals. Their positive perception toward the lack of domination from powerful actors in the MSF reflects their confidence in the forum's ability to provide an equal arena for different actors (see Getha-Taylor et al. 2019). This supports Ran and Qi's (2019) description of the relationship between power and trust in which trust could be a basis for, or influence, power-sharing in collaborative spaces.

Further, the implications of trust can be seen in participants' positive views of the MSF as a safe arena for knowledge and data exchange. Transparency, according to Rapp (2020), is an aspect that influences an individual's trust toward a system, forming a basis for procedural trust in which they believe the MSF promotes equity and transparency.

Lessons learned

An MSF is not just a platform for people to meet and gather but can also be a way to achieve transformational change (Sarmiento Barletti et al. 2020). Likewise, implementation of MSFs does not automatically lead to achievement of solutions. Certain efforts need to be taken to make a change (in this case, strategies to reduce deforestation and achieve sustainable forest management); these can be different depending on the contexts. These efforts, in turn, influenced how participants' trust/distrust was formed. Analysis has shown some factors that contributed to development of different types of trust and influenced the MSF process and achievements.

The MSF's fair procedure, with facilitators perceived as neutral coupled with informal relationships, indeed contributed to the development of rational, affinitive, and procedural trust in which participants evaluated the MSF system and their peers positively (Figure 3.6). In this context, those who had no past collaboration history may have based their judgments on the perceived procedural fairness in which inclusion of stakeholders at different levels was promoted. Informal dialogue and continuous discussion also contributed toward the development of shared values (affinitive trust) and over time, participants could base their trust on rational evaluation of the process and their peers.

Collaboration as a result of this positive evaluation contributed to the common agreement to protect the customary forest. Customary forest protection has continued even when the surrounding individual plots were converted into oil palm plantation. This shows how joint decisions and inclusivity contributed to increased participants' commitment in implementing MSF decisions. It also shows the importance of adhering to the initial decisions taken by the community members themselves, when sustainability of effort is required, as in this case.

The state recognition received in 2017, perhaps, helped a great deal in this. But we can't deny the communities' active role in protecting and defending their customary forest from private investors (e.g., oil palm companies) since the ACM project ended and before the 2017 state recognition. Trust in this context has

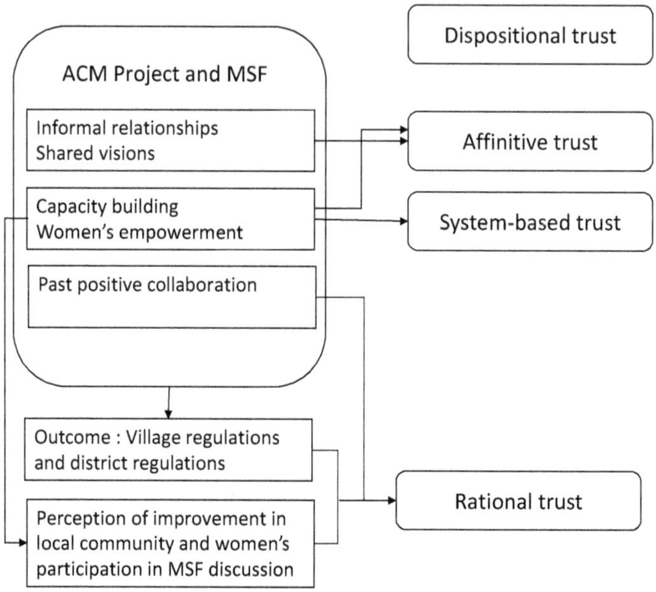

Figure 3.6 Trust development in ACM project and MSF

become the underlying factor of stakeholders' cooperation. Participants do not emphasize the importance of trust, but, as defined here, it has had an effect not only on the MSF process and achievements but also on participants' sense of belonging within the MSF group.

The ACM Jambi project and the MSF were closely related; both were driven by similar ideas of democracy, justice, and empowerment. What happened in the project influenced the MSF process where participants from different sectors met and deliberated. At the same time, what happened in the MSF process directly and indirectly influenced the achievement of the ACM project. Trust which was built along the way took turns both in shaping relationship dynamics within the MSF and outside the MSF in which agreements were implemented (see also Komarudin et al. 2011, on factors that influenced trust building). In the end, legal recognition of the customary community (the district regulations) was perceived as an ultimate achievement by both the ACM MSF and project participants.

Conclusion

The ACM forum in Jambi existed as the result of an effort from several organizations who had previously worked together and had similar objectives and goals. The forum's understanding of common needs synergized the different programs and resources, with mutual acknowledgment of the important role played by each stakeholder. As a result, unlike many collaborative processes in Indonesia (and elsewhere), nobody claimed that the success of the MSF relied only on a particular stakeholder. Mutual recognition shaped cooperation among the participants involved.

Likewise, the cooperation built was also related to trust building in the MSF process. Several efforts had been made in the ACM forum process that may have influenced how trust was formed, also supported by relevant literature on trust. First, the ACM MSF placed greater emphasis on capacity development for previously less powerful actors such as local communities. Not only that, women in this case were also the targets of capacity development activities. As a result, they became more empowered and confident in voicing their interests in public (see also Sari 2007). More importantly, they had meaningful participation in the forum and a sense of ownership toward the MSF process and outcomes.

A facilitator who is perceived to be relatively unbiased is also needed in the MSF, in which facilitators take neutral positions and avoid showing favoritism toward certain stakeholders. Participants' perspectives on the presence of an unbiased facilitator led to more trust building toward the process, where they saw the system as fair. In this context, a relatively unbiased facilitator can support the provision of a safe and equal arena for all stakeholders to discuss and negotiate. Second, the informal relationship among participants coupled with past positive collaboration led to a positive expectation over how the relationship among participants would evolve within the MSF. This means that one perceived others to "behave" and cooperate in a way that was similar to what had happened prior to and outside of the MSF.

Finally, the development of a shared vision contributed to the emergence of trust in the MSF process. Through face-to-face discussion and informal conversation, each could understand the vision and interest of others as well as find common ground. These kinds of processes, so supportive of meaningful collective action, particularly at this intermediate scale, will be crucial as we move forward addressing climate change and other global challenges.

Looking at the trust framework we learnt from the ACM forum process, this forum has helped in building not only rational trust but also affinitive and system-based (procedural) trust. The ACM forum has been perceived as successful by its participants and achieved meaningful outcomes. Further, when the ACM project ended and the forum was no longer as active as in the past, local communities were still holding on to the ACM legacy, especially in implementing rules and sanctions regarding the customary forest. Our evidence also shows their positive perceptions of the ACM project and forum processes. The achievements of Baru Pelepat in obtaining legality of the village and district regulations were due in part to MSF efforts, including the investment of time and human and financial resources. Furthermore, building trust is one of the key successes in the MSF process.

Indeed, there's no blueprint for a successful MSF, and trust building processes can differ depending on the MSF context. However, good MSFs are likely to need the success factors identified herein: the development of trust, excellent "neutral" facilitation (see Hagmann, this volume), and inclusion of marginalized groups – the absence of which have hindered the achievement of many MSF objectives elsewhere.

Acknowledgements

Any views expressed in this publication are those of the authors. They do not necessarily represent the views of CIFOR, the editors, the authors' institutions, the financial sponsors, or the reviewers. This work was carried out as part of the Center for International Forestry Research's Global Comparative Study on REDD+ (https://www.cifor-icraf.org/gcs/). Funding partners that have supported this research include the Norwegian Agency for Development Cooperation (Norad, Grant No. QZA- 21/0124), International Climate Initiative (IKI) of the German Federal Ministry for the Environment, Nature Conservation and Nuclear Safety (BMU, Grant No. 20_III_108), and CGIAR Research Programme on Forests, Trees and Agroforestry with financial support from CGIAR Fund donors.

Notes

1 The Transmigration Program was an initiative of the Dutch colonial government, later continued by the Government of Indonesia (GoI). It aimed to move landless people from densely populated areas of Indonesia to less populous areas of the country.
2 Although Baru Pelepat is one village, we speak here of multiple communities, because of the ethnic differences and geographical distribution among hamlets.
3 Village Regulation No. 2 of 2005 on customary forest management and utilization.
4 This regulation legally recognizes Baru Pelepat as a customary community.
5 The degree to which this decision was one taken autonomously by the community (à la Yuliani et al.) or determined through the efforts of broader scale actors (à la Liswanti et al.) remains a question.

References

Adnan, Hasantoha, Djuhendi Tadjudin, E. Linda Yuliani, Heru Komarudin, Dicky Lopulalan, Yuliana L. Siagian, and Dani Wahyu Munggoro. 2008. *Belajar Dari Bungo: Mengelola Sumberdaya Alam di Era Desentralisasi*. Bogor, Indonesia: CIFOR. https://doi.org/10.17528/cifor/002357.

Alfantoukh, Lina, Yefeng Ruan, and Arjan Durresi. 2018. "Multi-Stakeholder Consensus Decision-Making Framework Based on Trust: A Generic Framework." In 2018 IEEE 4th International Conference on Collaboration and Internet Computing (CIC), 472–479. IEEE. Philadelphia. https://doi.org/10.1109/CIC.2018.00012.

Ansell, Chris, and Alison Gash. 2008. "Collaborative Governance in Theory and Practice." *Journal of Public Administration Research and Theory* 18, no. 4: 543–571. https://doi.org/10.1093/jopart/mum032.

Armitage, Derek R., Ryan Plummer, Fikret Berkes, Robert I. Arthur, Anthony T. Charles, Iain J. Davidson-Hunt, Alan P. Diduck, et al. 2009. "Adaptive Co-Management for Social-Ecological Complexity." *Frontiers in Ecology and the Environment* 7, no. 2: 95–102. https://doi.org/10.1890/070089.

Bäckstrand, K. 2006. "Multi-Stakeholder Partnerships for Sustainable Development: Rethinking and Effectiveness." *European Environment* 16: 290–306. https://doi.org/10.1002/eet.425.

Badan Pusat Statistik Kabupaten Bungo. 2020. *Kabupaten Bungo dalam Angka 2020* (Bungo District in Figures). Katalog No. 1102001.1509, Jambi, Sumatra, Indonesia.

Berkes, Fikret. 2009. "Evolution of Co-Management: Role of Knowledge Generation, Bridging Organizations and Social Learning." *Environmental Management* 90, no. 5: 1692–1702. https://doi.org/10.1016/j.jenvman.2008.12.001.

Berkes, Fikret. 2010. "Devolution of Environment and Resources Governance: Trends and Future." *Environmental Conservation* 37, no. 4: 489–500. https://doi.org/10.1017/S037689291000072X.

Bourgoin, Jeremy, Jean Christophe Castella, David Pullar, Guillaume Lestrelin, and Bounthanom Bouahom. 2012. "Toward a Land Zoning Negotiation Support Platform: 'Tips and Tricks' for Participatory Land Use Planning in Laos." *Landscape and Urban Planning* 104, no. 2: 270–278. https://doi.org/10.1016/j.landurbplan.2011.11.008.

Butler, William H., and Courtney A. Schultz, eds. 2019. *A New Era for Collaborative Forest Management: Policy and Practice Insights from the Collaborative Forest Landscape Restoration Program*. London: Earthscan and Routledge.

Coleman, Kimberly, and Marc J. Stern. 2018. "Boundary Spanners as Trust Ambassadors in Collaborative Natural Resource Management." *Journal of Environmental Planning and Management* 61, no. 2: 291–308. https://doi.org/10.1080/09640568.2017.1303462.

Cronkleton, Peter, Kristen Evans, and Anne Larson. 2022. "Capacity Building for ACM: Lessons Learned from Training in Distinct Contexts." In *Adaptive Collaborative Management of Forest Landscapes: Villagers, Bureaucrats and Civil Society*, edited by Carol J. Pierce Colfer, Ravi Prabhu, and Anne Larson, 134–157. London: Earthscan and Routledge.

Davies, Althea L., and Rehema M. White. 2012. "Collaboration in Natural Resource Governance: Reconciling Stakeholder Expectations in Deer Management in Scotland." *Journal of Environmental Management* 112: 160–169. https://doi.org/10.1016/j.jenvman.2012.07.032.

de Vries, Jasper R. 2014. *Understanding Trust: Longitudinal Studies on Trust Dynamics in Governance Interactions*. Wageningen, The Netherlands: Wageningen University.

Edelenbos, Jurian, and Erik Hans Klijn. 2007. "Trust in Complex Decision-Making Networks: A Theoretical and Empirical Exploration." *Administration and Society* 39, no. 1: 25–50. https://doi.org/10.1177/0095399706294460.

Edmunds, David, and Eva Wollenberg. 2002. "Disadvantaged Groups in Multistakeholder Negotiations." CIFOR Programme Report, Bogor, Indonesia.

Emborg, Jens, Steven E. Daniels, and Gregg B. Walker. 2020. "A Framework for Exploring Trust and Distrust in Natural Resource Management." *Frontiers in Communication* 5 (April). https://doi.org/10.3389/fcomm.2020.00013.

Emerson, Kirk, Tina Nabatchi, and Stephen Balogh. 2011. "An Integrative Framework for Collaborative Governance." *Journal of Public Administration Research and Theory* 22, no. 1: 1–29. https://doi.org/10.1093/jopart/mur011.

Fisher, Micah R., Moira Moeliono, Agus Mulyana, E. Linda Yuliani, Andi Adriadi, Kamaluddin, Junaid Judda, and Muhammad Alif K. Sahide. 2018. "Assessing the New Social Forestry Project in Indonesia: Recognition, Livelihood and Conservation?" *International Forestry Review* 20, no. 3: 346–361. https://doi.org/10.1505/146554818824063014.

Fisher, Roger, and Scott Brown. 1989. *Getting Together: Building Relationships as We Negotiate*. 1st edition. New York: Penguin Books.

Fulmer, Ashley C., and Michele J. Gelfand. 2012. "At What Level (and in Whom) We Trust: Trust Across Multiple Organizational Levels." *Journal of Management* 38: 1167–1230.

Getha-Taylor, Heather, Misty J. Grayer, Robin J. Kempf, and Rosemary O'Leary. 2019. "Collaborating in the Absence of Trust? What Collaborative Governance

Theory and Practice Can Learn from the Literatures of Conflict Resolution, Psychology, and Law." *American Review of Public Administration* 49, no. 1: 51–64. https://doi.org/10.1177/0275074018773089.

Gray, Steven, Rachael Shwom, and Rebecca Jordan. 2012. "Understanding Factors that Influence Stakeholder Trust of Natural Resource Science and Institutions." *Environmental Management* 49, no. 3: 663–674. https://doi.org/10.1007/s00267-011-9800-7.

Hamm, Joseph A. 2017. "Trust, Trustworthiness, and Motivation in the Natural Resource Management Context." *Society and Natural Resources* 30, no. 8: 919–933. https://doi.org/10.1080/08941920.2016.1273419.

Juerges, Nataly, Alisa Viedma, Jessica Leahy, and Jens Newig. 2017. "The Role of Trust in Natural Resource Management Conflicts: A Forestry Case Study from Germany." *Forest Science* 64, no. 3: 330–339. https://doi.org/10.5849/FS-2016-050.

Kearney, John, Fikret Berkes, Anthony Charles, Evelyn Pinkerton, and Melanie Wiber. 2007. "The Role of Participatory Governance and Community-Based Management in Integrated Coastal and Ocean Management in Canada." *Coastal Management* 35, no. 1: 79–104. https://doi.org/10.1080/10.1080/08920750600970511.

Komarudin, Heru, Yuliana L. Siagian, Carol J. Pierce Colfer, Neldysavrino, Yentirizal, Syamsuddin, and Deddy Irawan. 2011. "The Role of Collective Action in Securing Property Rights for the Poor: A Case Study in Jambi Province, Indonesia." In *Collective Action and Property Rights for Poverty Reduction: Insights from Africa and Asia*, edited by E. Mwangi, H. Markelova and R. Meinzen-Dick, 235–269. Philadelphia, PA: University of Pennsylvania Press. https://doi.org/10.9783/9780812207873.235.

Kusters, K., L. Buck, M. de Graaf, et al. 2018. "Participatory Planning, Monitoring and Evaluation of Multi-Stakeholder Platforms in Integrated Landscape Initiatives." *Environmental Management* 62: 170–181. https://doi.org/10.1007/s00267-017-0847-y.

McDougall, Cynthia Y., and Hemant Ojha. 2022. "How Adaptive Collaborative Management Can Leverage Changes in Power: Insights from Social Theory." In *Adaptive Collaborative Management of Forest Landscapes: Villagers, Bureaucrats and Civil Society*, edited by Carol J. Pierce Colfer, Ravi Prabhu, and Anne Larson, 186–216. London: Earthscan and Routledge.

Mena, Sébastien, and Guido Palazzo. 2012. "Input and Output Legitimacy of Multi-Stakeholder Initiatives." *Business Ethics Quarterly* 22, no. 3: 527–556. https://doi.org/10.5840/beq201222333.

Ran, Bing, and Huiting Qi. 2019. "The Entangled Twins: Power and Trust in Collaborative Governance." *Administration and Society* 51, no. 4: 607–636. https://doi.org/10.1177/0095399718801000.

Rapp, Claire. 2020. "Hypothesis and Theory: Collaborative Governance, Natural Resource Management, and the Trust Environment." *Frontiers in Communication* 5, no. 28: 1–12.

Reed, Mark S. 2008. "Stakeholder Participation for Environmental Management: A Literature Review." *Biological Conservation* 141, no. 10: 2417–2431. https://doi.org/10.1016/j.biocon.2008.07.014.

Reed, Mark S., Anil Graves, Norman Dandy, Helena Posthumus, Klaus Hubacek, Joe Morris, Christina Prell, Claire H. Quinn, and Lindsay C. Stringer. 2009. "Who's in and Why? A Typology of Stakeholder Analysis Methods for Natural Resource Management." *Journal of Environmental Management* 90, no. 5: 1933–1349. https://doi.org/10.1016/j.jenvman.2009.01.001.

Riley, Shawn J., J. Kevin Ford, Heather A. Triezenberg, and Patrick E. Lederle. 2018. "Stakeholder Trust in a State Wildlife Agency." *Journal of Wildlife Management* 82, no. 7: 1528–1535. https://doi.org/10.1002/jwmg.21501.

Rondinelli, Dennis A., and Ted London. 2003. "How Corporations and Environmental Groups Cooperate: Assessing Cross Sector Alliances and Collaborations." *The Academy of Management Perspectives* 17, no. 1: 61–76.

Sari, Effi Permata. 2007. "Kembali Ke Ladang Menggapai Asa." In *Dari Desa Ke Desa: Dinamika Gender Dan Pengelolaan Kekayaan Alam (From Village to Village: Gender Dynamics and the Management of Nature's Wealth)*, edited by Yayan Indriatmoko, Elizabeth Linda Yuliani, Yunety Tarigan, Farid Gaban, Firkan Maulana, Dani Wahyu Munggoro, Dicky Lopulalan, and Hasantoha Adnan, 71–82. Bogor, Indonesia: Center for International Forestry Research (CIFOR).

Sarmiento Barletti, Juan Pablo. 2022. "Learning from Adaptive Collaborative Management: A Participatory Tool to Support Adaptive and Reflective Learning in Multistakeholder Forums." In *Adaptive Collaborative Management of Forest Landscapes: Villagers, Bureaucrats and Civil Society*, edited by Carol J. Pierce Colfer, Ravi Prabhu, and Anne Larson, 159–184. London: Earthscan and Routledge.

Sarmiento Barletti, Juan Pablo, and Anne M. Larson. 2019a. "The Role of Multi-Stakeholder Forums in Subnational Jurisdictions: Framing Literature Review for in-Depth Field Research." *CIFOR Occasional Paper* 194. https://doi.org/10.17528/cifor/007150.

Sarmiento Barletti, Juan Pablo, and Anne M. Larson. 2019b. *The Role of Multi-Stakeholder Forums in Subnational Jurisdictions: Methods Training Manual and Tools for in-Depth Research*. Bogor, Indonesia: CIFOR. https://doi.org/10.17528/cifor/007149.

Sarmiento Barletti, Juan Pablo, Anne M. Larson, Christopher Hewlett, and Deborah Delgado. 2020. "Designing for Engagement: A Realist Synthesis Review of How Context Affects the Outcomes of Multi-Stakeholder Forums on Land Use and/or Land-Use Change." *World Development* 127: 104753. https://doi.org/10.1016/j.worlddev.2019.104753.

Sayer, Jeffrey, Terry Sunderland, Jaboury Ghazoul, Jean Laurent Pfund, Douglas Sheil, Erik Meijaard, Michelle Venter, et al. 2013. "Ten Principles for a Landscape Approach to Reconciling Agriculture, Conservation, and Other Competing Land Uses." *Proceedings of the National Academy of Sciences of the United States of America* 110, no. 21: 8349–8356. https://doi.org/10.1073/pnas.1210595110.

Selsky, John W., and Barbara Parker. 2005. "Cross-Sector Partnerships to Address Social Issues: Challenges to Theory and Practice." *Journal of Management* 31, no. 6: 849–873. https://doi.org/10.1177/0149206305279601.

Smith, J., K. Obidzinski, Subarudi, and I. Suramenggala. 2003. "Types of Corruption and Their Dynamics During Periods of Political Transition." *International Forestry Review* 5, no. 3: 293–302. http://www.cifor.org/publications/pdf_files/articles/asmith0301.pdf.

Stern, Marc J., and Kimberly J. Coleman. 2015. "The Multidimensionality of Trust: Applications in Collaborative Natural Resource Management." *Society and Natural Resources* 28, no. 2: 117–132. https://doi.org/10.1080/08941920.2014.945062.

Stern, Marc J., and Timothy D. Baird. 2015. "Trust Ecology and the Resilience of Natural Resource Management Institutions." *Ecology and Society* 20, no. 2: 14. https://doi.org/10.5751/ES-07248-200214.

Suyanto, S., and Keijiro Otsuka. 2001. "From Deforestation to Development of Agroforests in Customary Land Tenure Areas of Sumatra." *Asian Economic Journal* 15, no. 1: 1–17.

Suyanto, S., T.P. Tomich, and K. Otsuka. 2001. "Land Tenure and Farm Management Efficiency: The Case of Smallholder Rubber Production in Customary Land Areas of Sumatra." *Agroforestry Systems* 50, no. 2: 145–160. https://doi.org/10.1023/A:1010625019030.

Vangen, Siv, and Chris Huxham. 2003. "Building Trust in Interorganizational Collaboration." *Journal of Applied Behavioral Science* 39, no. 1: 5–31. https://doi.org/10.1177/0021886303253179.

Introduction to Chapter 4

This chapter builds on the work reported in Chapters 2 and 3. It is the only one in this collection to explicitly address the 'wickedness' of societal problems – as highlighted in Chapter 1. Kusumanto et al. evaluate the appropriateness of ACM as a tool to address them. The authors first outline the increasing severity of Jakarta's flooding problem, showing its links to rising sea levels, climate change, high and growing population densities, and upriver forest and land use change. They then build on Kusumanto's long-term experiential involvement in the Jambi site, combining that with the diverse set of knowledges of her co-authors, including the more technical angle, extracting elements of both that seem applicable, and identifying likely challenges as they conduct a 'thought experiment'.

This chapter attacks the issue of expanding an ACM approach to broader levels head on. The authors lay out the complexity of stakeholders – from various government sectors and levels to NGOs to academics to local communities and more – and their varying interests. They build on our sense that solutions to the flooding problem (and other such wicked problems) will only come when these varying actors and their interests are incorporated meaningfully into policymaking and planning. Although this is definitely *not* the case now, there is encouraging movement in the Indonesian national policy arena that could allow some more people- and learning-oriented approaches to blossom.

One of the most (imaginative) of the chapters in this book, it also moves us forward as we consider whether and how to move 'upward' and 'outward' with ACM approaches – a topic also discussed explicitly in Chapters 10 and 11.

DOI: 10.4324/9781003325932-6

4 ACM as a pathway to mitigate Jakarta's flood impacts in a changing climate

Trikurnianti (Yanti) Kusumanto, Gusti Ayu Ketut Surtiari, Chris Zevenbergen, Annisa Triyanti, D. Ary A. Samsura, Tristam Pascal Moeliono and Yus Budiyono

Introduction

It was over two decades ago that Kusumanto began CIFOR's field research to apply and investigate the adaptive collaborative management (ACM) approach to forest management in Baru Pelepat village in Jambi Province, central Sumatra (see Chapters 2 and 3, this volume). The endeavour was a partnership between CIFOR, a Jambi-based nongovernmental organisation, and Jambi University and was supported by the Bungo District Government. The team's involvement as process facilitators and action researchers lasted from 2000 until 2006 and has left traces of social, relational, and institutional changes at community and forest landscape levels.

CIFOR's ACM conceptual underpinnings (Prabhu, McDougall, and Fisher 2007) have been imperative for generating the above as well as other outcomes and are inspired by Paulo Freire's philosophy of 'reflection and action upon the world in order to transform it', Holling's ideas on adaptive management of large ecosystems, Habermas' 'communicative action', the social theory of Giddens, Kolb's learning in development, and social learning in contexts of forest and natural resources (Maarleveld and Dangbegnon 1999; Wollenberg et al. 2001).[1] ACM's focus is on transforming social-ecological systems by employing participatory action research (PAR) as a framework for engaging stakeholders. It is an approach that potentially offers a pathway for dealing with larger scale, so-called 'wicked problems'. Underlying values and causes of such problems are typically ambiguous and contested (Lönngren and van Poeck 2020). They are in essence unsolvable and addressing them can at best be done by attempting to improve the situation and learn from the effort (Sol et al. 2018). The governance of natural resource management is characterised by mutual dependencies between the many actors, each with different interests, perspectives, and values, having a stake in the problem. The circumstances in which a given problem occurs can include social plurality, lack of trust among stakeholders, environmental change, scientific disagreement, inadequate legal tools, and varied policy framing. Examples of societal problems with a wicked attribute include environmental degradation, economic crises, or failing educational systems (Rittel and Webber 1973).

DOI: 10.4324/9781003325932-7

Here, we build on thoughts expressed earlier in ACM's history: "Although climate change issues were just a faraway twinkle in the eyes of the original ACM researchers, the relevance of ACM results to climate change adaptation has become increasingly clear" (CIFOR Website, accessed in June 2022). Now we think it is time that the potential of ACM is assessed as a pathway to address the flooding problem of Greater Jakarta, Indonesia – a notorious wicked problem which calls for adaptation with multiscale and equitable participation, learning, and innovation. ACM can arguably offer a path forward in a world subject to a changing climate and other global environmental change. Our present team is particularly interested to see if ACM offers avenues for transboundary collaboration and transformation at various levels and scales in the Jakarta region. Lessons from our investigation may nonetheless also benefit other global regions with similar wicked problems. The urban setting is a relatively new arena for ACM; only a few studies have been conducted on this approach in such a context. Furthermore, our assessment entails a cross-border endeavour which necessarily reaches out to forest-related realms, and geographically to the upstream parts of the Jakarta delta.

In this chapter we assess ACM's applicability by means of a thought experiment to identify, explore, and develop alternative approaches to better understand and hopefully better manage Jakarta's flooding problem. We believe that these are direly needed. The chapter is not based on an in-depth analysis of empirical work but an exercise whereby our team of experts envisions the application of ACM in Greater Jakarta.

Our multidisciplinary team represents diverse science and development disciplines, comprising human ecology, social learning, flood resilience, urban and regional planning, environmental governance, spatial planning and environmental law, and system dynamics. The lead author specialises in social learning and inclusivity in sustainability governance; she co-led CIFOR's ACM research in Jambi in 2000–2006.

In the following sections, we first describe the background of Jakarta's flooding problem. We then discuss in general terms the wicked problem concept in connection to water governance. A discussion of the ACM concept and applicability follows, including a brief account of why ACM, as our Jambi team applied and experienced it, was successful in delivering positive outcomes. The chapter proceeds with pinpointing the objective of our study and presenting its methodology which, as mentioned above, is essentially a thought experiment. It allows us to draw on the lessons from ACM application in Jambi and link these with our team's expertise, while amalgamating with relevant literature. The chapter continues by framing Jakarta's flooding as a wicked problem and subsequently discussing the results of our thought experiment. Concluding remarks highlight the significance and potential of ACM as a pathway for mitigating the impacts of Greater Jakarta's flooding in the context of climate change.

Background

Greater Jakarta covers a land area of 7,062 km^2 (Kamarzuki 2020) and stretches over the province of the Special Capital Region of Jakarta (DKI Jakarta) and parts of West Java and Banten provinces (Figure 4.1). Major parts of the metropolitan area are the five satellite cities of Depok, Bekasi, Bogor, Tangerang, and South-Tangerang; and the regencies of Bogor, Bekasi, and Tangerang. The low-lying delta is known as one of the most flood-prone metropoles globally (Marfai, Sekaranom, and Ward 2015), through which 13 rivers and two canals flow for the discharge of water into Jakarta Bay (Budiyono et al. 2017). In 2020, Greater Jakarta was home for 35.5 million people (BPS 2021) and is projected to accommodate around 75.6 million in 2039 (Florczyk et al. 2019). The region already experienced floods during the ancient Hindu Kingdom Tarumanegara[2] (4th–7th centuries CE) and they have persisted through colonial Batavia until today's super city.[3] Yet, it has been only since the 1970s that, due to urbanisation and rapid economic growth, land use-land cover change (LULCC) has become a key driving factor of flooding (Rustiadi et al. 2015).

LULCC potentially reduces the area for water catchment and adversely affects drainage systems. Especially LULCC due to urbanisation can importantly influence hydrological behaviour by reducing surface infiltration and increasing surface runoff and flow volumes (Goudie 2018; Rogger et al. 2017). Jakarta's urbanisation is marked by a fast increase in built-up area,[4] thereby rapidly reducing green space surface area and hence also the region's water retention capacity (Maheng, Pathirana, and Zevenbergen 2021). Furthermore, disturbance of the area's hydrology has been due to the loss of upstream forests and of water catchment and urban forest areas in more downstream localities (Afriyanie et al. 2022).[5]

In the case of Jakarta, land subsidence has been another key driving factor of flooding (Budiyono et al. 2016; Yan et al. 2020). The megacity is sinking fast, crucially driven by excessive extraction of deep groundwater which has occurred since the mid-1970s and has resulted in subsidence up to four metres in parts of northern Jakarta (Kooi and Yuherdha 2018). Forty per cent of DKI Jakarta is under sea level (Koto and Negara 2017). Land subsidence has also been caused by soil compaction due to loads from infrastructural construction and buildings associated with urban development (Hasanuddin et al. 2011).

Climate change is posing Greater Jakarta with yet another challenge. Sea level rise, intense rainfall, and extended wet monsoons induced by climate change have increasingly become causal factors that drive the occurrence of flooding and ensuing social and economic disasters. In the occasional case where high volumes of water flow down from the upstream rivers and high rainfall locally together meet up with (tidal) water coming from the sea, these result in disastrous flood levels.[6] Given LULCC, land subsidence, and climate change in the foreseeable future, Jakarta's flood hazards are expected to intensify.

The increasingly frequent and severe flooding has been among the arguments of the incumbent government to relocate the country's capital to East Kalimantan (on Borneo island) by 2024 (Van de Vuurst and Escobar 2020; Yusriyah et al. 2020).[7] Since the 2007 flood in Greater Jakarta – the largest flood over the last two and a half decades (see Figure 4.1), the region has been stricken by devastating floods in 2013, 2015, 2018, and 2020. Yet, while moving the capital may help to evade potential loss and damage associated with flooding (Januariyadi et al. 2020), it could merely mean a transfer of the problems confronting Jakarta to the new

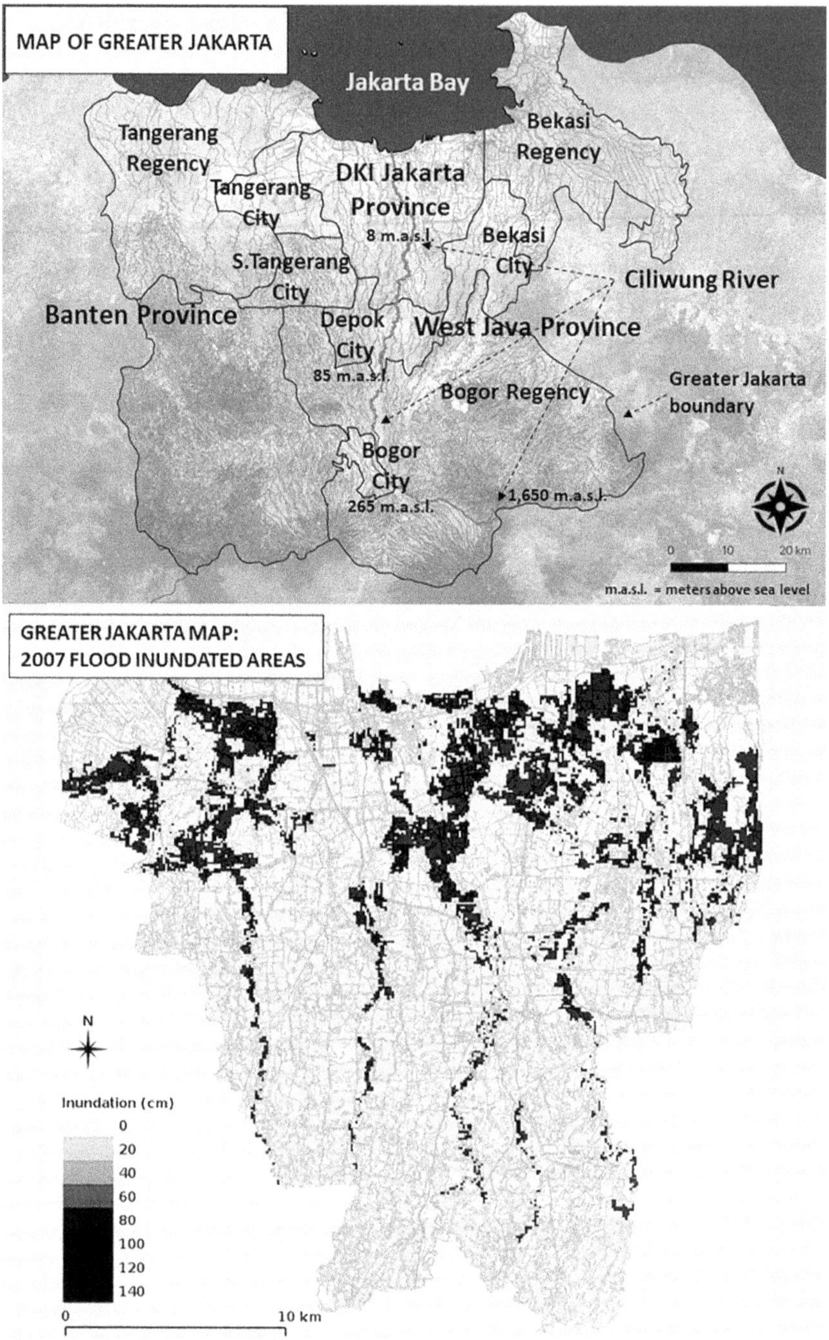

Figure 4.1 Maps of Greater Jakarta showing the different jurisdictions and its most flood-prone Ciliwung River, with inundated areas at village administrative level due to the 2007 flood

capital (Van de Vuurst and Escobar 2020). Moreover, without any doubt, the current capital – and Greater Jakarta – must still be protected from future flooding.

Early flood mitigation measures date back to 17th-century Dutch colonial times (Caljouw, Nas, and Pratiwo 2009; Garschagen, Surtiari, and Harben 2018; Octavianti and Charles 2018). With a view to obtaining control over the city's hydrology, for 400 years, infrastructure-focused engineering solutions – e.g., structured canal systems, flood reservoirs, and the giant sea wall – have been the mainstream paradigm in flood management (Octavianti and Charles 2018). Limited attention has been paid to the anthropogenic root causes of flooding and recent studies revealed that responses to Jakarta's flooding problem should be sought beyond the engineering and technocratic realms (Asdak, Supiah, and Subiyanto 2018; Cao et al. 2021; Garschagen, Surtiari, and Harben 2018).

'Wicked problems' and water governance

Since Rittel and Webber introduced the wicked problem concept in 1973, literature on this idea has grown exponentially, dispersed across a wide variety of scientific disciplines with distinct epistemological assumptions (Lönngren and van Poeck 2020). We position our investigation in a multidisciplinary context which is informed by a social constructivist understanding of wicked problems. From this perspective, there is no 'true' definition of a given wicked problem; it is rather a social construct. Following sociological literature, theoretical concepts function in research as descriptive/analytic tools, sensitising/creative tools, or critical/emancipatory tools (ibid.). The utility of the wicked problem concept as a sensitising tool, particularly, fits well in our assessment of ACM to explore the approach's applicability to Jakarta's flooding. Because of its multifaceted and suggestive character, the concept creates space for reflection, creativity, and surprises in developing understanding of the flooding phenomenon and, importantly, in finding decentralised and equitable pathways for dealing with it.

In general terms, a wicked problem is a situation of high complexity, uncertainty, and divergence which involves multiple stakeholders with distinct needs, values, interests, knowledges, and expectations (Kharel, Romsdahl, and Kirilenko 2018). Following Sol et al. (2018), the most realistic effort for addressing such problems is to make improvements in problematic situations and learn from such efforts. With wicked problems, optimising prevailing practices, routines, and systems – that is, 'doing better the things we do' – will not help much (see Prabhu and Colfer, this volume); rather, the values and assumptions on which actions are grounded are reconsidered so that we 'do better things' (Sol et al. 2018, 1385). This implies a need for alternative policy pathways, relationship building, ways of thinking and perceptions, behaviours and lifestyles, and at the heart of all these: novel processes for the production of knowledge and social learning.

The multiplicity of jurisdictions often associated with wicked problems implies that the solutions as preferred by the various stakeholders may diverge or even be contradictory. Only an unbounded time frame offers chances to appraise the efficacy and consequences of potential solutions (Kharel, Romsdahl, and Kirilenko

2018), learn from our attempts, and accordingly improve these (Sol et al. 2018). For solutions are consequential and can create new problems. Flexibility and adaptability are therefore warranted through learning or trial-and-error (ibid.; Rittel and Webber 1973). Governmental planning tends nevertheless to have limited room for adaptations (Kharel, Romsdahl, and Kirilenko 2018) and largely responds to short-term time horizons (Adams-Schoen 2016).

Water governance problems are arguably ill-defined (Kharel, Romsdahl, and Kirilenko 2018). They greatly vary in scope and nature in terms of causes and consequences and resolving them tends to rely on elusive political judgement. While conventional expert-driven and administrative routines can solve many of the problems, they may provide little solace in solving wicked problems and can create stalemates for policymakers and flood managers. Among the challenges confronted when dealing with water-related wicked problems are (ibid.; Adams-Schoen 2016) (i) the transboundary nature across jurisdictions, sectors, and institutions; (ii) ignorance in governmental planning of hydrological processes at watershed scales; (iii) limited scope and enforcement of policies to resolve newly emerging water management problems, such as those caused by climate change; and (iv) water conflicts arising between economic, environmental, and social objectives.

Concept and applicability of adaptive collaborative management

When CIFOR coined the term 'adaptive collaborative management' around 1997, it was intended to investigate adaptive management (Lee 1999) in varying social (stakeholder) contexts (Plummer et al. 2012). Since the early stages of ACM scholarship, its focus has gradually broadened to also comprise collaborative management in connection to complexity science and resilience thinking in social-ecological systems (Armitage, Marschke, and Plummer 2007), including in connection to climate change (CIFOR Website). Hence, ACM is conceptually a convergence of adaptive management and collaborative management. A substantial body of literature between 1997 and 2010 views ACM as "an emergent governance approach for complex social-ecological systems that connects the *learning function* (experimental and experiential) of adaptive management with the *linking function* (vertically and horizontally) of co-management" (Plummer et al. 2012; emphasis added).

ACM's instrumental rationale is twofold. Equipped with the learning and linking functions as modalities, the approach is expected to "deal with the complexity of interdependent social-ecological systems and enhance the fit between ecosystem dynamics and governance systems" (Olsson, Bodin, and Folke 2010, 263). Furthermore, ACM is also postulated as a *continual and iterative process of action and reflection* whereby outcomes shape pre-conditions for the process to continue (Colfer 2005a; Plummer et al. 2017).

ACM has received much attention from scientists and practitioners alike, including criticists. The relationship between learning and outcomes, in particular, is often referred to as troublesome. ACM is a relatively young field and the absence of a framework to organise the many considerations, definitions, variables, driving factors, and outcomes makes it difficult to analyse plausible relations between process, social interactions, and outcomes of an empirical case (Plummer et al. 2012). CIFOR's extensive investigation asserts that ACM is highly contextual and hardly any variable is deterministic (Colfer 2005a). Yet, various empirical works have evidently established positive relationships between ACM process and outcomes – i.e., ecological and livelihood effects (Colfer 2005b; Guijt 2007; Plummer et al. 2017). Interestingly, some scholars attach ACM's value to its evocative nature which suggests pathways for transitionary changes in attaining desirable resource and environmental governance objectives, rather than that it provides particular benchmarks (Huitema et al. 2009).

While not always explicitly labelled 'adaptive collaborative management', the approach has been widely applied to addressing various resource and environmental management and governance challenges (e.g., concerning agriculture, water management, or restoration). Plummer's ACM literature review (2012) reveals that the approach is predominantly applied in 'typical' common-pool resources, such as forestry, water resources, and fisheries. Not much can be found about the extent to which ACM potentially contributes to climate change adaptation and mitigation, nor to risk reduction of climate-induced disasters.

Crucially, the utility of ACM in research and practice for dealing with the diverse challenges lies particularly in the convergence of collaboration and knowledge-oriented processes. In Jambi's empirical work (Adnan et al. 2008; Colfer 2005a; Diaw and Kusumanto 2005; Indriatmoko 2002; Kusumanto 2006, 2007a, 2007b; Kusumanto et al. 2005), learning and linking processes were crucially connected to the pragmatic character of our team's role as action researchers and process facilitators. The team encouraged linking and learning to be organised and maintained between diverse community stakeholders, and vertical linking and learning between community stakeholders and village institutions, as well as between community representatives and the district government. The use of PAR as a framework crucially enabled the collaborative and learning processes to take place: (i) *substantively* (by way of locally prioritised issues); (ii) *structurally and relationally* (through PAR's joint plan-act-reflect iterations); and (iii) *via transdisciplinary/transboundary* learning between different social and institutional entities – including our Jambi ACM team. In this way, ACM's facilitated intervention is essentially a blended assortment of smaller-scale interventions grounded in systems thinking; smaller interventions were linked horizontally and vertically and nested in ACM in its entirety.[8]

Investigation by thought experiment: objective and methodology

The objective of our investigation is twofold. First, we position the flooding problem of Greater Jakarta within a wicked problem framing, which allows us to develop understanding about the way flooding governance has evolved over time in response to a changing flooding context. Second, we assess the applicability of

ACM to addressing Jakarta's flooding as a wicked problem in the climate change context so as to deliver recommendations with long-term objectives.

We employ a methodology referred to as a 'thought experiment', which originates in philosophy – essentially relying on human intuition and imagination – and its use has gradually expanded to the natural and social science disciplines (Brown and Fehige 2022). The methodology can be applied to investigate phenomena with the purpose of thinking through a hypothetical situation and its probable human and societal consequences. In our case, we obtain new insights by using already known information based on previous ACM empirical work from the Jambi research and rearranging this information from the new perspective of Jakarta's flooding context. Prior lessons that draw on ACM Jambi research are combined with insights from our team's expertise and the literature.

Our choice for a thought experiment has a practical and conceptual rationale. Not an empirical investigation, our study uses prior ACM Jambi research outcomes while relying on the team's expertise and empirical works by others in a different locale and context. An actual ACM investigation at the scale of Greater Jakarta would at present not be affordable. Nevertheless, it should be cautioned that the cost of a 'business as usual' approach to Jakarta's flooding could presumably be significantly higher. The conceptual rationale of our thought experiment is that outcomes derived from our study can offer insights for handling Jakarta's flooding problem, applying ACM, or scaling-up initiatives using ACM or similar approaches.

Framing Greater Jakarta's flooding as a wicked problem

The flooding problem of Greater Jakarta is characterised in the literature as complex, uncertain, and multijurisdictional (Cao et al. 2021; Dwirahmadi et al. 2019; Simarmata and Surtiari 2020). The problem cannot easily be defined. As discussed previously, these attributes are typical of wicked problems, which have seriously challenged Jakarta's flood policymakers.

Viewing the metropolitan area as a typically deltaic megacity in the Global South can cast some light on the issue. Population growth, urbanisation, and urban sprawl in the megacities situated in deltas of this global region tend to manifest themselves in the conversion of waterways to other uses and in an expansion of informal settlements in flood-prone areas, both of which complicate flood management (Cao et al. 2021). In such areas, water supply for household use, as well as for urban and industrial development, typically relies on the over-extraction of groundwater, resulting in land subsidence. An increased risk of coastal flooding is often the consequence. Greater Jakarta is among the delta megacities with the most severe flood risk in the future (Cao et al. 2021; Garschagen, Surtiari, and Harben 2018; Marfai, Sekaranom, and Ward 2015; Rukmana 2021).

Different stakeholders perceive Jakarta's flooding problem differently and envisage therefore distinct solutions. Governments and other stakeholders of downstream flood-prone parts of the region – particularly low-lying areas in DKI Jakarta Province – view Jakarta's flood problem primarily as an outlet problem. They seek solutions for the protection of people, resources, and infrastructures through flood mitigation – which means diverting flood water away from flood-prone zones and/or relocating communities living there to flood-safe localities. For vulnerable people living in informal settlements, flooding can affect water quality and hinder a safe, healthy, and productive life. Many of these groups consider these localities as key to making a living, and government relocation programmes mean to them a loss of livelihood (Dovey, Cook, and Achmadi 2019; Simarmata and Surtiari 2020). For these groups, leaving these areas does not resolve the flooding problem and they seek solutions in informal, small-scale flood adaptation measures, such as raising the floor of dwellings or building simple water barriers (ibid.; Cao et al. 2021). Downstream stakeholders have often scapegoated upstream stakeholders, such as the government of Bogor Regency, upstream farmers, or plantation holders for flooding occurring downstream. Greater Jakarta's flooding is an intricate transboundary issue that involves multiple jurisdictions, sectors, and institutions, each with its own preferred solutions, which can lead to conflicts.

The 'wickedness' of Jakarta's flooding problem is obvious as well from the nature of the diverse adaptation pathways followed by stakeholders in response to changing flooding contexts. We follow Cao et al. (2021) who define adaptation pathways as "sequences of measures that can be implemented to reduce the impacts of changes in environmental conditions" (88). The discussion that follows, makes clear that whatever pathway is pursued by the different stakeholders, most proposed solutions tend, borrowing Rittel and Webber's (1973) wicked problem terminology, to be a 'one-shot operation'. The tendency exists that solutions are expected to immediately resolve the flooding problem. However, negative consequences or side-effects of solutions may not be reversible and new problems are likely to arise. Furthermore, it is difficult to determine how long Greater Jakarta's flooding will continue. This temporal uncertainty – yet another attribute of wicked problems – implies that 'there is no stopping rule' (ibid.): there is no point in time that establishes that the handling of a problem is complete. This temporal aspect is exacerbated in Greater Jakarta by the emerging challenges of land subsidence and climate change. While Jakarta's current flooding governance may tame the flooding problem temporarily, the risk is real that future problems are much more severe. Below, we describe how the 'wickedness' of adaptation pathways in Jakarta's flooding case has manifested itself in contemporary times.

Although in response to changing flooding contexts, Jakarta's flood governance has been continually adapted over time, yet the core paradigm has remained largely focused on 'taming nature' by attempting to control hydrology (Garschagen, Surtiari, and Harben 2018). Floods have been perceived as an annual recurrence linked to the monsoon cycle, hence requiring tactical, short-term responses and a focus on controlling flow from outlets. Consequently, a canal and drainage system connecting the city's waterways, initially developed by the Dutch in the 17th century, has been the main flood management strategy (Caljouw, Nas, and Pratiwo 2009). It diverts flood discharge of the Ciliwung – the region's largest and most flood-prone river – and other waterways to the city's peripheries and further into Jakarta Bay.

The devastating flood in 2007, shown in Figure 4.1, led to a significant adaptation of the outlet-based flood governance. Extreme precipitation accumulations in the metropolitan area met with water coming from the sea pushed by an extremely high tide. This event triggered the government to expand its flood governance to also include coastal protection and the enhancement of the city's water retention capacity (Garschagen, Surtiari, and Harben 2018). In 2011, the Jakarta Coastal Defense Strategy (JCDS) was adopted by the Indonesian government in collaboration with the government and experts from The Netherlands. Subsequent revisions and expansion of JCDS delivered in 2014 the National Capital Integrated Coastal Development (NCICD) Masterplan with the 46-kilometre Giant Seawall that closes off Jakarta Bay from the sea as a main component, complemented by a large pumping system for the metropolitan area's flood drainage.[9] The various infrastructural works – i.e., dredging canals, dykes, and the giant sea wall – have demanded space for which around 4,000 households were relocated between 2015 and 2018 (Simarmata and Surtiari 2020), sometimes involving coercion (ibid; Padawangi and Douglass 2015).

Despite the changing course of flood governance, hard engineering-infrastructural solutions have remained central. Yet, Cao et al. (2021) remark that NCICD's main component – the sealing off of Jakarta Bay from the sea – is not going to solve the fundamental cause of Jakarta's flooding, namely land subsidence.[10] New problems and critics have also emerged from civil society and from within the government itself – namely, the Ministry of Maritime Affairs and Fisheries. These officials foresaw significant environmental degradation in Jakarta Bay, including changes in local currents, a decrease in fish stocks, and rapid sedimentation of the seabed, thereby adversely affecting the bay's ecosystem and putting the livelihoods of fishing communities at risk.

It was not until 2017 that the government drastically refocused its flood governance in response to mounting criticism on, particularly, the relocation of communities and exclusion of residents from decision-making and planning. As envisioned in the 2017–2022 medium-term development plan of DKI Jakarta, flood risk management measures should include the strengthening of water institutions and human capital, besides a stricter policy on the use of groundwater (Simarmata and Surtiari 2020). Obviously, in Jakarta, the dominant formal adaptation pathway applied by the government is not really connected to the informal adaptation trajectories of many, often-times vulnerable, local communities and in some instances has even hampered their capacities to adapt (Cao et al. 2021).

Results of thought experiment and discussion

Viewing Greater Jakarta's flooding as a wicked problem leads to the question whether ACM would be applicable in this context and, if so, what outcomes could be expected from applying the approach. In the sections below, we assess ACM's applicability and focus on the following three interconnected points of discussion: (i) can ACM be applied, given Greater Jakarta's flood governance structure; (ii) will ACM's social learning work in Jakarta's flooding context; and (iii) if ACM were applicable to the case of Jakarta, what operational indicators could be used.

Can ACM be applied, given Jakarta's flood governance structure?

Assessment of ACM's applicability

Our central thesis is that Jakarta's flooding risks can effectively be managed *if* adaptation is strategic – which we see as encompassing and connecting sufficiently large spatial and temporal horizons. This means that the governance of adaptation must include and interconnect all necessary nested levels and scales of decision-making. We have shown, based on our past ACM research in Jambi's forest environs, that a multilayer, nested governance structure, organised around interdependent formal and informal decision-making nodes at various levels and scales where stakeholders are represented, is key to effective adaptation in complex and uncertain forest settings (Diaw and Kusumanto 2005; Kusumanto 2007a). Such a structure allows for a more balanced power distribution in the social-ecological system, transboundary learning, and stakeholder communication, resulting in the construction of shared values and knowledges. Our ACM research made clear that a polycentric governance structure for attaining sustainability in the forest system is crucial, necessarily comprising formal as well as informal structures and mechanisms. In Jambi, it became clear to us that informal decision-making in polycentric systems was a critically important, and often missing, adjunct to formal structures. Literature on collaborative governance also underlines the importance of informal structures and mechanism (Emerson, Nabatchi, and Balogh 2011) but the link with polycentric decision-making is usually less explicit than in our Jambi case.

Huitema et al. (2009) have similarly underlined the importance of polycentric decision-making for fostering adaptation and collaboration in social-ecological systems. Translating this into the context of water governance, a polycentric structure implies that the lowest possible jurisdictional level should hold decision-making authority for the implementation of flood policies, spatial plans, and flood disaster protocols (Becker, Huitema, and Aerts 2015). The central government would thereby be responsible for oversight of legal procedures, with well-functioning coordination between different levels as a prerequisite (ibid.).

Bringing these insights to bear in the case of Greater Jakarta's governance structure, we see a rather weak resemblance to a polycentric governance system, potentially hindering effective implementation of flood policy measures. This situation is further complicated because of the multiplicity of jurisdictions – namely, the different administrative areas (Figure 4.1) and sectoral mandates (Samsura, Kusumanto, and Triyanti 2022). In current decentralised Indonesia, decision-making authority in the water and land sectors is held by regional governments – to wit, provincial, and municipality/regency governments (Simanjuntak et al.

2012). In reality, however, flood risk management and strategic authority tend to remain concentrated at the central level (Rukmana 2016). The Ministry of Public Works and Public Housing has essentially the sole authority over the most flood-prone river of the region, the Ciliwung, including efforts to improve its discharge and retention capacity. Besides, the upstream parts of the Ciliwung watershed have remained the responsibility of the Ministry of Environment and Forestry,[11] leaving limited space for decision-making by local governments. Furthermore, local governments are involved in public infrastructure development and services only when permitted by the Ministry of Public Works and Public Housing (Simanjuntak et al. 2012). All of this is further compounded by poor inter-sectoral and inter-agency coordination and collaboration (ibid.; Samsura, Kusumanto, and Triyanti 2022; Simarmata and Surtiari 2020).[12]

Hence, we assess that Jakarta's flood governance structure is weakly appropriate for shaping the enabling conditions for adaptation that is strategic. As noted previously, we consider adaptation to be strategic, if it incorporates and interconnects sufficiently large spatial and temporal horizons and, as such, could effectively address Jakarta's flooding. Nevertheless, we have identified a unique opportunity for improving Greater Jakarta's flooding governance structure and in that way develop the necessary conditions to apply ACM to Jakarta.

Window of opportunity for improving Jakarta's flood governance structure

Presidential Regulation No. 60 of 2020 concerning the Greater Jakarta Urban Area Spatial Plan offers room for improving Jakarta's flood governance structure. With the main aim to transform the metropolitan area into an economic stronghold for industrial development, trade and service delivery (Kusumanto et al. 2022), it directs development in the region by regulating spatial patterns and area utilisation (Afriyanie et al. 2022). A recently established regional body affiliated with the Ministry of Agrarian and Spatial Planning holds the mandate for implementation. The regulation, however, does not include climate change impact considerations and despite its flood mitigation and adaptation directives, little guidance is provided on how spatial planning should be linked to flood risk management under the pressing conditions of economic growth, urban development, and rapid land use change.

We view these missing elements in the new regulation as a window of opportunity for applying ACM and, simultaneously, carrying through the necessary adjustments in the current flood governance structure. The central aim would thereby be transforming the governance structure into a polycentric system. By using ACM as a framework, we argue that restructuring would be operational with limited need for investments in financial and human resources. Structural adjustments would be implemented under the new spatial plan regulation and specific flood policies could be developed that pertain to subsets of relevant jurisdictions. Jurisdictions should be given sufficient space for self-governance and decision-authority at relevant scales. The new coordinating regional body, previously mentioned, would be responsible for supervision over legal procedures and be mandated to establish a legal framework and formal regulations, as well as to improve inter-agency coordination and simplify governance mechanisms. The engagement of civil society groups in the interactions and communication with local stakeholders are both crucial, facilitating local participation in decision-making.

Time and effort would be needed to institutionalise the adjustments to Jakarta's flood governance structure as discussed above. In the following section, we seek to understand whether social learning – at the heart of ACM – would work for bringing these about and, if so, what role it could play, and with what outcomes.

Would social learning in ACM work in Jakarta's flooding context?

Assessment of ACM's social learning

By definition, social learning in ACM connects collaboration with the adaptive process by encouraging a continuous reflection and revisits of plans, relationships, knowledges, and worldviews, fostering agency and transforming structures and social systems (Prabhu, McDougall, and Fisher 2007). Hence, as a concept, ACM should be well-equipped for addressing Jakarta's wicked flooding problem because of its multiple perspectives, systematic and iterative protocol, and sustainability (long-term) objectives. The question remains, however, if social learning in ACM could effectively address the complexities inherent in the megacity's flooding problem. We assess below ACM's applicability to dealing with this wicked problem, focusing on the central role which social learning is expected to play in fostering institutional adaptations, collaboration, and policy change.

It is obvious that the region's flood governance structure, as discussed previously, is rather weak for creating a culture of social learning and a collaborative and adaptive environment. As discussed, Jakarta's water and flood governance are relatively centralised and fragmented over different ministries, government units, and implementing agencies with disparate viewpoints and weak horizontal and vertical communication. Where collaboration across institutions, sectors, and jurisdictions is required, existing communication mechanisms tend to be along hierarchical lines. Hence, space is limited for learning or the exchange of values and knowledges. Inadequate social learning between government and communities is common as well, except for some cases where civil society organisations facilitate participatory interactions between the two (Padawangi and Douglass 2015; Rukmana 2016) or public agencies are endowed with the necessary communication capacity.

We argue here that social learning in ACM can only work in Jakarta's flooding context *if* transboundary communication channels and mechanisms are put in place for a systematic and iterative exchange of values, perspectives, and knowledges. In CIFOR's ACM, PAR was central for building these. Viewing Jakarta's flooding as a wicked problem as previously discussed, communication channels and mechanisms would ideally encourage a continual appraisal of possible consequences of joint plans and actions as well as of implemented solutions. As also noted above, it is important that informal mechanisms complement formal structures and processes. Furthermore, crucial as well are effective facilitation for learning at the boundaries between different stakeholders and how to ensure that all relevant stakeholders are justly represented in the learning and collaboration. It is also obvious that sufficient financial, human, and time resources, as well as political support are key prerequisites.

It has become clear at this point that for social learning to work in Jakarta's context, the core issue is how to develop communication channels and mechanisms for a systematic, participatory, and just exchange of perspectives, values, and knowledges. In other words, how to shape the enabling conditions for transboundary learning that encourages collaboration and adaptation. To this end, below we envision the initiation of *nested platforms for transboundary learning and collaboration* using ACM as a framework.

Shaping opportunities for transboundary learning and collaboration in nested platforms

For this particular initiative, we define

> a transboundary learning platform as consciously constructed opportunities for multiple stakeholders to jointly learn about a particular flood policy issue by iteratively exploring, implementing, and appraising flood adaptation policy and measures and their consequences, improving circumstances accordingly, and learning from the efforts.

Following Buck, Wollenberg, and Edmunds (2001), we consider that unlike stakeholder meetings or forums, the way platforms are designed and facilitated influences what is perceived and experienced by platform participants and what can be achieved. Platforms involve human as well as non-human entities, such as technology, a given resource, and data (Steins 2002). Over time, meanings, perceptions, and social experience are reshaped through collective human action with the non-human entities making part of the collective action itself. Uncertainty is inherent in complex problem-solving as the full implications of neither the process, nor the outcomes of the activity can be known (Aarts and Van Woerkum 2002). Rationality is one of the key emergent properties from platform processes (Steins 2002) and in circumstances of uncertainty, collective decisions should be made by platform participants each of whom has his or her own starting rationality (Aarts and Van Woerkum 2002). A carefully designed platform can assist participants in dealing with uncertainties more creatively and effectively. The role of facilitation is thereby pivotal (see, e.g., Hagmann et al., this volume). Uncertainty external to the platform (or for that matter, to any collaborative endeavour), such as uncertainty of flood risk or of possible impacts of climate change, can act as a '*driver*' (Emerson, Nabatchi, and Balogh 2011) or '*trigger*' (Kusumanto et al. 2005) for different stakeholders to seek collaboration for managing the problem at hand.

To ensure legal and policy support, the initiative would be implemented within the legal framework of Greater Jakarta's urban spatial plan issued in 2020, mentioned previously. The new regulation is helpful instrumentally in three ways: first, high-level institutional coordination has been put in place, headed by the Minister of Agrarian Affairs and Spatial Planning and run by heads of government at provincial and municipality/regency levels, creating opportunities for fostering institutional adaptations and collaboration at the various levels. Second, the regulation provides an excellent learning opportunity for collaborative governance between state agencies, as well as between government, community, the private sector, and civil society. Third, the regulation provides a basis from which cognition can be improved and awareness enhanced among policymakers about the links between Jakarta's course of development and increasing flood risk, potentially complicated by climate change, and how to curtail potential impacts of flooding.

The overall objective of the proposed nested platforms for transboundary learning and collaboration is two-fold: (i) improve compatibility between river basin ecosystems and the institutions that manage human activities affecting these; and (ii) foster Jakarta's regional transition towards collaborative water and flood governance and management. The adjective 'nested' emphasises the incorporation of multiple jurisdictional and social-ecological landscapes into official policy and plans. A given platform constitutes cross-cutting jurisdictions specialising in or affected by a specific flood policy issue. We discuss below an example of envisioned transboundary learning platforms.

CILIWUNG RIVER ACM NESTED PLATFORMS

The Ciliwung River ACM initiative is envisioned to shape learning conditions that foster creativity and discovery towards new problem framings and alternative solutions. The platforms provide a learning track which functions in parallel to and delivers policy options to the formal policy track. We discuss the four core components of the initiative: key stakeholders; challenges; platform design; and facilitation of transboundary learning.

Following the World Meteorological Organization and Global Water Partnership (WMO and GWP 2009), **key stakeholders** for an effective urban flood risk management and basin development include the following categories: government ministries, departments, and agencies; communities in flood-prone areas; other basin communities; research institutions; NGOs; and voluntary organisations. The 119-km long Ciliwung River crosses DKI Jakarta Province, West Java Province, Bogor Regency, Bogor City, and Depok City. Box 4.1 lists the key stakeholders of the Ciliwung ACM platforms.

Box 4.1 Key stakeholders of Ciliwung River's proposed ACM nested platform*

- **Government** includes relevant departments and agencies of the Ministry of Agrarian Affairs and Spatial Planning (ATR/BPN), Ministry of Environment and Forestry, Ministry of Public Works and Housing, National Development Planning Agency, Regional Development Planning Board, Ciliwung-Cisadane River Basin Agency**; Ciliwung-Citarum Watershed Management and Protection Forest Agency**; Indonesian Agency for Climatology, Meteorology, and Geophysics; National Disaster Management Agency; Provincial and relevant Municipality Governments of DKI Jakarta Province, Provincial Government of West Java, Bogor Municipality Government, Bogor Regency Government, Depok Municipality Government.
- **Communities in flood-prone areas** along the Ciliwung, Citarum, and Cisadane rivers include those in DKI Jakarta areas such as Manggarai, Tanah Abang, Tomang, Jembatan Lima, Pluit, Duren Sawit, Pondok Kopi, and Cakung; and in Depok City areas such as Sukmawijaya, Pancoran Mas, Cipayung, and Sawangan.
- **Other river basin communities** include upstream Ciliwung, Puncak sub-regency, and Bogor Regency, and Action Consortium for Saving Upstream Ciliwung communities.
- **NGOs** include Telapak, Ciliwung Merdeka (Free Ciliwung), and Friends of Ciliwung.
- **Voluntary organisations** include Gerakan Ciliwung Bersih (Clean Ciliwung Movement) and Komunitas Peduli Ciliwung (Community Concerned about Ciliwung).
- **Research institutions** include Indonesian National Research and Innovation Agency (BRIN), Institute of Technology Bandung, Bogor Agricultural University, and University of Indonesia.
- **Business** includes Indofood, Indonesia Power, Perusahan Listrik Negara (PLN, National Electricity Company), and Perusahaan Air Minum (PAM Jaya, Drinking Water Company).

*Stakeholder list not exhaustive.

**Located in Greater Jakarta, the Cisadane watershed is west and adjacent to the Ciliwung watershed; they and two smaller watersheds make up the Ciliwung-Cisadane river basin area (Arifin, pers. communication, 24 June 2022). The middle and downstream parts of the Citarum watershed are located in Greater Jakarta; these are east and adjacent to the Ciliwung watershed (Julian, pers. communication, 24 June 2022); the Citarum watershed is part of the Citarum river basin area (Website Major Office of Citarum River Basin).

A testing **challenge** which the platform design should particularly take into account is the value-laden policy and institutional context of Jakarta's flooding. In such a setting, the platforms can encounter a compounded challenge beyond the common lack of a social learning culture and collaborative environment in public institutions, as previously mentioned. In our Jambi ACM research, when social learning occurred at the formal level and community stakeholders interacted with district/regency officials, learning was at points influenced, even hindered by prevailing bureaucratic and institutional requirements (Kusumanto 2006).[13] We anticipate a similar but more extreme challenge in the Ciliwung platforms because of their specific policy and institutional setting.

Ward et al. (2013) note that in global delta cities, including Jakarta and Rotterdam (The Netherlands), changing flood governance or paradigms is not easy because of institutional path-dependency and deep-rooted policy beliefs. In interactive policymaking, policymakers must work with contradictory views and interests, but approaches that encourage multiple perspectives are not common in policy practice (Wagemans 2002). The incorporation of multiple perspectives, however, is the bottom line for any policy to effectively address wicked problems. The policy system can be a learning barrier for those with a formal position, even when exposed to new perspectives. To a relatively limited degree, officials are likely to be receptive to alternative problem framings and new solutions that are brought onto the platforms by new stakeholders. It is unrealistic to expect public officials to ignore institutional mandates and responsibilities because of risks of being sanctioned institutionally. Due to this impasse, policy processes tend to reduce diverse perspectives to a single perspective that is acceptable from a formal standpoint and conforms with existing legal and policy frameworks. Nonetheless, the wicked nature of Jakarta's flooding problem requires a form of learning that allows for the incorporation of new values and multiple problem perceptions and concepts. Breaking through this policy and institutional deadlock is a trying task. In this context, the platforms must offer participants a new way of learning which they derive from collaboration and experimenting, without bearing the risk of becoming punished institutionally.

Bearing in mind this challenge, the **platform design** constitutes three central elements: its core idea, structure, and learning focus. The **core idea** of the design of the platforms is to shape learning conditions that stimulate creativity and discovery outside the policy system and, hence, independent from existing flood policy. Platform participants are assigned to collaboratively experiment with new problem framings and solutions, monitor experiences, and learn from the efforts. Official platform participants are not expected to evaluate and approve policy innovations in accordance with legal and policy frameworks. Successful flood policy options or alternative policy measures resulting from the platform processes are eventually mainstreamed in the existing flood policy. As such, the platforms provide a learning track or learning pathway which functions in parallel to, is independent from, yet delivers policy options that feed into the formal policy track.

Because the platforms proceed in parallel to the formal policy process, the **platform structure** should provide supports to the multilayer, multiscale, and nested polycentric governance system. In analogy with the way we structured system-wide ACM learning in Jambi, the platforms' transboundary learning comprises a blend of smaller-scale subsystems nested in wider-scale subsystems. The lowest level jurisdiction is framed by wider subsystems – e.g., a spatial plan legislation or land tenure policy. Importantly, the use of PAR as a framework for bringing structure into platform activities effectively encourages learning and collaboration. Through PAR's joint plan-act-reflect iterations, relations become more structured and actor networks within platforms take shape organically. Figure 4.2 illustrates the nested structure of the Ciliwung River ACM platforms, as we propose.

For proposing the **focus of learning and collaboration** of the platforms, we argue that attempts to solve Jakarta's flooding have exhausted legal instruments and policy measures currently available. Given the wickedness of the flooding problem and the fragmentation of water and flood governance over different sectors, institutions, and jurisdictions as previously described, the platform should use a basin-wide, cross-sectoral programmatic approach. This approach can become an impetus for different sectors, institutions, and jurisdictions to integrate spatial planning more effectively with the multiple social, economic, and environmental values. At the same time, they will be able to maintain implementation of their projects and formal duties as well as, where relevant, deliver public services (e.g., improving drainage, sanitation, or waste management). Slightly adapted from the Global Environment Facility's definition (GEF 2009, 7), we view a 'programmatic approach' as "a long-term and strategic arrangement of individual or sectoral yet interlinked projects aimed at achieving large-scale impacts on the (global) environment". Learning and collaboration from the platforms would deliver policy measures in connection to the above-mentioned Greater Jakarta spatial plan. The

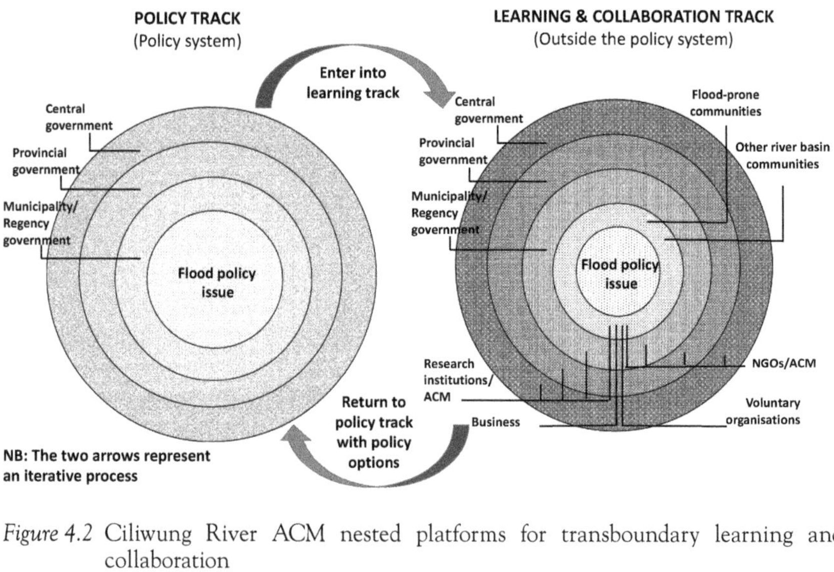

Figure 4.2 Ciliwung River ACM nested platforms for transboundary learning and collaboration

programme should allow connecting problems that are currently addressed by different policies, sectors, and/or institutions.

The **facilitation of transboundary learning** is challenging. Some learning points from past ACM research can be of benefit here. In handling the flooding problem, it is imperative that the social diversity of the platform be made salient to platform participants (Diaw and Kusumanto 2005). The researcher-facilitator is engaged in real-world processes and should maintain stakeholder boundaries as they are, up to the point that the stakeholders themselves decide to enter into cooperation. The building of trust between 'would-be collaborators' is thereby one of her/his key tasks (see Chapters 2 and 3, this volume). 'Collaboration' and 'adaptation' are platform outcomes that must fully be in the interests of platform members. The facilitator's role in this is no more and no less than helping them in attaching meaning to these concepts from their own perspective (see Hagmann et al., this volume).

The facilitation of system-wide learning across boundaries of (stakeholder) subsystems goes beyond the mere convening of stakeholders: from a wicked problem perspective, the facilitator mobilises multiple stakeholders, each with their own problem framing and preferred solutions; mediates or advocates where these are contradictory; facilitates the collaborative development of platform ground rules; and safeguards platform processes (rather than taking decisive action). Facilitation, therefore, also includes the following, approaching, and enrolling of actors, especially those who are less visible. This requires the facilitator's engagement rather than neutrality, sensitivity for hidden relationships and needs, and openness to acknowledge research or facilitation biases. It is clear that the facilitator must be credible, equitable, and considered authoritative by platform participants, including on technical matters. The role of an ACM researcher-facilitator could also be played by the so-called boundary organisations, in the literature referred to as those enabling collaboration between stakeholders by intervening structurally and cognitively (Perkmann 2016).

The facilitation of system-wide change in connection to Jakarta's flooding has inevitable drawbacks because facilitation would almost always need to work through mechanisms of stakeholder representation. However, in circumstances of stakeholder conflict or gridlock, a representation system may offer a way out from stalemates. Although working through stakeholder representatives may not be the ideal, there is much to gain from carefully designed representation mechanisms. While this is beyond this chapter's scope, a major issue that we need to pinpoint here is the contingent nature of representation. In our Jambi case, some representatives perceived their involvement in ACM activities as a privilege rather than as a mandate that was provided by the constituents they represented (Kusumanto 2006). It is therefore crucial that system-wide facilitation ensures that learning processes encompass the representatives, the stakeholders at the subsystem level, and, most crucially, that representation checks and balances are put in place.

Hagmann et al., this volume, discuss some ACM facilitation methods. Facilitation does not necessarily mean that platform participants should meet in person;

it can, in some cases, make good use of, for example, computer modelling (as we did for ACM in forest contexts, see, e.g., Purnomo, Mendoza, and Prabhu 2004; for flooding contexts, see, e.g., Teng, Jakeman, and Vaze 2017) or satellite imaging for facilitating collaborative planning and monitoring exercises. Yet, when stakeholders need to meet in close encounters, the facilitator should make sure that conventional dominance patterns of communication are not replicated (Sarmiento Barletti 2022).

What operational indicators could be used for applying ACM to Greater Jakarta's wicked flooding problem?

Our thought experiment on ACM's applicability to Greater Jakarta's flooding problem has resulted in the following key points: first, ACM can be applied to Jakarta's flooding if adjustments are made to the flooding governance structure and in this way enabling conditions are created for a strategic adaptation that can effectively address the problem. Second, time and effort would be needed to institutionalise the envisaged structural adjustments. To this end, ACM can be of benefit to accelerate the process by shaping conditions for transboundary learning, collaboration, and adaptation in handling the wicked problem. In Box 4.2, we provide operational indicators that reflect the structural adjustments to flooding governance as well as those for transboundary learning, collaboration, and adaptation.

BOX 4.2 Operational indicators for applying ACM to Greater Jakarta's flooding*

- Substantive authority in flooding governance at the municipality/regency level.
- National and provincial governments responsible for supervision over legal procedures.
- Solving problems is the responsibility of multiple parties, some of whom compete with one another.
- Mechanisms for the discussion of novel approaches exist at the national and sub-national levels.
- Space for diversity of problem framing and experimentation, including innovative competition, cross-fertilisation.
- Formal and informal water management and flooding governance structures go hand in hand so as to create space for learning and interactions.
- Formal and informal structures and mechanisms allow for redundancy and overlaps in the system.

- Horizontal and vertical stakeholder representative structures and processes are in place and effective.
- To some extent formalised, (facilitated) PAR is deployed as a foundation for learning (adaptation) and linking (collaboration).
- Government, nongovernmental organisations, educational and research institutions' officials, staff and field facilitators are trained in PAR.

* Based on Jambi ACM research in 2000–2006 and Huitema et al. (2009)

Concluding remarks

In this chapter, we reflect on the ACM approach as previously applied to investigate its ecological and livelihood effects in a local forest setting in Sumatra. Our reflections form the basis of a thought experiment to assess ACM's applicability as a pathway to address a much larger, complex, long-lived, and even more multiscale 'wicked problem', namely the flooding problem of Greater Jakarta. In this context, our central thesis is that Jakarta's flooding risks can effectively be managed *if* adaptation is strategic, which we see as encompassing and connecting sufficiently large spatial and temporal horizons. In addition, Jakarta's future flood risks may be greatly exacerbated by climate change, for which long-term projections and effective pathways to a climate proofing future are urgently needed.

Our assessment could not have been carried out at a better time. Globally, flood management is undergoing a shift from approaches focusing on flood control towards adaptive approaches aimed at reducing the impacts of floods. This trend can be observed in Greater Jakarta as well. Yet, in spite of the current ambitions to apply adaptive approaches to flood management, this chapter shows that some adaptation measures have led to 'maladaptation' – a term we borrow from the Intergovernmental Panel on Climate Change (IPCC 2022).[14] This has manifested itself in increased vulnerabilities of flood-prone communities and of the region's economies and ecosystems. Hence, the governance of adaptation for solving Jakarta's flooding problem has obviously fallen short. In responding to changing flooding contexts, adaptation governance has shown a predominantly *ad hoc* character while lacking a long-term vision and being locked into institutional frameworks, ingrained policy beliefs, and a technological engineering paradigm.

We see this stalemate as offering a window of opportunity for Greater Jakarta to play a leading role in the global quest for and application of flood adaptation approaches with long-term sustainability and, importantly, one that would be less likely to deliver 'maladaptive' outcomes. Jakarta's past and current infrastructural flood mitigation projects clearly show that the Indonesian government has been decisive in taking bold steps in the pursuit of the adaptation measures it deems necessary. This has been the case, regardless of the large investments in finance and technology needed. We anticipate an urgent need for a new form of leadership and recommend that Greater Jakarta (the Indonesian Government) take up this leadership role. The multiple gridlock, discussed in this chapter, can become a stimulus for the government to be a leader in this by leaving behind well-worn paths and entering new avenues.

We recommend as the first step of the pathway using an ACM framework that this new leadership encourage the proposed adjustments to the current flood governance structure. An adjusted adaptation governance should take into account ecological and societal impacts, stakeholder engagement, long-term effectiveness, and climate resilience. The adaptation governance we recommend has the following characteristics: a long-term goal that accommodates short-term interests and needs; a multilayer, multiscale, more balanced basin-wide distribution of decision-making authority with the lowest jurisdiction level (municipality/district level) holding full substantive authority; space for a diversity of problem framing and experimentation; and cross-boundary formal and informal structures for stakeholder collaboration and communication, thereby fostering redundancy and system-wide learning.

The recommended structural adjustments can be implemented under Presidential Regulation No. 60 of 2020 concerning the Greater Jakarta Urban Spatial Plan. The regulation provides unique opportunities for the integration of social, environmental, and economic values into the region's spatial plans and as such feeds into the region's climate-resilient development and flood risk policies.

Despite observed formal ambitions in exploring alternative adaptation trajectories in the face of flooding and climate change, adjusting the current flood governance structure cannot happen overnight. Political-will and open minds will not suffice for breaking open Jakarta's lock-ins; and time will be needed before the adjustments in governance structures are institutionalised and enable processes of change. Yet, our assessment makes clear that the main hurdles for change processes at the formal level are a weak social learning culture and the lack of a collaborative and adaptive environment. In interactive policymaking, the traditional policy system can be a learning barrier for policymakers: they may be less likely to be receptive to alternative problem framings and new solutions, hindered by strict institutional mandates and responsibilities.

As the second step of the pathway with an ACM framework, we recommend the initiation of nested platforms for shaping excellent learning conditions that stimulate creativity in and discovery of new problem framings and solutions outside the policy system and, hence, independent from existing flood policy. Official platform participants do not need to evaluate and approve policy innovations in accordance with legal and policy frameworks. Flood policy options or alternative policy measures resulting from the platform processes are eventually mainstreamed in the existing flood policy. As such, the platform provides a learning track or learning pathway which functions in parallel to, is independent from, yet delivers policy options that feed into the formal policy track. A basin-wide, cross-sectoral programmatic approach can become the basis for different sectors, institutions, and jurisdictions to integrate spatial planning more effectively with multiple social, economic, and environmental values. At the same time, participants will be able to maintain implementation of their own projects, formal duties and mandates as well as, where relevant, deliver public services. The basin-wide and cross-sectoral approach will allow the linking of problems that are currently addressed by different policies, sectors, and/or institutions.

This second step should be implemented under the regional spatial plan presidential regulation, referred to above. As such, learning and collaborative platform processes would deliver policy options for the benefit of Greater Jakarta's spatial planning policy process and support the region's climate-resilient development.

In this chapter, we acknowledge that the challenges that would confront ACM when addressing the flooding problem of Greater Jakarta are considerable. At the same time, we emphasise that the megacity's flooding problem has reached a crisis stage and, therefore, there is a pressing need for approaches that can break the cycle of long-established paradigms and maladaptive path-dependency processes. We hope this chapter encourages further discussion, with on-the-ground action, examining the potential as well as the drawbacks of ACM and similar approaches for mitigating the impacts of the flooding of Jakarta and elsewhere and, for that matter also, the impacts of climate change.

Acknowledgements

The authors would like to express special thanks of gratitude to Carol J. Pierce Colfer and Ravi Prabhu for the opportunity to undertake this exciting journey, the results of which are presented here. Their continued encouragement, critical feedback, and patience were invaluable in experimenting with ideas, writing the work, and completing the project. We also wish to acknowledge Dr. Hendricus Andy Simarmata, of the University of Indonesia, with whom at the start of our endeavour we discussed possible foci of our exploration. He was, however, unavailable to join us in the writing. Finally, we would like to thank Y.H. (Wiwi) Tjiook and Y. Yulia of the Liveable Cities Task Force of the Indonesian Diaspora Network in The Netherlands for providing us with a basic map of Greater Jakarta, which we modified for presentation purposes.

Notes

1 See also Colfer, Prabhu, and Larson (2022), for a systematic rendering of ACM's theoretical and conceptual forebears.
2 See Wessing (2011) and Tarumanagara – Wikipedia.
3 See National Geographic Indonesia, 27 February (2019) and Ward et al. (2013).
4 Between 1972 and 2012, the region's built-up area increased from 65 to 2,015 km^2 (Rustiadi et al. 2015).
5 Fifty-seven hundred hectares of forest in upstream Puncak vanished between 2000 and 2016 and 3,925 hectares of water catchment and urban forest areas in DKI Jakarta were lost between 1985 and 2006 (Afriyanie et al. 2022).
6 The single-day rainfall of 377 mm on January 1, 2020 led to devastating floods, and was preceded by a strong trans-equatorial monsoon flow (Yesi et al. 2021). Under current climatic conditions, the probability of rainfall extremities, including increasing

surface temperature, is already higher than 100 years ago and still higher risk is likely in the future.

7 Other factors behind the relocation of Indonesia's Capital are environmental degradation, rapid urbanisation, economic disparity and other societal problems, and traffic congestion in Greater Jakarta.

8 The embeddedness of smaller interventions in one larger ACM intervention can be referred to as a nested system approach (Groot et al. 2002).

9 At time of writing (June 2022), 12.6 kilometres of the giant dyke had been built.

10 Present solutions for Jakarta's land subsidence – predominantly due to groundwater extraction for drinking water – are sought in improving water management and/or sea dyke construction (Yan et al. 2020). A renewed water management design is planned for accomplishment over a ten-year period. Main challenges include a decrease in surface water resources due to pollution of the 13 rivers flowing through Jakarta; as well as a decrease of water retention areas resulting from massive land acquisition for buildings and infrastructures. Rain harvesting is one solution, which is at present beyond the agenda.

11 By means of the Watershed Management and Protection Forest Agency (*Balai Pengelolaan Daerah Aliran Sungai dan Hutan Lindung*, BPDAS-HL), which is a technical unit of the ministry at central level or an agency under the ministry at lower government levels (Pambudi and Kusumanto, in press).

12 Marshall Murphree calls this 'a socially constructed stalemate', which occurs when external agencies impose their agendas upon local populations. Interestingly, Murphree argues that such a stalemate can be broken when local communities are given the authority and responsibility necessary to create 'internally legitimate regimes', Local level scenario planning, iterative assessment and adaptive management : final technical report, July 2006 to November 2011 (https://agris.fao.org/agris-search/search.do?recordID=QD2021004709)

13 Nonetheless, learning occurred at the individual level with some public officials. For instance, officials felt encouraged to improve government programmes and sought new, creative ways to go about this – hence, *triple loop* learning clearly occurred here (Kusumanto 2006).

14 IPCC (2022) defines "*maladaptation*" as

> actions that may lead to increased risk of adverse climate-related outcomes, including via increased greenhouse gas emissions, increased or shifted vulnerability to climate change, more inequitable outcomes, or diminished welfare, now or in the future. Most often, maladaptation is an unintended consequence.

References

Aarts, N., and C. van Woerkum. 2002. "Dealing with Uncertainty in Solving Complex Problems." In *Wheelbarrows Full of Frogs. Social Learning in Rural Resource Management*, edited by C. Leeuwis and R. Pyburn Leeuwis, 421–435. Assen, The Netherlands: Koninklijke van Gorcum.

Adams-Schoen, S. 2016. "Taming the Super-Wicked Problem of Waterfront Hazard Mitigation Planning: The Role of Municipal Communication Strategies." In *Contemporary Issues in Climate Change Law and Policy: Essays Inspired by The IPCC 123*, edited by Robin K. Craig and Stephen R. Miller, 123–141. New York: Touro Law Center.

Adnan, H., D. Tadjudin, E.L. Yuliani, H. Komarudin, D. Lopulalan, Y.L. Siagian, and D.W. Munggoro (eds.). 2008. *Belajar dari Bungo: Mengelola Sumber Daya Alam di Era Desentralisasi.* Bogor, Indonesia: CIFOR.

Afriyanie, D., T.P. Moeliono, A. Triyanti, and T. Octavianti. 2022. "Policy Brief 3. Working towards an Adaptive Spatial Planning Policy for Flood Risk Management in Greater Jakarta." In *Policy Brief Series. Dealing with Greater Jakarta Floods in Times of Climate*

Change, edited by T. Kusumanto, A. Triyanti, and Y.H. Tjiook 39–48. Delft, The Netherlands; Utrecht, The Netherlands; Jakarta, Indonesia: TYK research & action consulting, Liveable Cities, IDN-NL, Utrecht University, and Indonesian National Research and Innovation Agency (BRIN).

Armitage, I., M. Marschke, and R. Plummer. 2007. "Adaptive Co-management and the Paradox of Learning." *Global Environmental Change* 18, no. 1: 86–98. doi: 10.1016/j.gloenvcha.2007.07.002.

Asdak, C., S. Supiah, and Subiyanto. 2018. "Watershed Management Strategies for Flood Mitigation: A Case Study of Jakarta's Flooding." *Weather and Climate Extremes* 21: 117–122. doi: 10.1016/j.wace.2018.08.002.

Becker, G., D. Huitema, and J.C.J.H. Aerts. 2015. "Prescriptions for Adaptive Comanagement: The Case of Flood Management in the German Rhine basin." *Ecology and Society* 20, no. 3: 1. doi: 10.5751/ES-07562-200301.

BPS. 2021. *Statistik Indonesia 2021* (*Statistical Yearbook of Indonesia 2021*). Jakarta, Indonesia: Badan Pusat Statistik (Central Bureau of Statistics).

Brown, J.R., and Y. Fehige. 2022. "Thought Experiments." In *The Stanford Encyclopedia of Philosophy*, edited by Spring 2022. Stanford, CA: Stanford University. https://plato.stanford.edu/cgi-bin/encyclopedia/archinfo.cgi?entry=thought-experiment

Buck, L., E. Wollenberg, and D. Edmunds. 2001. "Social Learning in the Collaborative Management of Community Forests: Lessons from the Field." In *Social Learning in Community Forests*, edited by D. Edmunds, E. Wollenberg, L. Buck, J. Fox, and S. Brodt, 1–20. Bogor: Indonesia: CIFOR.

Budiyono, Y., J.C.J.H. Aerts, D. Tollenaar, and P.J. Ward. 2016. "River Flood Risk in Jakarta under Scenarios of Future Change." *Natural Hazards and Earth System Sciences* 16, no. 3: 757–774. doi: 10.5194/nhess-16-757-2016.

Budiyono, Y., M.A. Marfai, J. Aerts, H. de Moel, and P.J. Ward. 2017. "Flood Risk in Polder Systems in Jakarta: Present and Future Analyses." In *Disaster Risk Reduction in Indonesia. Progress, Challenges, and Issues*, edited by M. Garschagen, R. Djalante, F. Thomalla, and R. Shaw, 517–537. Cham, Switzerland: Springer.

Caljouw, M., P.J.M. Nas, and M.R. Pratiwo. 2009. "Flooding in Jakarta: Towards a Blue City with Improved Water Management." *Bijdragen tot de Taal- Land- en Volkenkunde/Journal of the Humanities and Social Sciences of Southeast Asia* 161, no. 4: 454–484. doi: 10.1163/22134379-90003704.

Cao, A., M. Esteban, V.P.B. Valenzuela, M. Onuki, H. Takagi, N.D. Thao, and N. Tsuchiya. 2021. "Future of Asian Deltaic Megacities under Sea Level Rise and Land Subsidence: Current Adaptation Pathways for Tokyo, Jakarta, Manila, and Ho Chi Minh City." *In Environmental Sustainability* 2021. *Themed Issue on Slow Onset Events Related to Climate Change*, edited by S. Adamo, R. Djalante, P.G.D. Chakrabarti, F. Renaud, A.W. Yalew, D. Stabinsky, and Z. Zommers, 87–97. Amsterdam: Elsevier B.V. doi: 10.1016/j.cosust.2021.02.010.

Colfer, C.J.P. (ed.). 2005a. *The Equitable Forest: Diversity, Community and Resource Management*. Washington, DC: RFF and CIFOR.

Colfer, C.J.P. 2005b. *The Complex Forest: Communities, Uncertainty, Adaptive Collaborative Management*. Washington, DC: RFF and CIFOR.

Colfer, C.J.P., R. Prabhu, and A. Larson. (eds.). 2022. *Adaptive Collaborative Management in Forest Landscapes: Villagers, Bureaucrats, and Civil Society*. London: Earthscan and Routledge.

Diaw, M.C., and T. Kusumanto. 2005. "Scientists in Social Encounters. The Case for an Engaged Practice of Science." In *The Equitable Forest: Diversity, Community and Resource Management*, edited by C.J.P. Colfer, 72–109. Washington, DC: RFF and CIFOR.

Dovey, K., B. Cook, and A. Achmadi. 2019. "Contested Riverscapes in Jakarta: Flooding, Forced Eviction and Urban Image." *Space and Polity* 23, no. 3: 265–282. doi: 10.1080/13562576.2019.1667764.

Dwirahmadi, F., S. Rutherford, D. Phung, and C. Chu. 2019. "Understanding the Operational Concept of a Flood-Resilient Urban Community in Jakarta, Indonesia, from the Perspectives of Disaster Risk Reduction, Climate Change Adaptation, and Development Agencies." *International Journal of Environmental and Public Health* 16, no. 20: 3993. doi: 10.3390/ijerph16203993.

Emerson, K., T. Nabatchi, and S. Balogh. 2011. "An Integrative Framework for Collaborative Governance." *Journal of Public Administration Research and Theory* 22: 1–29. doi: 10.1093/jopart/mur011.

Florczyk, A.J., M. Melchiorri, C. Corbane, M. Schiavina, M. Maffenini, M. Pesaresi, P. Politis, S. Sabo, S. Freire, D. Ehrlich, T. Kemper, P. Tommasi, D. Airaghi, and L. Zanchetta. 2019. *Description of the GHS Urban Centre Database 2015*. Luxembourg: Publications Office of the European Union.

Garschagen, M., G.A.K. Surtiari, and M. Harben. 2018. "Is Jakarta's New Flood Risk Reduction Strategy Transformational?" *Sustainability* 10, no. 8: 2934. doi: 10.3390/su10082934.

GEF. 2009. *Adding Value and Promoting Higher Impact through the GEF's Programmatic Approach*. Washington, DC: Global Environment Facility.

Goudie, A.S. 2018. *Human Impact on the Natural Environment*. New York: John Wiley & Sons.

Groot, A., N. van Dijk, J. Jiggins, and M. Maarleveld. 2002. "Three Challenges in the Facilitation of System-wide Change." In *Wheelbarrows Full of Frogs. Social Learning in Rural Resource Management*, edited by C. Leeuwis and R. Pyburn, 199–213. Assen, The Netherlands: Koninklijke van Gorcum.

Guijt, I. (ed.). 2007. *Negotiated Learning: Collaborative Monitoring in Forest Resource Management*. Washington, DC: RFF and CIFOR.

Hasanuddin, Z.A., I. Gumilar, Y. Pohan, H. Andreas, Y. Fukuda, and T. Deguchi. 2011. "Land Subsidence of Jakarta (Indonesia) and Its Relation with Urban Development." *Natural Hazards* 59, no. 3: 1753–1771.

Huitema, D., E. Mostert, W. Egas, S. Moellenkamp, C. Pahl-Wostl, and R. Yalci. 2009. "Adaptive Water Governance: Assessing the Institutional Prescription of Adaptive (Co-) management from a Governance Perspective and Defining a Research Agenda." *Ecology and Society* 14, no. 1: 1–19.

Indriatmoko, Y. 2002. "Penelitian Aksi Bersama secara Adaptif dan Interaksi berbagai Domain Pengetahuan: Pengalaman dari Desa Baru Pelepat, Kabupaten Bungo, Jambi." Presented at The 2nd International Symposium of Journal Antropologi Indonesia, Padang, West Sumatra.

IPCC. 2022. "Climate Change 2022. Impacts, Adaptation and Vulnerability. Summary for Policymakers." Working Group II Contribution to the Sixth Assessment Report of the Intergovernmental Panel on Climate Change, Geneva, Switzerland.

Januariyadi, N.F., S. Kazama, I.R. Moe, and S. Kure. 2020. "Effectiveness of Structural and Nonstructural Measures on the Magnitude and Uncertainty of Future Flood Risks." *Journal of Water Resource and Protection* 12, no. 5: 401–415.

Kamarzuki, A. 2020. *Jabodetabekpunjur Spatial Policy Direction*. Presented as 1st Webinar of Regional Development Studies Webinar Series, Jabodetabek Study Forum, P4W-LPPM IPB University. 11 June 2020. IPB University: Bogor, Indonesia. Jabodetabek Study Forum (ipb.ac.id)

Kharel, G., R.J. Romsdahl, and A.P. Kirilenko. 2018. "Managing the Wicked Problem of Devils Lake Flooding along the US–Canada Border." *International Journal of Water Resources Development* 35, no. 6: 938–958. doi: 10.1080/07900627.2018.1523050.

Kooi, H., and A.T. Yuherdha. 2018. "Updated Subsidence Scenarios Jakarta." In *MODFLOW SUB-CR Calculations for Sunter, Daan Mogot and Marunda*. Delft, The Netherlands: Deltares.

Koto, J., and R.B. Negara. 2017. "Review Study on Rainfall Characteristics in DKI Jakarta." *Journal of Aeronautical Science and Engineering* 9, no. 1: 1–16.

Kusumanto, T. 2006. Social Learning, Policymaking and Customary Forest: Lessons from Jambi (Central Sumatra). Unpublished Report. Bogor, Indonesia: CIFOR.

Kusumanto, T. 2007a. "Shaping Opportunity for Improving Forest Quality and Community Livelihoods in Central Sumatra and East Kalimantan, Indonesia." In *Adaptive Collaborative Management of Community Forests in Asia. Experiences from Nepal, Indonesia and the Philippines*, edited by R. Prabhu, R. Fisher, and C. McDougall, 93–133. Bogor, Indonesia: CIFOR.

Kusumanto, T. 2007b. "Learning to Monitor Political Processes for Fairness in Jambi, Indonesia." In *Negotiated Learning: Collaborative Monitoring in Forest Resources Management*, edited by I. Guijt, 124–134. Washington, DC: RFF and CIFOR.

Kusumanto, T., G.A.K. Surtiari, S. van Bemmelen, P. Wijayanti, T.P. Moeliono, and D. Elhasani. 2022. "Policy Brief 4. Dealing with Greater Jakarta Flooding Inclusively from Social, Environmental and Economic Perspectives." In *Policy Brief Series: Dealing with Greater Jakarta Floods in Times of Climate Change*, edited by T. Kusumanto, A. Triyanti, and Y.H. Tjiook, 51–63. Delft, The Netherlands; Utrecht, The Netherlands; Jakarta, Indonesia: TYK research & action consulting, Liveable Cities, IDN-NL, Utrecht University, and Indonesian National Research and Innovation Agency (BRIN).

Kusumanto, T., E.L. Yuliani, P. Macoun, Y. Indriatmoko, and H. Adnan. 2005. *Learning to Adapt. Managing Forests Together in Indonesia*. Bogor, Indonesia: CIFOR.

Lee, K.N. 1999. "Appraising Adaptive Management." *Conservation Ecology* 3, no. 2: 3.

Lönngren, J., and K. van Poeck. 2020. "Wicked Problems: A Mapping Review of the Literature." *The International Journal of Sustainable Development and World Ecology* 28, no. 6: 481–502. doi: 10.1080/13504509.2020.1859415.

Maarleveld, M., and C. Dangbegnon. 1999. "Managing Natural Resources: A Social Learning Perspective." *Agriculture and Human Values* 16, no. 3: 267–280.

Maheng, D., A. Pathirana, and C. Zevenbergen. 2021. "Study on the Impact of Landscape Pattern Changes Due to Urbanization: Case Study of Jakarta, Indonesia." *Land* 10, no. 2: 218. doi: 10.3390/land10020218.

Marfai, M.A., A.B. Sekaranom, and P. Ward. 2015. "Community Responses and Adaptation Strategies toward Flood Hazard in Jakarta, Indonesia." *Natural Hazards* 75, no. 2: 1127–1144.

Octavianti, T., and K. Charles. 2018. "The Evolution of Jakarta's Flood Policy over the Past 400 Years: The Lock-in of Infrastructural Solutions." *Environment and Planning C: Politics and Space* 37, no. 6: 1102–1125. doi: 10.1177/2399654418813578.

Olsson, P., Ö. Bodin, and C. Folke. 2010. "Transformative Capacity for Ecosystem Stewardship in Social–Ecological Systems." In *Adaptive Capacity and Environmental Governance*, edited by D. Armitage and R. Plummer, 263–385. Berlin: Springer-Verlag.

Padawangi, R., and M. Douglass. 2015. "Water, Water Everywhere: Toward Participatory Solutions to Chronic Urban Flooding in Jakarta." *Pacific Affairs* 88, no. 3: 517–550.

Pambudi, A.S., and T. Kusumanto. In press. "Water Resources Governance towards Environmental Sustainability along with Social and Economic Development." In *Environmental Governance in Indonesia*, edited by A. Triyanti, R. Djalante, M.A. Marfai, M. Indrawan, and L. Nurhidayah. Cham, Switzerland: Springer. doi: 10.1007/978-3-031-15904-6.

Perkmann, M. 2016. "How Boundary Organizations Facilitate Collaboration across Diverse Communities." In *Managing Knowledge Integration Across Boundaries*, edited by C. Berggren, F. Tell, S. Brusoni, and A. van de Ven, 155–170. Oxford: Oxford University Press.

Plummer, R., B. Crona, D.R. Armitage, P. Olsson, M. Tengö, and O. Yudina. 2012. "Adaptive Comanagement: A Systematic Review and Analysis." *Ecology and Society* 17, no. 3: 11. doi: 10.5751/ES-04952-170311.

Plummer, R., J. Baird, A. Dzyundzyak, D. Armitage, Ö. Bodin, and L. Schultz. 2017. "Is Adaptive Co-management Delivering? Examining Relationships between Collaboration, Learning and Outcomes in UNESCO Biosphere Reserves." *Ecological Economics* 140: 79–88. doi: 10.1016/j.ecolecon.2017.04.028.

Prabhu, R., C. McDougall, and R. Fisher. 2007. "Adaptive Collaborative Management: A Conceptual Model." In *Adaptive Collaborative Management of Community Forests in Asia. Experiences from Nepal, Indonesia and the Philippines*, edited by R. Prabhu, R. Fisher, and C. McDougall, 6–49. Bogor, Indonesia: CIFOR.

Purnomo, H., G.A. Mendoza, and R. Prabhu. 2004. "Model for Collaborative Planning of Community-Managed Resources based on Qualitative Soft Systems Approach." *Journal of Tropical Forest Science* 16, no. 1: 106–131.

Rittel, H.W.J., and M.M. Webber. 1973. "Dilemmas in a General Theory of Planning." *Policy Sciences* 2: 155–169. doi: 10.1007/BF01405730.

Rogger, M., M. Agnoletti, A. Alaoui, J.C. Bathurst, G. Bodner, M. Borga, V. Chaplot, F. Gallart, G. Glatzel, J. Hall, J. Holden, L. Holko, R. Horn, A Kiss, S. Kohnová, G. Leitinger, B. Lennartz, J. Parajka, R. Perdigão, S. Peth, L. Plavcová, J.N. Quinton, M. Robinson, J.L Salinas, A. Santoro, J. Szolgay, S. Tron, J.J.H. van den Akker, A. Viglione, and G. Blöschl. 2017. "Land Use Change Impacts on Floods at the Catchment Scale: Challenges and Opportunities for Future Research." *Water Resources Research* 53, no. 7: 5209–5219.

Rukmana, D. 2016. "Flood Governance in Jakarta: The Role of CBOs in Mitigating Annual Floods." In *Essay Series Governing Megacities in the MENA and Asia*. Washington, DC: Middle East Institute (mei.edu).

Rukmana, D. 2021. "Participatory Governance in Mitigating Annual Floods in Jakarta." In *The Routledge Handbook of Planning Megacities in the Global South*, edited by D. Rukmana, 317–326. Abingdon, UK: Routledge.

Rustiadi, E., D.O. Pribadi, A.E. Privatisari, G.S. Indraprahastra, and L.S. Imam. 2015. "Jabodetabek Megacity: From City Development Toward Urban Complex Management System." In *Urban Development Challenges, Risks and Resilience in Asian Mega Cities*, edited by R.B. Singh, 421–445. Tokyo, Japan: Springer.

Samsura, D.A.A., T. Kusumanto, and A. Triyanti. 2022. "Policy Brief 5. Multi-Stakeholder Collaboration for Integrating Flood Risk Management with Spatial Planning in Greater Jakarta." In *Policy Brief Series. Dealing with Greater Jakarta Floods in Times of Climate Change*, edited by T. Kusumanto, A. Triyanti, and Y.H. Tjiook, 67–79. Delft, The Netherlands; Utrecht, The Netherlands; Jakarta, Indonesia: TYK research & action consulting, Liveable Cities, IDN-NL, Utrecht University, and Indonesian National Research and Innovation Agency (BRIN).

Sarmiento Barletti, J.P. 2022. "Learning from Adaptive Collaborative Management. A Participatory Tool to Support Adaptive and Reflective Learning in Multi-stakeholder Forums." In *Adaptive Collaborative Management in Forest Landscapes. Villagers, Bureaucrats, and Civil Society*, edited by C.J.P. Colfer, R. Prabhu, and A. Larson, 159–184. Abingdon, UK: Earthscan and Routledge.

Simanjuntak, I., N. Frantzeskaki, B. Enserink, and W. Ravesteijn. 2012. "Evaluating Jakarta's Flood Defence Governance: The Impact of Political and Institutional Reforms." *Water Policy* 14: 561–580. doi: 10.2166/wp.2012.119.

Simarmata, H.A., and G.A.K. Surtiari. 2020. "Adaptation to Climate Change: Decision Making and Opportunities for Transformation in Jakarta, Indonesia." Research Paper 2020–3, prepared for the UNRISD Project Transformative Adaptation to Climate Change in Southeast Asian Coastal Cities. United Nations Research Institute for Social Development (UNRISD) and Rosa Luxemburg Stiftung, Geneva, Switzerland.

Sol, J., M.M. van der Wal, P.J. Beers, and A.E.J. Wals. 2018. "Reframing the Future: The Role of Reflexivity in Governance Networks in Sustainability Transitions." *Environmental Education Research* 24, no. 9: 1383–1405. doi: 10.1080/13504622.2017.1402171.

Steins, N. 2002. "Understanding Platform Processes through Actor-network Theory." In *Wheelbarrows Full of Frogs. Social Learning in Rural Resource Management*, edited by C. Leeuwis and R. Pyburn, 407–419. Assen, The Netherlands: Koninklijke van Gorcum.

Teng, J., A. Jakeman, and J. Vaze. 2017. "Flood Inundation Modelling: A Review of Methods, Recent Advances and Uncertainty Analysis." *Environmental Modelling and Software* 90: 201–216.

Van de Vuurst, P., and L.E. Escobar. 2020. "Perspective: Climate Change and the Relocation of Indonesia's Capital to Borneo." *Frontiers in Earth Science* 8: 5. doi: 10.3389/feart.2020.00005.

Wagemans, M. 2002. "Institutional Conditions for Transformations. A Plea for Policy Making from the Perspective of Constructivism." In *Wheelbarrows Full of Frogs. Social Learning in Rural Resource Management*, edited by C. Leeuwis and R. Pyburn, 245–258. Assen, The Netherlands: Koninklijke van Gorcum.

Ward, P.J., W.P. Pauw, M.W. van Buuren, and M.A. Marfai. 2013. "Governance of Flood Risk Management in a Time of Climate Change: The Cases of Jakarta and Rotterdam." *Environmental Politics* 22, no. 3: 518–536. doi: 10.1080/09644016.2012.683155.

Wessing, R. 2011. "Tarumanegara: What's in a Name?" *Journal of Southeast Asian Studies* 42, no. 2: 325–337. doi: 10.1017/S0022463411000075.

WMO, and GWP (World Meteorological Organization and Global Water Partnership). 2009. "Urban Flood Risk Management: A Tool for Integrated Flood Management." In *Associated Programme on Flood Management (APFM) Technical Document* 11, Flood Management Tools Series. Geneva, Switzerland; Stockholm, Sweden: WMO and GWP.

Wollenberg, E., D. Edmunds, L. Buck, J. Fox, and S. Brodt (eds.). 2001. *Social Learning in Community Forests*. Bogor, Indonesia: CIFOR.

Yan, X., T. Yang, Y. Xu, L. Tosi, E. Stouthamer, H. Andreas, P. Minderhoud, A. Ladawadee, R. Hanssen, G. Erkens, P. Teatini, J. Lin, R. Boni, J. Chimpalee, X. Huang, C. Da Lio, C. Meisina, and F. Zucca. 2020. "Advances and Practices on the Research, Prevention and Control of Land Subsidence in Coastal Cities." *Acta Geologica Sinica* (English Edition) 94, no. 1: 162–175. doi: 10.1111/1755–6724.14403.

Yesi, U.S., Siswanto, H. Andhika, A. Tamima, Marzuki, E. Trihadi, Nasrullah, and Herizal. 2021. "Extreme Rainfall Causes 2020 Flood in Jakarta: A Climate Change Perspective." Presented at: The 2nd International Conference on Tropical Meteorology and Atmospheric Sciences (ICTMAS), Virtual Conference, 23–25 March 2021.

Yusriyah, K., S. Sudaryanto, A. Fatoni, and M.A. Mansyur. 2020. "Communication Networks Analysis on Information Dissemination of the Moving of Capital City from Jakarta to East Kalimantan." *Aspiration Journal* 1, no. 1: 31–55.

Introduction to Chapter 5

This chapter takes us to another Indonesian island, Sulawesi; and to a second generation of ACM-style research and action. Fisher et al. describe their experience using participatory action research – a key ACM method – primarily at an intermediate scale, in the mid-2010s. They echo some of the findings of Liswanti, Tamara and Arwida (Chapter 3). These authors too demonstrate the longevity of the multistakeholder group that they initially facilitated, and which continues to function, albeit with changing topical foci (most recently focused on climate change). The successful devolution of rights to manage a local forest, now recognized more broadly, also echoes the findings of Yuliani et al., in Jambi, Sumatra (Chapter 2).

The multistakeholder groups these authors revisited (taskforces), after successfully gaining formal rights for the Kajang to manage their 313-ha customary forest (previously part of Indonesia's national forest estate), have taken on two new topics.[1] The first (the taskforce's 'Life 2') took as its mandate to develop landscape-scale implementation guidelines for the district (*Kabupaten*), with special attention to watershed management. The excitement and sense of urgency that had characterized the first, 'Life 1', had waned, and the work had taken on a more mundane, routine aspect but the communication and shared concerns of the taskforce remained, as they switched their focus to a new task.

In 'Life 3', the task force turned its attention, most recently, to the national, indeed global, concern (among those we focus on here): climate change adaptation and how to contribute to that within their own district – specifically in three bounded watersheds that draw water from the original Kajang sacred forest.

These authors look to the future, examining what design elements change when opportunities appear and a group addresses new challenges. They examine important landscape principles and how a PAR process can contribute to environmental sustainability, observing also the link between PAR and multiple stakeholders' hopes for greater accountability. One important source of insights in this chapter is the authors' analysis and account of their own roles, their own conflicts as they strove to facilitate in a relatively neutral manner.

Following the evolution of an intermediate level, PAR-initiated taskforce through its three lives, this chapter shows how taskforce emphasis changed over time, responding to new opportunities, such as broader concerns with landscapes

DOI: 10.4324/9781003325932-8

and climate change (and also to donor concerns, highlighted in Egunyu, Chapter 9, as a danger). These Indonesian examples provide additional evidence of the enduring nature of the bonds of trust and communication that a PAR process can initiate – showing also its feasibility at an intermediate scale.

Note

1 The experience of Fisher et al., as CAI [critical action intellectuals] themselves and members of these taskforces, is reminiscent of what Ojha et al. (2022) report: "In both cases [Nepal and Kenya], the first-generation CAI contributed to environmental policy reforms, while the second generation engaged in defending or refining those policies as they were implemented" (p. 631).

Reference

Ojha, H., A. J. Nightingale, N. Gonda, B. O. Muok, S. Eriksen, D. Khatri, et al. 2022. "Transforming Environmental Governance: Critical Action Intellectuals and their Praxis in the Field." *Sustainability Science* 17: 621–635. DOI: 10.1007/s11625-022-01108-z

5 The power of possibility in landscape governance

Multiple lives of participatory action research in Kajang, Sulawesi

Micah Fisher, Agus Mulyana, Ardi Labarani, Kamaluddin, E. Linda Yuliani and Moira Moeliono

Introduction

There have been remarkable advances over the past decade for establishing regulatory frameworks to support rural land tenure policy in Indonesia. Many of the regulations in place would have been unimaginable by the initial proponents of forest and land tenure rights in the 1990s. As of 2022, there are now regulatory guidelines in place to recognize Indigenous land rights, as well as formal government commitments to provide tenure arrangements to rural communities in and around state forests in the form of social forestry permits. These developments gained significant momentum in 2013. At that time, a series of court cases laid out the legal basis for returning state forests to the control and authority of Indigenous/customary (henceforth, *adat*) communities (Myers et al. 2017), followed by presidential commitments in 2014 aimed to put 10% of Indonesia's vast state forests under social forestry schemes supporting access and livelihood opportunities to local communities (Fisher et al. 2018; Fisher, Dhiaulhaq, and Sahide 2019). In 2016, the first set of *adat* land titles began to materialize and were symbolically handed over to local communities, and by 2021, about one-third of the social forestry permit goal of 12.7 million hectares were in place. One of the persistent questions beyond the legal dimensions of recognizing land rights for rural communities has been how to go about doing it, particularly in contexts of historical state and corporate land enclosures. Given the policy developments sympathetic to conferring local tenure and rights in Indonesia, how would formal institutions go about working with local communities to produce these and maintain them for the longer term.

The most symbolic and precedent-setting example for operationalizing *adat* forest rights recognition was the Kajang community of Bulukumba, South Sulawesi (Fisher and Muur 2020). Although the state forest area implicated for *adat* recognition was small, at only 313 hectares, the case proved a legal mechanism to secure the most robust form of recognition over state forests, and was celebrated by President Widodo in a ceremony to deliver land title to local Kajang leadership. The implications are significant, given that two-thirds of Indonesia's territorial area is designated state forests. Such precedence opens opportunities

DOI: 10.4324/9781003325932-9

for Indigenous Peoples and other local communities across Indonesia to formally reclaim land based on their historical customary tenure systems. What gained particular attention by policymakers, activist groups, and international observers alike, with regard to land rights recognition in Kajang, was the participatory means by which land rights had been secured. Applying participatory action research (PAR) principles involved joint fact-finding across multi-stakeholder teams from the government, activists, local communities, and other Taskforce members. The process was hailed for its ability to convene and collaborate across historically contentious members and institutions.

This PAR has been well documented (Colfer et al. 2015; Fisher et al. 2017; Kemitraan 2015; Workman et al. 2015). In this chapter, we examine what happened to PAR after the objective of *adat* land rights recognition was secured. After gaining recognition, PAR stakeholders began to think about how to approach planning processes, allocate and commit resources, and sustain landscape governance under an Adaptive Collaborative Management (ACM) framework. PAR efforts were initially established largely for making the case for recognition, but once formalized land rights were secured, the more challenging work of sustaining engagement across stakeholder groups on land management outcomes began. Although the PAR team penned new commitments, the basis for external support began to wane, influencing various elements of participation. Nevertheless, the day-to-day goals related to environmental governance continued beyond these legal achievements. This chapter focuses on processes that helped sustain collaborative landscape governance goals.

In what follows, we examine three different lives of distinct PAR initiatives at the same site, taking a close look at what happens when mandates and convening frameworks shift, goals change, and new questions of representation emerge. We begin by situating our broader understanding and overall engagement with ideas of collaborative initiatives in landscape governance, particularly in Indonesia. We then turn to documenting the process that led to the establishment of PAR in Kajang (Life 1). Next, we return to local landscape-scale PAR initiatives that sought to "implement" land rights recognition after the spotlight of national and international attention refocused elsewhere (Life 2). We extend questions and critiques around stakeholder representation throughout and evaluate the extent to which PAR delivers on its initial promises in this context. Finally, as momentum has surfaced for programmatic initiatives to re-enliven PAR processes around climate change adaptation (Life 3), we not only identify the legacy of preceding PARs but also consider what works, to what ends, and what might be done differently to deepen meaningful, longer lasting public participation on complex socio-environmental challenges.

Collaborative management and research in Indonesian landscapes

As a tradition, collaborative research and management initiatives have sought to combine multi-stakeholder interests with other experts interested in seeking out sustainable outcomes (Lee 1994). Particularly in the form of ACM, shared learning, experimentation, adaptation, and inclusivity are fundamental principles (Colfer, Prabhu, and Larson 2022). As a practice, collaborative research initiatives have long explored ways to integrate both the method and the outcomes of research initiatives addressing the complexity of landscapes (Werkman et al. 2011). With PAR initiatives, researchers and key actors can take on various roles, including networking across stakeholders, mobilizing for greater representation, and being actively involved to influence outcomes. There are also different reasons that stakeholders convene to collaborate over a wicked landscape governance issue or work together on a research question. They usually do so to drive processes of institutional change or policymaking, to resolve conflicts, or to rebalance power relations and advocate for representation. Much of the formal research and writing about collaboration in landscapes, however, tends to be explored as externally driven processes supported by a time-bound initiative implemented over a confined area. For this reason, research explicitly aimed at understanding facets of participation in landscape governance has had difficulty linking what came before, with the implications of what happens afterward. In other words, it can be challenging to explain the extent to which collaborative research initiatives succeed or fail, and how.

The notion of PAR in Indonesia also has its own deep history, one that connects to networks of thinkers and practitioners committed to community engagement. Movements for rapid rural appraisal and participatory rural appraisal run deep in landscape governance initiatives since at least the late 1970s, particularly as a subversive and ostensibly apolitical form of civil society intervention due to the complex institutional arrangements that were not then conducive to stakeholder representation (Leurs 1996). Indeed, rural development programming and especially the establishment of national parks often meant landscape governance on forest landscapes translated into eviction and resettlement (Fox et al. 2009). Those promoting the early versions of PAR in Indonesia were a network of researchers, activists, and practitioners, committed to going to the field and understanding issues from local perspectives. Often, they were seeking to create visibility around local concerns and highlighting issues at different scales and contexts as a form of advocacy. Cadres of researchers and practitioners inspired networks not only to seek out knowledge from books but also to view knowledge from the source, among the communities whose lives and livelihoods were situated within the landscapes (Mulyana et al. 2008).

Those that were drawn to and subsequently shaped ideas of PAR came from these networks and also began to find greater currency in large research institutions. Headquartered in Indonesia, the Center for International Forest Research (CIFOR) had a robust group of practitioners eager to apply research that worked for the communities they were working with in rural and forested landscapes. However, PAR and ACM proponents also faced pressure from the strong biases in favor of experimental and quantitative research present in such research institutions. These researchers made the case for studying collaborative research in comparative contexts, identifying governance principles across sites, applicable even when locations may seem to present such diverse ecological and political contexts (Colfer et al. 2011; Colfer, Prabhu, and Larson 2022). The visibility ACM teams provided regarding stakeholder representation has hence driven ongoing commitment within some segments of government to pursue community-driven forest and land tenure reforms like social forestry. But with the status quo of bureaucratic process, stated commitments for participation have also weakened participation in practice. Over time, participation and collaboration have sometimes been treated as dirty words, synonymous with broken promises, corruption, manipulation, and coercion (Arnstein 1969). Nevertheless, there is growing acknowledgment that sound collaboration is fundamental to carrying out sustainability initiatives in complex landscapes. Almost all external initiatives involving forests continue to require a variety of safeguards embedded in principles of consultation and collaboration (Arhin 2014).

In Indonesia, the political economy of development and social movements for human rights and the environment are important contextual factors to any participatory engagement. Industrial-scale plantation development shaped much of modern Indonesia's political economy, beginning at the turn of the 19th century with the Dutch colonial model, expanding and intensifying in the 1970s and 1980s through the New Order regime (President Soeharto's three-decade long authoritarian rule from 1966 to 1998 (Anderson 1983)). In post-*reformasi* (reform) Indonesia, the period associated with governing frameworks after Soeharto, political economy drivers favoring plantation development sparked new violent evictions across rural Indonesia (Lund 2021). Political reforms under the framework of a democratically decentralized nation-state, however, helped expand civil society rights and shaped new forms of rural development activism. These were often expressed through frames of dispossessed peasants and local Indigenous communities (Muur 2018). Conflicts skyrocketed across Indonesia through emboldened people reclaiming and repossessing lands that had been enclosed by the state. Numerous demonstrations, litigation, and conflict resolution initiatives ensued.

Social movement organizations consolidated their voice most prominently through the international visibility of a 2007 global climate change conference convened in Bali, Indonesia. This highly anticipated Conference of Parties (COP 13) was among the first to feature the emerging issue of Reduced Emissions from Deforestation and Forest Degradation (REDD+). Advocacy groups strategically centered discussions on the politics of Indigenous rights sweeping across Indonesia, calling for a new language of recognition that began to hold a high degree of political legitimacy (Afiff 2016). The PAR initiatives discussed in this chapter initially emerged out of an agroforestry program to support rural communities that are inseparable from these political trends. The governance initiatives, in particular, aligned with the high political stakes of seeking to recognize *adat* land rights.

Before getting into each of the stages of engagement, a note on method. This chapter is written by a confluence of the teams that facilitated and formed the Bulukumba Taskforce(s) over three distinct phases. The first included the CIFOR agroforestry initiative, the second included a follow-up CIFOR initiative on integrated watershed management, and the third did not directly involve CIFOR, namely, an Adaptation Fund initiative supporting local NGOs to convene PAR. As a result, the views expressed are those of the facilitation team from across these three initiatives and reflect a set of views of stakeholders that include local government leadership from the forestry agency, a network of NGOs tasked with carrying out assessments and convening meetings, and also reflect the viewpoints of key local leadership and community members over a time frame that spans nearly a decade. Committed to a reflective practice, we always aimed to be self-critical (Schön 2017). This chapter is based on revisiting perspectives from key proponents of the Taskforce. However, we also take a more critical view to explore questions about what succeeded for whom, and how PAR initiatives changed over time to reflect its multi-stakeholder representation. We set out to reflect on the extent to which proponents understood its successes, failures, and limitations. In short, was it a good thing – how and why not – and what were some of the unanticipated outcomes?

Life 1: the Bulukumba policymaking taskforce

The context behind PAR

Although discussions about community-based resource management are longstanding in Indonesia (Fisher, Dhiaulhaq, and Sahide 2019), the 2007 COP presented a renewed opportunity to showcase the important role of communities in conserving forests in the context of climate change (Afiff 2016). This resulted in a broad coalition of NGOs advocating for Indigenous People's rights and land rights recognition, social forestry, and agrarian reform (Bettinger, Fisher, and Miles 2014). Given connections to international movements, the rights of Indigenous Peoples were first to gain explicit political attention, and by 2013 the Constitutional Court ruled over the illegal enclosures of *adat* lands, showcasing a groundswell of support for local participation in policymaking on forest land tenure (McCarthy and Robinson 2016; Myers et al. 2017). The court's ruling established a legal basis for millions of hectares to be "returned" to Indigenous and local authority. Legal decisions subsequently shifted to the more operational elements of implementing regulations and practices, a challenging prospect given the history of dispossession and relocation across rural Indonesia. Questions emerged whether there would be a run on resources, or whether local communities would be left without the support to manage forests (Erbaugh 2019), while others pointed to the potential for elite capture and new forms of exclusions to unfold at different scales (Fisher and van der Muur 2020).

The Kajang case offered a unique situation, serving as a precedent-setting case for handing over land rights. First, it was a small community, and crucially, there was no question about the extent of local adherence to Kajang cultural norms, given the clear outward appearances of customary practice (Maarif 2012). The inner zones, *ilalang embaya*, were particularly strict in their adherence to the moral code (*pasang*), with residents wearing all black to symbolize values of modesty. While there are several thousand people that identify ethnically as Kajang across the subdistrict of Kajang in Bulukumba, strict adherents of the *pasang* are largely confined to the village of Tanah Toa. *Ilalang Embaya* spatially overlaps with much of Tanah Toa village, which also extends into the Kajang sacred forest, an area that according to state administrative documents had been enclosed under the designation of *limited production forests*. This state forest, however, was mostly a state forest on paper, given the very strict rules by the Kajang on entry into their sacred forest, where important local rituals to restore balance of human-environment relations take place (*andingingi*) (Workman et al. 2015). As we will highlight later, however, the role of the state in this production forest had developed a relationship with local customary leaders by hiring local rangers to help protect its boundaries as a sacred forest. Legally, this helped ease the transfer of a bounded state forest area to an Indigenous forest, as the change did not disrupt any existing land uses, which are usually complicated elsewhere by the layers of claims over cultivation rights.

Furthermore, there were existing sympathies and precedent by the Bulukumba local government to designate Kajang forests as *adat* forests. Several years prior to the Constitutional Court ruling in 2013, as early as 2008, Hasanuddin University convened discussions to formally designate the limited production forest in Kajang as *adat* forests but these faltered due to the lack of higher level regulatory support. The growing regional and national attention to identify and designate *adat* forests served to reinforce acceptance of formal recognition among the local government in Bulukumba, especially as it would make them a standout in promoting ideas about good governance over forest land tenure. Nevertheless, there were many competing reasons for recognition among the different government agencies of the local government. Some leaders in the Bulukumba district government believed that Kajang forests should be recognized in order to restore justice over past violent dispossession, while others believed that recognition provided tourism development opportunities. Others still believed that recognition could help direct resources to address poverty and support for local economic development, while others were concerned about the extent to which local recognition would divert resources to Kajang over other subdistricts in Bulukumba.

Complicating the process of recognition however was the historical conflict over nearby enclosures of rubber plantations (Muur 2018). Of the total 22,593 hectares of the Kajang region, over 5,000 hectares were enclosed by the large rubber plantations of PT Lonsum (Tyson 2009). The company continues to argue that its land concessions date back to documentation in 1919. But it also expanded significantly in area during the 1970s–1990s, and again began to enclose additional lands in 2003. These most recent attempts at enclosure led to violent clashes between company and state security forces against local farmer and Indigenous alliances that led to three deaths, accompanied by other forms of violence carried out toward local demonstrators (SNUB [Solidaritas Nasional Untuk Bulukumba] 2003). Conflict resolution efforts ensued, with surprising settlements for farmer groups to reclaim land. Unusual for grievances against a plantation, some smallholders even received compensation but local farmers still largely felt that it did not address grievances or make up for the enclosures of their family groves. Regular public demonstrations against the plantation have continued to take place in Bulukumba ever since.

In 2011, a consortium between ICRAF and CIFOR began to implement an international development project on agroforestry and forestry livelihoods and governance in South Sulawesi.[1] Although the initiative focused in large part on working with farmers across the province, in Bulukumba, governance interests around *adat* land rights recognition shifted to addressing the policy moment of formally returning land and facilitating conflict resolution. A CIFOR governance team helped to facilitate dialogues, and with buy-in from the Bulukumba district government, the District Head established a policymaking Taskforce, assigned to draft a local regulation recognizing the existence and empowering the rights of the Kajang. In a unique approach to policymaking in Indonesia, leadership within the local government invited explicit representation of local community stakeholders in Kajang, and even included historically contentious advocacy groups like the Alliance for Indigenous Peoples' of the Archipelago (AMAN).

Researcher positionality

For the CIFOR research/governance team, we were acutely aware of entering the context of land rights, particularly the historical incidences of violence and dispossession (SNUB 2003). Furthermore, being invited by the Bulukumba local government to participate in facilitating a policy initiative on *adat* recognition posed many challenges, particularly when many national advocates began to suggest that Kajang could serve as a precedent-setting case for *adat* land rights recognition elsewhere. To navigate this process, we first discussed our role as researchers in the context of facilitating a PAR initiative, as well as our role as facilitators to support in securing normative outcomes. Indeed, the impetus for this chapter emerged because we continue to assess and re-examine the extent to which efforts improved the quality of governance relations between stakeholders.

At the outset, we asked ourselves if we were able to perform that role well. We considered whether our role as facilitators to a PAR process would be helpful or even necessary. Throughout the PAR initiatives, we continued to explore whether our facilitation supported local authority and local smallholders to convene over questions of access to land and protecting forested landscapes. Beyond the formal involvement of the CIFOR team involved in PAR from 2011 to 2016, a new set of teams among us continue to return to these overarching goals in the two subsequent PAR initiatives that took place in 2017–2018, and an initiative currently underway in 2022. The subsequent narrative follows through the design and implementation of each of these phases.

The Kajang landscape and the taskforce for rights recognition

After the 2013 Constitutional Court ruling, many questions turned to how to operationalize returning *adat* forests to local communities. There were guidelines put in place on how to do this that drew from activist legal analysis and regulatory interpretations by the Forestry Ministry (Arizona 2013; Mancayo and Firmansyah 2014). These guidelines explained that justification for *adat* recognition required establishing evidence of adherence to *adat* norms and a territorial area of continued land and resource management. The research team was well aware that one of the well-documented challenges for Indigenous land recognition elsewhere around the world involved the many contradictions of reinstating cultural authority in locations previously excluded, neglected by the state, or purposefully dispossessed from land (Erazo 2013). Such legal and policy approaches are especially complex when re-establishing institutions in complex political economies that have undergone many changes over time (Fisher and van der Muur 2020).

In Kajang, there were several institutions with authority. Over an area of approximately 22,000 hectares, considered part of the historical area of influence of the Kajang, about one-quarter is under cultivation by PT Lonsum, the rubber plantation corporation. Much of the remainder of that landscape consists of the settlements of about 25 villages administered across four subdistricts, with a small portion deemed to cross over into the neighboring district of Sinjai (important because it is outside the jurisdiction of the Bulukumba local government). Outside of the village settlement areas that make up a small footprint, much of the rural landscape is divided in three, between smallholder plots of wet rice paddy cultivation, commodity crops and agroforestry groves. Rice paddies and some plots of corn are for subsistence, while the other large portions of the landscape are covered with smallholder rubber, cloves, black pepper, with some intermixed cropping systems, and some smaller forests and sacred sites. In this variegated landscape, the Forestry Ministry claimed limited production forest as overlapping with the Kajang sacred forests, while village governments and other government agencies claim jurisdiction over various development functions. Meanwhile, the Kajang cultural institutions also hold much influence, particularly in the traditional zone *of Ilalang Embaya*. A polycentric form of governance has taken place through the process of naming elected village leaders, which also confers parallel authority for Kajang leaders. For example, the elected head of the village of Tanah Toa also gains the customary Kajang leadership title as *Galla Lombo*' that comes with its own set of responsibilities.

Across this landscape, Bulukumba local government officials were keen to operationalize national interests to recognize *adat* land rights. For this reason, much of the policy discussions narrowed to the 331-hectare sacred forest and neighboring traditional hamlets of *Ilalang Embaya*. This forest also neatly overlapped with the Forestry Ministry's jurisdiction, making for a strategic opportunity to transfer rights without actually enclosing any livelihoods. Indeed, almost too conveniently for national *adat* recognition of state lands, the sacred forest was very much still under the control of the Kajang, where rights of entry are closely guarded, various ritual sites are located, and rituals continue to be performed. In the context of historical conflict with the plantation company and the social movements protesting local authorities, Bulukumba government authorities knew that there needed to be a process to convene groups around a common goal and begin to build mutual understanding. Several sympathetic government offices had already been working to recognize local Kajang authority over forests, especially to the areas that continue to function as sacred groves. But others in government also had a negative view of the tactics that rights advocacy groups had employed to implicate them.

The facilitator team from CIFOR consulted with the local government over governance elements of the agroforestry initiative in target locations across Sulawesi. With formal Bulukumba government buy-in, CIFOR sought to convene local and regional activist NGOs with government stakeholders sympathetic to Kajang recognition. One key element was to gather on site in Kajang, sitting with local cultural leadership, conducting focus groups to identify key priorities and approaches to pursuing an initiative for Kajang rights recognition. Early questions focused on local consultation, exploring the openness of Kajang cultural leaders about how they would like to go about it. Initial understanding began to emerge across groups about the ways that Kajang cultural authority were de facto recognized by the government but that these were not formally in place. Various discussions also proceeded on the merits and implications of formal recognition. Gatherings involved several training initiatives on PAR principles and approaches to conflict resolution that several local NGOs (Balang Institute, and later OASE) began to internalize. Cronkleton, Evans, and Larson (2022) describe the need for facilitators to better understand PAR processes and develop more appropriate ways for listening and integrating villager perspectives. The Bulukumba Forestry Agency also committed to shifting their approach from discussion to greater intentionality in the form of *dialogues* and expressed a unique willingness to listen and involve non-government agencies (*Kemitraan* 2015).

Once there was buy-in from the local government to proceed, a policymaking process was initiated through the establishment of a Taskforce for the recognition of the Kajang *adat* community.[2] A local regulation would then pave the way for the Forestry Ministry to have a basis to release land management authority. Drafting regulations in Indonesia is approached by establishing a team of relevant government agencies. The process usually requires a process whereby an academic or expert team is contracted to conduct a study *(naskah akademik)* that serves as a technical and legal analysis to form the basis of a policy. Links to similar policy and planning-directed PARs with local government have been documented elsewhere in Indonesia (Adnan et al. 2008, also Yuliani et al., this volume). In a unique development, the local government promoted and signed an order to establish the policymaking Taskforce by also including representation of NGOs and local Kajang members. This Taskforce was formally tasked with developing mechanisms to pass the first *adat* land rights recognition to challenge national forest territorial designation.

At this point, the CIFOR research team also began to facilitate discussions about PAR. As a method, PAR offered an opportunity to address a policy problem with different stakeholders and competing interests in a way that identifies key questions, deepens engagement on emerging findings collectively, and guides the policy drafting process based on those findings. Following from the national guidelines on *adat* recognition, the Taskforce set out to identify two key questions, namely the extent of *adat* adherence in the area and the territorial scope of recognition. Research teams conducted 150 household interviews sampled across almost half of all the villages to better understand local livelihoods and cultural practices. The enumerators consisted of multi-stakeholder staff teams from the Taskforce, which eagerly participated in the household interviews, and expressed this to be a unique learning experience to view issues from local perspectives. In mixed teams, they entered Kajang villager households to deliver the questionnaires. A series of focus groups with Kajang customary elders were crafted based on preliminary findings, which assisted in listing out all of the complex Kajang leadership functions and customary authority, as well as documenting the role of different ritual sites. These details were incorporated directly into the policy draft to reflect this bottom-up process.

Mapping Kajang territory

In another unique show of trust, both government agencies and NGOs agreed to share all spatial data within the Taskforce and identified other spatial data that would be necessary to answer the question of territorial scope. Very quickly, efforts to obtain a map of the PT Lonsum forest concession met difficulties, which led to NGOs promoting alternative efforts to map the extent of plantation land, using satellite remote sensing (see also Yuliani et al., this volume). This is a common practice in participatory mapping employed by Indonesia, which continues to grow in sophistication (Radjawali and Pye 2015). In addition, key inconsistencies in the forestry agency maps were also identified. AMAN's regional office also shared maps produced on earlier initiatives to identify oral histories on the extent of Kajang influence in the region. These were further consulted with local Kajang leadership, triangulating for consistency and policy translation. Inaccurate village maps were rectified by walking all boundary zones, whereby individual parcels were also mapped over any disputed boundaries. Local trainings on the use of GIS technology were provided to the local NGOs, who quickly learned to render overlapping maps, creating presentations that served as facilitation tools to help answer the collectively guided PAR questions. The NGO members eagerly went to the field to collect GPS points, including protocols with local Kajang, always making sure to consult with cultural authorities before collecting spatial data. This was deemed especially important, given the Kajang beliefs about the use of technology, inappropriate in certain territories.

Participatory mapping can produce uneven benefits and unintended consequences (Fox et al. 2009). Changing perceptions about landscape and new articulations of politics through mapping processes have also been documented in Kajang (Fisher 2021). The transparency of data allowed for new opportunities for dialogue throughout the PAR process, especially in terms of accountability over questions of land ownership and access. On the one hand, mapping helped to highlight areas of traditional ancestral authority that were enclosed by the plantation, which heightened the legitimacy of Kajang authority over those lands. But on the other hand, remapped boundaries of villages in the process also produced discussions about households to be included or excluded from areas of traditional authority, which some wanted, and others did not. This affected different households as to how they are allowed to build homes, cultivate lands; it also reshaped power and authority locally. Specifically, village heads were eager to use the mapping data to develop village plans that interacted with local decision-making processes and influenced the allocation of resources.

Life 2: "implementing" recognition in a new taskforce

Once the local regulation was passed in 2015 recognizing Kajang land rights, national advocacy groups and government stakeholders celebrated the event. It was described as the actualization of the Constitutional legal victories that promised to place Indigeneity back into the landscape. Representatives from Bulukumba and Kajang, both formal and customary leaders, were invited to the national palace to meet with the President and were formally given title to the land that was previously part of state forests (as in Baru Pelepat, see Figure 2.3). Local government representatives expressed that this was the longest and most intense district regulation they had ever produced, and described the deep commitments that were required to produce such a policy document across non-traditional policy-making stakeholder groups. Many PAR and ACM processes are critiqued for the lengthy time frames and heavy time commitments to process. This was also a critique among Taskforce members, most of whom generally expressed a sense that time commitments were beneficial and necessary to build common ground and produce an agreeable draft across all stakeholder groups. At the national level, the success was hailed as a precedent-setting event, connecting with national and regional Indigenous social movement aims, stating that this was the beginning of a possibility to reclaim 40 million hectares of Indigenous lands across Indonesia (Gaol and Dahlia 2017).

Back in Kajang and Bulukumba, however, there was a sense that the work was still incomplete. Without the initial formal mandate to draft a regulation, there were no longer any resources or grounds for stakeholders, particularly government agencies, to convene. Furthermore, officials in certain offices were promoted or rotated to other agencies, a recurrent problem in efforts to upscale ACM processes. The implications of land rights recognition in Kajang also needed explaining to those that were not part of the Taskforce but had taken up positions that had the corresponding authority to follow up on earlier commitments. Indeed, competing interpretations began to emerge about the role and outcomes of recognition, which Fisher (2019) documents as the "double edge of recognition."

First, within the Kajang territory, the idea that the signed regulation meant that the sacred forest was no longer under the jurisdiction of the Forestry Ministry began to embolden some actors to claim individual ancestral rights to land within the boundaries of the sacred forest. Village administrators also began to view these lands under their jurisdictions for potential development projects. Learning of these incidents in the forestry agency, one ranger felt compelled to ask: without funds to support the rangers to protect the boundaries of the sacred forest, who would defend it against future encroachment (Workman et al. 2015)? Second, some concluded that since the 331 hectares of sacred forest were returned to the Kajang, it meant that remaining lands of the 22,000 hectares were outside of Kajang jurisdiction. This argument was especially politically motivated. Some considered that one implication of this regulation was support for closure over claims by the plantation corporation, suggesting that the regulation constituted a major concession of land rights given to the Kajang. Third, the regulation was viewed as an effort to develop cultural tourism opportunities, which gained formal traction in the regional development planning processes, and also led to outside contractors benefitting from these plans. Finally, many people in Kajang were also unsure of the benefits of recognition. Everyday Kajang smallholder farmers felt that they gained no additional cultivable land that would benefit them, and some land-poor households felt that the regulation served to further empower and direct resources to local elite authorities with ties to the Bulukumba district government. This also extended a sense of distrust among local farmers toward NGO meetings and discussions about Indigenous land rights. Farmers concluded that a lot of resources were being allocated to fund gatherings without any benefits accruing to address their day-to-day livelihood challenges.

Given many of these competing narratives, more senior officials in Bulukumba advocated for the re-establishment of a Taskforce (henceforth, Taskforce 2) with the objectives of drafting implementing regulations for the District Head. This Taskforce 2 would help articulate the ways in which the regulation on recognition would be applied in practice. It would seek to lay out the ways in which local government programs would provide support to local livelihoods, addressing questions about disbursements of farm equipment, support to local women's groups, assistance for village development planning, and others. The CIFOR team also secured an additional grant to support initiatives for landscape governance framed around the concept of integrated water resource management and applied through a PAR approach.

The early Taskforce 2 meetings began by identifying the challenges noted above, as well as bridging local Kajang ideas on the relational elements of recognition between government and *adat* institutions. The meetings also considered questions about the idea of state citizenship in the context of *adat* recognition and how the responsibilities of governing subsequently changes with these new distinctions. This was especially important for the continued allocation of resources for forest rangers to protect any encroachment on the forest. Taskforce members were also forthright about accountability questions, such as possibilities for elite capture of any development assistance to the regions, particularly due to the high political stakes of local village elections and the patronage systems in place. Taskforce 2 remained committed to multi-stakeholder representation between government, NGO, and Kajang *adat* membership. Meetings continued to rotate between different hosts and sites, and also highlighted women's unusually strong representation within Taskforce leadership (Fisher et al. 2018). Nevertheless, the community meetings in Kajang generally reflected much greater representation of men, reflective of dynamics in formal settings in Bulukumba, though women do have unique standing in household decisions given their control of finances (Colfer et al. 2015).

In the field, a geographic element had also changed the discussions. While much of the focus in the past had been around the sacred forests and the main gate to enter the inner Kajang at *Ilalang Embaya*, interests had shifted to questions of landscape-scale management more broadly. There were efforts to extend mapping efforts to identify water resources and highlight the interconnectivity of the forest as a water resource with the numerous rice fields that depend on these water resources across three different watersheds (Fisher 2019). Village planning initiatives also began to take place and Taskforce 2 teams helped to develop proposals under the new Village Law.

Over time, energy around Taskforce 2 waned without the sense of urgency that was present in its previous manifestation. The first Taskforce had a tangible and high-stakes goal of drafting a regulation for recognition that had significant implications at a national level. The forestry agency, refashioned as a forest and environment agency in step with changing national regulations, continues to fulfill its commitments through implementing budgetary and work planning commitments that emerged from Taskforce 2 discussions. For example, the agency still assists in implementing a joint monitoring program with cultural leaders that helps to ensure continued protection over sacred forests. With the village planning initiatives, the water resources discussions helped to identify mutual interests for protecting water, given key concerns over irrigation resources, the loss of drinking water resources, and the flooding that has taken place due to land use change. Other initiatives were implemented for a time but there was also a sense that the momentum did not develop into sustained landscape-scale initiatives.

Life 3: a new framing around climate adaptation

Although there have been various watershed management initiatives undertaken toward the aims of climate change adaptation in South Sulawesi, most of this work has taken place under top-down technical initiatives focused on analysis at scale from a provincial or district level. The NGO proponents of the program in Kajang were well regarded by the eventual donors – the Adaptation Fund – because the project was similarly framed as integrated watershed management in support of identifying opportunities for climate change adaptation (see also Kusumanto et al., this volume, which explicitly connects ACM and PAR relevance to flood governance and climate change). More specifically, the initiative focused on bottom-up initiatives using innovative approaches for engaging with villages and households. The fact that plans would be developed by local *adat* farmer groups and institutions also provided an attractive grounding for external support organizations. The ability to convene the multi-stakeholder groups grew out of the past two Taskforce initiatives, with proponents pointing to the ability to quickly convene and establish networks. Furthermore, the proposal writing process was locally driven and grew out of the interests of the young NGO members who had been so eager to implement fieldwork and pursue baseline assessments during Taskforce 1.

The program was framed around the lack of responsive integrated watershed initiatives across key stakeholder groups, especially in ways that could support the resilience of local livelihoods and environmental sustainability. Drawing from past PAR approaches, proponents hope to derive legitimacy from the experience of the previous stakeholder groups successfully working together to develop regulatory outcomes. Such a policy goal will be operationalized through drafting of a roadmap and action plan for development planning, which will be integrated across district-level agencies and partner organizations. Doing so would require agencies and other stakeholders to develop specific tasks, including assigning key roles and resources, and establishing the budgetary means for supporting efforts to maximize resources. These would be formalized through an umbrella regulation to be signed by the District Head. Taskforce 3, in this case, would work across stakeholders to identify planning and regulatory commitments, and other forms of support for greater flexibility in climate change adaptation initiatives that would guide institutional commitments. Stakeholders included the key planning and implementing agencies at the Bulukumba district level as before but also included more implementation-minded bodies such as the water utility. Meanwhile, the core of the work would engage with local farmers on identifying and supporting resilient cropping and livelihoods systems, as well as exploring ways to transition to and buttress more sustainable inputs and pest control systems.

The geographic scope of the program follows three bounded watersheds that draw their water resources from the Kajang sacred forest. Taskforce 3 would convene around questions of addressing household water supply, drought, flooding, irrigation, and other water considerations determined to make up the existing vulnerabilities of the region, but which could be exacerbated by climate change (see also Kusumanto et al., this volume). One of the initiatives that emerged was to map out and identify spring water resources and protect them from conversion. Another included identifying already changing cropping systems due to pronounced drought that would shift agriculture to more drought-resistant systems. Overall, the program involves a set of 14 villages that overlap with the three priority watersheds in the upstream, midstream, and downstream areas. By establishing multi-stakeholder groups, institutions would be able to work together to identify ways to develop adaptive landscape management efforts that directly support local livelihoods. This work is still ongoing but the precedence of process, trust, and guiding frameworks for PAR settings that convene multi-stakeholder groups in familiar ways had been set in motion long ago.

Discussion

For proponents of land rights recognition in Indonesia's vast state forests, Kajang was a convenient success. Initially, it seemed like an intractable conflict between historical enclosures of state-supported plantation expansion versus Indigenous smallholders. But when discussions focused on the growing national-level interest to reclassify state forests as *adat* lands, it served as a strategic site for passing a regulation over a sacred forest already acknowledged *de facto* by the local government. Nevertheless, drafting a regulation without precedence was by no means automatic. PAR offered an opportunity to bring stakeholders together to begin building common ground that led to continuing forms of collaboration in the future. These manifested in two subsequent Taskforces aimed at some form of landscape-scale governance involving rights, livelihoods, and conservation.

Key features that were present in the process revolved around spatially explicit ways of knowing, clarifying objectives and fact-finding. These offered an opportunity to clarify, ensure transparency, and build accountability around often misleading questions of territory. Spatial representations also created new tensions about what activities were allowable culturally or jurisdictionally in certain spaces, and both allowed for contestations against the plantation while also legitimating its presence on the landscape. It led to new normative approaches to problem solving and created the basis for new imaginaries about the future of Kajang. While the case supported downward accountability in some ways, it also privileged those with links to state institutional authority, raising questions about the benefits of Indigenous recognition for day-to-day and land-poor farmers (Fisher and Muur 2019). Although we tend to think of PAR process as having a higher purpose, often couched in the language of "win-wins," decisions are inseparable from land and power relations. Throughout our process, we needed to be able to understand the implications of decisions from a regional scale, but also from the hoe-level. While external observers focused on the symbolic opportunity from releasing state forests to sacred forests, the facilitation team undertook a committed process to better understand underlying drivers of land conflicts.

State recognition over *adat* forests through policy regulation also faced the aftereffects of its message of triumph in environmental justice. These were tempered by the reality of trying to articulate governance actions around what comes next after *adat* recognition. Given that the first Taskforce had succeeded in a policy outcome, a new Taskforce bridging multi-stakeholder representation was established, with strong organic leadership across stakeholder groups. Nevertheless, while the initial regulation provided symbolic possibility for *adat* communities elsewhere, Taskforce 2 quickly had to contend with normative governing challenges, such as spatial planning regulations and agency budgeting processes. These were divided across development interests and fragmented across local politics. Taskforce 2 aimed to remain rooted in commitments to meaningful local representation but initiatives lost momentum in its larger aims of building a concerted approach to landscape-scale management.

Years later, a new initiative arrived. It was catalyzed by an unrelated international donor grant to convene stakeholders. This time framed under climate change adaptation, the language of landscape and watersheds remained, harking back to the unfinished goals of earlier PARs. Convening stakeholder groups in Taskforce 3 came naturally across the multi-stakeholder groups. The ability to quickly establish the networks of participants eager to engage in PAR is testament to the lingering commitments to process that PAR can afford (cf. Liswanti et al., this volume). Not only did past initiatives provide an institutional framework for advocating to elected leadership, it also made it easy to convene key agencies, with representation afforded to more technical agencies, such as the water utility. Furthermore, the principles of inclusivity, representation, and dialogue extended beyond formal departments to involve NGOs and local Kajang leadership. These were second nature to the process, a fact that certainly would have been unlikely a decade ago. The initiatives were rooted in bottom-up principles to seek out solutions at the village and household level, and prioritized local Indigenous knowledge and experiences. The continuity through Taskforce 3 was made possible by the initial NGO actors who had so eagerly taken on the task to develop the baselines and worked so closely with communities in Taskforce 1. This time, establishing local partnerships was easy, undertaking pragmatic steps of planning and convening meetings to establish common goals. Although the façade of legitimacy of external organizations and resources provides a key convening base, over time PAR principles were internalized in the individuals and institutions involved.

As a team of self-critical PAR facilitators participating in three separate lives of PAR in Kajang, we often felt as outsiders creating work for local institutions. We often questioned ourselves by asking whether PAR is a tool that makes busy people busier. We would come in to talk about our interests in being part of these activities and try to align them across interest groups, particularly as a group of researchers leaning toward social and environmental justice goals. This can potentially lead local stakeholders to prioritize our agendas as outsiders at the expense of their priorities. Bureaucrats are also easily intimidated by these ideas, giving a nod to meaningful participation as an idea, without genuine commitments in practice. Committed to reflective practice, we chose not to judge stakeholders by our own standards but only aimed to evaluate the quality and intent of our own processes.

From a broader systems perspective, throughout the process, we asked ourselves whether we as facilitators needed a redefinition of PAR. Could it be that our interests to assist could be manipulated into a tool to intimidate, silence, or reshape people's thinking in the name of PAR? This was often referred to by the research team through a combination of the words facilitation and manipulation: "facipulation" (or *fasipulasi* in Bahasa Indonesia). In this respect, to provide productive pathways forward, we discussed the inverse, namely can PAR serve as a tool to get people to be more involved or change their views about concerns that fell into our broader interests of environmental justice. To these ends, we suggest that the power of possibility over process that continues to bring people together in strategic ways presents a powerful symbol and a guiding compass. Nevertheless, the work of involving marginalized communities such as women and the landless will continue to be perplexing dilemmas to address in PAR implementation, particularly at this broader scale.

One of the reflections that emerged from our engagements was that each of the PAR processes was ad hoc and reactive to certain conditions, often dovetailing with political interests. We provided support through fact-finding initiatives in the form of participatory mapping and others, but also realized that a lot of PAR falls short of scrutinizing and researching the quality and extent to which the PAR objectives themselves were exercised and achieved. Many PAR researchers are usually focused on the more immediate concerns of our worlds as practitioners. Being able to be more rigorous in the practice of day-to-day facilitation and the learning about PAR practice in and of itself would help build profound new insights on process and outcome (see Hagmann et al., this volume). This is especially pertinent across Indonesia, as we encounter numerous initiatives to build forums embodying the principles of PAR. Commitment to such forums is important but more important is to ensure that the multi-stakeholder forums become operationalized in ways that respond to common goals of stakeholder groups.

The evolving process across three Taskforces involved a strategy of strengthening networks between bureaucrats, NGOs, and community members. It took place at multiple levels, from the grassroots level to national policy networks. As participatory a PAR process can be, however, it will always need to continue to consider creative approaches for establishing firmer roots with the community. This involves sustaining interest and momentum for stakeholders to continue to be involved and providing new avenues for just representation. In addition, given the rapid land uses transforming the region, the involvement of state institutions, civil society, organizations, and local concerns must also seek out common solutions that also make environmental sense.

Conclusion

Across three iterations of Taskforces convened under different pretexts, PAR continued to provide a language to convene more easily at each subsequent iteration. Each Taskforce had some form of landscape governance element. The impetus for the first Taskforce emerged due to broader concern for land rights, the second around implementing land rights recognition around a landscape and watershed approach, and a more recent third Taskforce was designed to apply principles of climate change adaptation. The framings around issues of natural resources and common property rooted in local knowledge and systems of authority were critical elements. Representation required active interest to sit for long meetings, and it was also bolstered or undermined by political contexts. Young staff members from NGOs and institutional memory in formal agencies, alongside strong networks with local communities based on trust, also establish generational potential to establish new governance trajectories. Though initial efforts at PAR can seem foreign, outlandish, onerous, or time-consuming, continued commitment to process can evolve into new discoveries that build upon each iteration. It empowers possibility for making improvements in multi-stakeholder landscape governance, of supreme importance in addressing climate change and other natural resources challenges.

Notes

1 The formal title of the project was "Agroforestry and Forestry in Sulawesi: Linking Knowledge to Action AgFor" Contribution Arrangement No. 7056890, Department of Foreign Affairs, Trade and Development (DFATD), Government of Canada (2011–2016). The authors of this research team were affiliated with this project.

2 Bulukumba Bupati Decision Letter No: 760/VII/2013 to form "The Taskforce for the designing and drafting of the local regulation on recognition of the Kajang indigenous communities of Bulukumba district."

References

Adnan, Hasantoha, Djuhendi Tadjudin, E. Linda Yuliani, Heru Komarudin, Dicky Lopulalan, Yuliana L. Siagian, and Dani Wahyu Munggoro, eds. 2008. *Belajar Dari Bungo Mengelola Sumberdaya Alam Di Era Desentralisasi*. Bogor, Indonesia: Center for International Forestry Research.

Afiff, Suraya A. 2016. "REDD, Land Management and the Politics of Forest and Land Tenure Reform with Special Reference to the Case of Central Kalimantan Province." In *Land and Development in Indonesia: Searching for the People's Sovereignty*, edited by John F. McCarthy and Kathryn Robinson, 113–140. Singapore: ISEAS – Yusof Ishak Institute.

Anderson, Benedict R. O'G. 1983. "Old State, New Society: Indonesia's New Order in Comparative Historical Perspective." *The Journal of Asian Studies* 42, no. 3: 477–496. Cambridge University Press.

Arhin, Albert Abraham. 2014. "Safeguards and Dangerguards: A Framework for Unpacking the Black Box of Safeguards for REDD+." *Forest Policy and Economics* 45: 24–31.

Arizona, Yance. 2013. "Masyarakat Adat Dalam Kontestasi Pembaruan Hukum." Presented at *Seminar Pemberdayaan Sosial Komunitas Adat: Upaya Peningkatan Efektivitas Pemberdayaan KAT Saat ini dan Pengembangan Kedepan*. Sponsored by Badan Perencanaan Pembangunan Nasional (BAPPENAS), Hotel Grand Sahid, Jakarta, Indonesia.

Arnstein, Sherry R. 1969. "A Ladder of Citizen Participation." *Journal of the American Institute of Planners* 35, no. 4: 216–224.

Bettinger, Keith, Micah Fisher, and Wendy Miles. 2014. "The Art of Contestation and Legitimacy: Environment, Customary Communities, and Activism in Indonesia." In *Occupy the Earth: Global Environmental Movements*, edited by Lim Leonard and Sya B. Kedzior, 195–224. West Yorkshire, England: Emerald Group Publishing Limited.

Colfer, Carol J. Pierce, Etienne Andriamampandry, Stella Asaha, Emmanuel Lyimo, Endri Martini, Jean Laurent Pfund, and John Watts. 2011. "Participatory Action Research for Catalyzing Adaptive Management: Analysis of a 'Fits and Starts' Process." *Journal of Environmental Science and Engineering* 5, no. 1: 28–43.

Colfer, Carol J. Pierce, Ramadhani Achdiawan, James M. Roshetko, Elok Mulyoutami, E. Linda Yuliani, Agus Mulyana, Moira Moeliono, Hasantoha Adnan, and Erni. 2015. "The Balance of Power in Household Decision-Making: Encouraging News on Gender in Southern Sulawesi." *World Development* 76: 147–164.

Colfer, Carol J. Pierce, Ravi Prabhu, and Anne M. Larson, eds. 2022. *Adaptive Collaborative Management in Forest Landscapes: Villagers, Bureaucrats and Civil Society*. London: Routledge and Earthscan.

Cronkleton, Peter, Kristen Evans, and Anne Larson. 2022. "Capacity Building for ACM." In *Adaptive Collaborative Management in Forest Landscapes: Villagers, Bureaucrats and Civil Society*, edited by C. Colfer, R. Prabhu, and A. Larson, 134–157. London: Routledge and Earthscan.

Erazo, Juliet S. 2013. *Governing Indigenous Territories*. Durham: Duke University Press.

Erbaugh, James T. 2019. "Responsibilization and Social Forestry in Indonesia." *Forest Policy and Economics* 109: 102019.

Fisher, Micah R. 2019. *Beyond Recognition: Indigenous Land Rights and Changing Landscapes in Indonesia*. Honolulu: University of Hawaii at Manoa.

Fisher, Micah R. 2021. "Engaging Voices in the Landscape: Participatory Geography in Indigenous Land Rights Recognition." In *Indigenous Peoples, Heritage and Landscape in the Asia Pacific*, edited by Stephen Acabado and Da-wei Kwan, 31–53. London: Routledge.

Fisher, Micah R., Ahmad Dhiaulhaq, and Muhammad Alif K. Sahide. 2019. "The Politics, Economies, and Ecologies of Indonesia's Third Generation of Social Forestry: An Introduction to the Special Section." *Forest and Society* 3, no. 1: 152–170. Makassar, Indonesia: Hasanuddin University.

Fisher, Micah R., Moira Moeliono, Agus Mulyana, E. Linda Yuliani, Andi Adriadi, Junaid Judda, and M. Aalif K. Sahide. 2018. "Assessing the New Social Forestry Project in Indonesia: Recognition, Livelihood and Conservation?" *International Forestry Review* 20, no. 3: 346–361.

Fisher, Micah R., Timothy Workman, Agus Mulyana, Institute Balang, Moira Moeliono, E. Linda Yuliani, and Carol J. Pierce Colfer. 2017. "Striving for PAR Excellence in Land Use Planning: Multi-Stakeholder Collaboration on Customary Forest Recognition in Bulukumba, South Sulawesi." *Land Use Policy* 99: 102997.

Fisher, Micah R., and Willem van der Muur. 2020. "Misleading Icons of Communal Lands in Indonesia: Implications of *Adat* Forest Recognition from a Model Site in Kajang, Sulawesi." *The Asia Pacific Journal of Anthropology* 21, no. 1: 55–76.

Fox, Jefferson, Yayoi Fujita, Dimbab Ngidang, Nancy Peluso, Lesley Potter, Niken Sakuntaladewi, Janet Sturgeon, and David Thomas. 2009. "Policies, Political-Economy, and Swidden in Southeast Asia." *Human Ecology* 37, no. 3: 305–322.

Gaol, Amy Lumban and Lia Dahlia. 2017. "Indonesian President Hands over Management of Forests to Indigenous People." February 15. http://blog.worldagroforestry.org/index.php/2017/02/15/indonesian-president-hands-management-forests-indigenous-people/#.WKVL0AVg670.facebook.

Kemitraan (Partnership). 2015. *Hutan Untuk Kesejahteraan: Cerita Dari Lapangan*. Jakarta, Indonesia: Kemitraan Partnership and The Royal Norwegian Embassy.

Lee, Kai N. 1994. *Compass and Gyroscope: Integrating Science and Politics for the Environment*. Washington, DC: Island Press.

Leurs, Robert. 1996. "Current Challenges Facing Participatory Rural Appraisal." *Public Administration and Development* 16, no. 1: 57–72.

Lund, Christian. 2021. *Nine-tenths of the Law: Enduring Dispossession in Indonesia*. New Haven, CT: Yale University Press.

Maarif, Samsul. 2012. *Dimensions of Religious Practice: The Ammatoans of Sulawesi, Indonesia*. Tempe: Arizona State University.

Mancayo, Andiko Sutan, and Nurul Firmansyah. 2014. *Mengenal Pilihan-Pilihan Hukum Daerah Untuk Pengakuan Masyarakat Adat: Kiat-Kiat Praktis Bagi Pendamping Hukum Rakyat (PHR), Masyarakat Sipil (Pelaku Advokasi) Dan Pemimpin Masyarakat Adat*. Jakarta, Indonesia: Perkumpulan HuMa Indonesia.

McCarthy, John F., and Kathryn Robinson (eds.). 2016. *Land and Development in Indonesia: Searching for the People's Sovereignty*. Singapore: ISEAS – Yusof Ishak Institute.

Mulyana, Agus, Hasantoha Adnan, Yayan Indriatmoko, Agus Priyono, and Moira Moeliono. 2008. *Belajar Sambil Mengajar: Menghadapi Perubahan Sosial Untuk Pengelolaan Sumberdaya Alam*. Bogor, Indonesia: CIFOR.

Muur, Willem van der. 2018. "Forest Conflicts and the Informal Nature of Realizing Indigenous Land Rights in Indonesia." *Citizenship Studies* 22, no. 2: 160–174.

Myers, Rodd, Dian Intarini, Martua Thomas Sirait, and Ahmad Maryudi. 2017. "Claiming the Forest: Inclusions and Exclusions under Indonesia's 'New' Forest Policies on Customary Forests." *Land Use Policy* 66: 205–213.

Radjawali, Irendra, and Oliver Pye. 2015 "Counter-mapping Land Grabs with Community Drones in Indonesia." In *Proceedings of the Land Grabbing, Conflict and Agrarian—Environmental Transformations: Perspectives from East and Southeast Asia*, 5–6 June. Chiang Mai, Thailand.

Schön, Donald A. 2017. *The Reflective Practitioner: How Professionals Think in Action*. London: Routledge.

SNUB (Solidaritas Nasional Untuk Bulukumba). 2003. "Kronologis: Kasus Penembakan Petani Dan Masyarakat Adat Kajang Bulukumba Propinsi Sulawesi Selatan Oleh Aparat POLDA Sulawesi Selatan Dan Polres Bulukumba 21 Juli 2003." Bulukumba, Sulawesi: SNUB.

Tyson, Adam D. 2009. "Still Striving for Modesty: Land, Spirits, and Rubber Production in Kajang, Indonesia." *The Asia Pacific Journal of Anthropology* 10, no. 3: 200–215.

Werkman, Renate, Jolanda van den Berg, Annemarie van Paassen, and Bette Harms. 2011. "What Is Collaborative Landscape Research About?" In *Knowledge in Action*, edited by Annemarie Van Paassen, Jolanda Van den Berg, Eveliene Steingröver, Renate Werkman, and Bas Pedroli, 41–56. Mansholt Publication Series, Vol. 11. Wageningen, The Netherlands: Wageningen Academic Publishers.

Workman, Timothy, Micah R. Fisher, Agus Mulyana, Moira Moeliono, E. Linda Yuliani, and Balang. 2015. "Out of the Lion's Den, Into the Crocodile's Jaws? Lessons from Policy Developments on Customary Forest in Bulukumba." ICRAF Brief No. 56. Bogor, Indonesia: World Agroforestry Centre - ICRAF Southeast Asia Regional Office.

Introduction to Chapter 6

This chapter provides a much-needed elaboration of what exactly is involved in excellent facilitation – a set of skills, attitudes and competences that are rarely spelled out and often go under-recognized. Any practitioner of ACM recognizes how central good facilitation is to bringing about a successful ACM process. This chapter deconstructs the process at both the community level and at broader scales, as community-identified needs spiral out and up to relevant hierarchies. Attempts to use ACM or similar approaches in the many arenas where we feel it can contribute will require much broader understanding of the complexities of effective facilitation.

Firmly grounded in respect for local people and their systems, whether values or subsistence or knowledge, Hagmann and his co-authors provide us with valuable guidance about how to understand such systems, work constructively with local people toward self-empowerment, and then how to mesh their systems and actions relatively seamlessly and sustainably with those at broader scales. A central element in the kind of facilitation these authors recommend is increasing self-awareness and examination of one's own values and biases. Such understanding is a precursor to the sensitivities required in understanding the values and biases of others.

Recognizing that a huge leap in knowledge is required for would-be facilitators (see e.g., Cronkleton, Evans and Larson 2022, on the learning typically required), Hagmann et al. also devote some space to explaining an effective way they have developed for training people in how to do it well. This approach is based on five facilitated workshops, between which participants explore and practice their new knowledge for several months in communities, returning for coaching, further teamwork, and planning the next phase at each subsequent workshop.

This thorough coverage of the topic of facilitation is much needed, as we consider using ACM-like approaches more commonly to address global problems like climate change, restoration, disaster relief, etc.

DOI: 10.4324/9781003325932-10

Reference

Cronkleton, Peter, Kristen Evans and Anne M. Larson. 2022. Capacity Building for ACM: Lessons Learned from Training in Distinct Contexts. In: *Adaptive Collaborative Management of Forest Landscapes: Villagers, Bureaucrats and Civil Society*, edited by Carol J. Pierce Colfer, Ravi Prabhu and Anne M. Larson, 134–157. London: Earthscan/Routledge.

6 Herding cats

Facilitation in social learning processes

Jürgen Hagmann, Edward Chuma, Joe Ramaru, Henning Peter, Kudakwashe Murwira, Paolo Ficarelli, Hlamalani Ngwenya and Klaus Krebs

Introduction

The Adaptive Collaborative Management (ACM) of Forest Landscapes programme was a fascinating experience in participatory action research and social learning. At the start of the programme, some of the concepts were not well defined, in terms of what exactly they meant in practice and how to operationalise them. Due to the main author's previous experience in action research and social change, he was asked to support the ACM team through facilitated meetings and workshops in order to move the conceptual thinking and practice forward and accompany the team in their learning process.

His first involvement was as a learning facilitator in a foundation workshop in early 2000, where concepts were clarified and first steps in implementation were set up. Later, he became an advisor and attended several meetings of the advisory group, interacted with the team and many individuals who were engaged in ACM; and at a later stage in 2005, he reviewed some of the ACM activities in Indonesia, Zimbabwe, Cameroon and Nepal. It was a fascinating journey the team had taken and he deeply appreciates having been involved.

The experiences on which this chapter on facilitation is grounded go way beyond the ACM programme. Many lessons and insights described here are based on long-term processes in community development and participatory learning and extension which we as PICOTEAM (Institute for People, Innovation and Change in Organisations) have gained since 1991 in Zimbabwe, South Africa, Tanzania, Cambodia and the Dominican Republic. Our team was developing the implementation capacity of government officers, NGO staff and community change agents to facilitate social learning processes, whom we accompanied for several years. This gave us the opportunity to deeply understand these processes, experiment with them and develop approaches and methodologies to facilitate processes and develop the necessary capacities. Together, we have facilitated more than a thousand events and processes at different levels – public and private organisations, multi-stakeholder processes, high-level consultations, as well as communities (www.picoteam.org). These varying levels often require similar principles but with a different architecture. This chapter mainly focuses on the community level. It is an experience-based analysis with myriads of insights and

DOI: 10.4324/9781003325932-11

conceptualisation from our team in different contexts. The chapter looks at facilitation in a comprehensive way from a practice perspective. For more information, see our website: www.picoteam.org.

As mentioned repeatedly in this book, facilitation is critical for the success of ACM. This chapter first looks at the concept of facilitation for change (F4C) as we understand it now, then it describes some pertinent issues in implementation with examples, after which we elaborate how to develop such complex competences among field staff in an effective way. We conclude with some critical issues to consider for ACM-type interventions.

When we talk of a facilitator/process manager, in most cases, it will be a facilitation team rather than one individual.

Facilitation – an overloaded term

In the 1980s, facilitation was understood in many development organisations as workshop moderation. In the 1990s when participatory development became more mainstream, it became clear that the role of facilitation needed to be deeper and more transformative. Its value was seen as going beyond workshops and it became a key concept in the implementation of participatory methodologies (see Chambers 2002). In the late 1990s and 2000s, facilitation evolved into a word that could mean almost anything related to participation. It became an overloaded term often with little defined meaning, and an over-emphasis on tools (Groot and Maarleveld 2000). In the present era, our understanding expanded towards facilitation of processes of change and development; it became clear that effective facilitation needs to be more than applying participatory tools and methods. However, the depth and breadth of the processes, such as its architecture, its psychological nature, the deeper psycho-social issues involved and the need for organisational development, have often been only minimally developed.

In our perspective, in the context of social learning processes and participatory action learning/research like ACM, we see facilitation as a process of guiding people in their own learning about their context, social systems and ways to create the future they hope for. Facilitation is an experience-based knowledge creation from a constructivist perspective, where people construct/change their own reality, rather than in a conventional teaching mode or knowledge transfer. It is about making people better understand their systems to change/re-create them, rather than an external understanding of their system by outsiders providing outside advice. Facilitation has a strong systemic dimension as change in systems is complex, dynamic and rarely succeeds with a linear approach (see 'soft systems' methodology by Checkland 1999). Interventions in facilitation oriented towards change are designed to create discomfort with the status quo, to trigger action and through this action, experiment and find out what works, what does not work and how the system functions. It follows the logic of Kurt Lewin, one of the main founders of action research who stated already in the 1940s that:

> If you want to understand the system, just try to change it!

Learning cycles are thus a fundamental architecture of such facilitation processes, and critical thinking and questioning are paramount characteristics of a facilitator. In mainstream interventions, a lot of focus was given to facilitation tools and methods (Bollinger and Zellweger 2007; Chambers 2002; Kaner 2007; Neuland 1998). These are all important, but were largely overrated. Without a solid process, these methods are likely to be disjointed and ineffective. In some contexts, facilitators acted like chairpersons and lecturers, which precluded the creation of a diverse dialogue that let people discover and experience rather than being lectured to.

The concept of facilitation for change

Building on the understanding above, this chapter will elaborate some key elements of the concept of F4C.

Purpose/goal of 'Facilitation for Change'

The fundamental question in any facilitation intervention is its purpose/goal: facilitation for what? There are three fundamentally different purposes to differentiate:

The first is the use of facilitation by outsiders to implement their interventions and ideas. Facilitation can be very powerful in convincing people of ideas, reducing resistance and smoothing processes with often pre-defined outputs. This form of facilitation can be rather manipulative and directive, an approach which many development interventions are choosing to achieve their set 'outsiders'' goals in projects. As long as there is sufficient monetary incentives and intervention, this can work, but normally falls apart the moment the external intervention is over.

The second purpose is fundamentally different as it starts from peoples' values and aspirations and addresses the changes needed to self-empower and emancipate people and communities to manage better their own world and resources and to pursue their aspirations. It is an intrinsic process of self-development of communities. This 'facilitation for change' process is by its very definition built on action and social learning, and social energy to create change. This is the kind of facilitation aimed for with ACM processes as they are described in this book.

The third purpose is the moderation of conferences and events which is also called facilitation. However, this form leans more towards chairing of a meeting than F4C. It is often accompanied by already pre-designed programmes with sessions that are neither necessarily linked nor built on each other. The power lies with the moderator and not necessarily the participants.

With the recent challenges of COVID-19 and related travel restrictions, we have seen an increase of digital platforms and the need for virtual moderation. We have not practised this with rural/forest communities yet, but we see a great potential for integrating it in future.

Different levels of facilitation

Facilitation happens at different levels and the processes that happen at community level are core, as that is where people who directly manage forests live. Facilitation processes at community level are central to the success of ACM-type processes. In almost all cases – after some successes at community level, the need arises to involve other levels of groups and organisational levels as they have roles in the process. For example, a forestry administration might need to learn about such processes if ACM-type processes should become a new way of working, policy makers might need to be engaged to adapt policies and regulations, service providers might need to adapt their services, etc. There are two requirements for outsiders to co-learn: involve them in the community processes and also facilitate their own learning and adaptation in their organisations. In action learning processes towards change, community processes spiral into other processes as actors throughout the whole systems need to adapt and learn. Facilitation interventions can become quite complex systemic multiple loop learning interventions as Cronkleton, Evans and Larson (2022) experienced in a similar way. None of the essential levels with possible 'sabotaging power' can be left out, and risk needs to be managed. The facilitation processes for these different levels might be different but the key principles are rather similar (Hagmann et al. 1999).

Architecture of process facilitation and management

ACM-type processes as described in this book are longer-term interventions which need a clear process model with facilitation and process management, which we call 'process architecture'. The two, facilitation and process management, are hard to separate as they will involve the same people/teams who need to see the whole and not just one workshop/event. Processes have their own dynamics which often appear successful in one step and after a next step may seem to be going backwards. It is a continuous up and down, so if one does not have a plan for the whole process, results of single events can be misleading. In the type of facilitation we encourage, single event/workshop facilitation is part of longer-term processes. These processes are not just multiple events; the activities between the workshops are equally important and part of the process design.

The fundamental change process design is based on action learning which occurs in phases and loops (described in this book in other chapters, see also Mukasa et al. 2022). The cycles normally follow five to six main phases:

1 **Initiating change:** developing trust and analysis of the situation towards a better self-understanding by the community and its goals, and ownership of their problems, challenges and opportunities.
2 **Searching for solutions/new ways**: exploring local and outsider solutions, learning about alternative ways.
3 **Planning and organising implementation**: planning how to move into action, strengthening local organisational capacity/overcoming weak organisation.
4 **Experimenting with new ideas/implementation of action**: enhancing creativity and learning by trying out new ideas.

5 **Reviewing outcomes and sharing ideas:** assessing new ideas and sharing with other community members.
6 **Reflection on lessons and replanning:** taking stock of the whole process, getting deeper in the analysis of issues, identifying new issues to deal with and adapting the whole intervention (Figure 6.1).

The role of the process facilitator/manager is to design and guide participants through such learning loops. It requires a meta-level analysis of what is happening and continuously analysing and adapting the next steps. Not a blueprint, each step is a logical consequence of the analysis of what happened in previous steps. Sometimes, steps might not be necessary; other times, they need to be intensified and a step back is needed to drill down deep enough. The main elements in such processes are as follows:

- Participatory/interactive meetings and workshops
- Coaching local teams
- Analytical work (e.g., institutional analysis in communities where an outsider role is advantageous)
- Team reflection sessions
- Conflict management processes
- Personal interaction with community members, agencies, bureaucrats, service providers, etc.

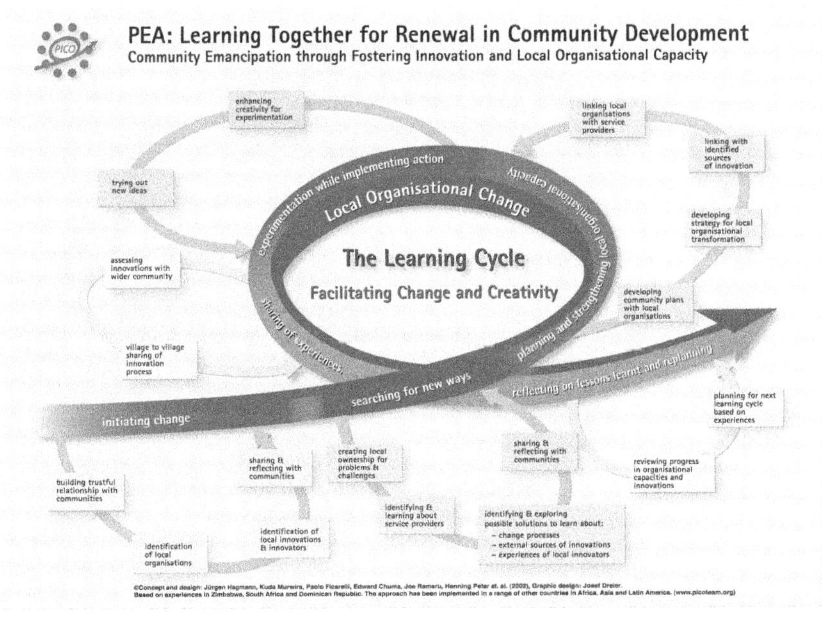

Figure 6.1 Example of a process architecture in the form of cyclical intervention loops in community development (Hagmann et al. 1998, 2002b)

In a rigorous action learning programme, there are several interlinked simultaneous loops. They serve as levels of learning if one monitors and reflects on each loop separately, bringing out the issues and next steps. Figure 6.2 shows an example from South Africa and Zimbabwe, where we had 5 overlapping loops – activities which are all in one bigger process but needed to be analysed separately.

The process facilitator/manager will facilitate the events in such processes, be a coach for local teams, a connector to outsiders and external knowledge, an advisor for critical areas, but at the same time learn from and adapt the greater process and multiple loops. This role is way more than a meeting facilitator and requires a deep understanding and vision of such processes and multiple skills. It is a challenging and high competence role which is notoriously underestimated, with the wrong people often engaged for the job – a recipe for poor outcomes of such processes. Actually, in our experience with myriads of processes across the world, this is the single most critical failure factor for action learning/social learning processes. It has often resulted in very low-quality process management and a lot of discouragement and disappointment about the value of facilitation.

Key elements of meeting/workshop facilitation processes

Workshop facilitation is one of the most crucial elements in the context of action learning processes. Therefore, we will describe it in more detail.

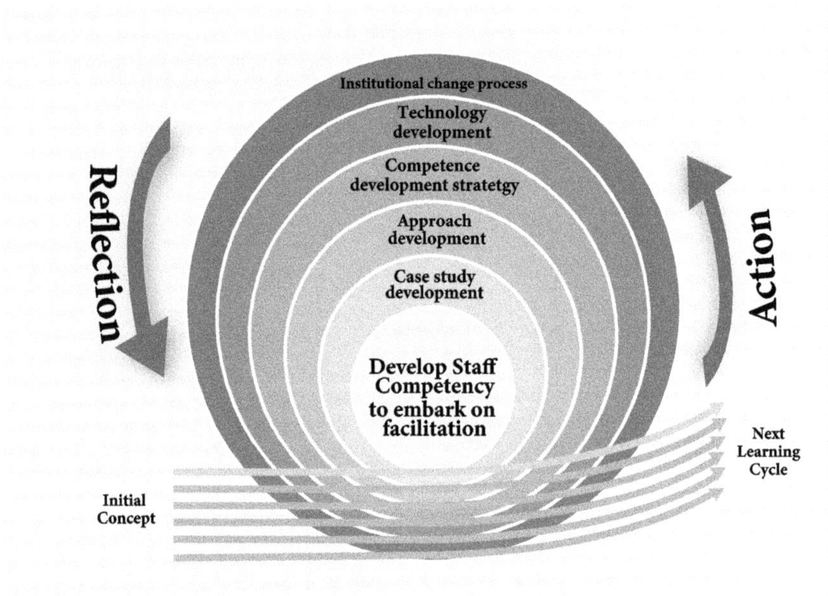

Figure 6.2 Five simultaneous levels (loops) of action and reflection towards development of participatory community development

Workshop facilitation is built around five cornerstones which are managed simultaneously to support a creative dialogue (see Figure 6.3).

The elements are:

1 **Guiding star, the logical flow/roadmap** for the meeting which guides the workshop process (the architecture of the meeting process) based on desired outcomes. The programme design represents a well thought through process that focuses on sequencing of items/sessions in a manner that strategically builds one step on another, rather than just agenda items as in conventional meetings.
2 **Facilitation techniques** which enable the understanding and clarity of the content (visualisation, consensus-based methods, participatory tools, etc.).
3 **Questioning for change** to trigger deep thinking and analysis, challenge the status quo and bring out creative thinking and innovation.
4 **Group dynamics and empathy** to understand and manage the needs of groups and bring everyone along in a socially inclusive way.

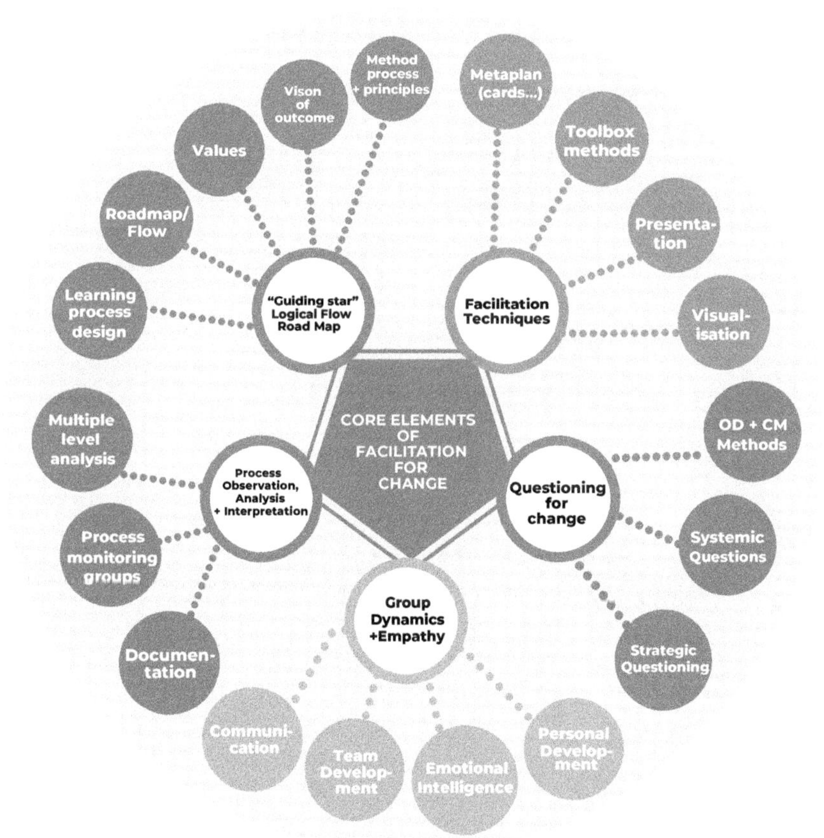

Figure 6.3 Key elements in workshop facilitation

5 **Process observation, analysis and interpretation** to understand what is happening; inform the next steps towards reaching the desired outcomes and unblocking deadlocks in the group. This is the navigation compass for the facilitator and comprises six levels of observation. A beginner will be fully absorbed by the content level and concentrate on this, while a highly skilled facilitator observes the dynamics of the interaction (behaviour of individuals who might dominate, the dynamics within the group, between the group and the facilitator and also the environment of the workshop). A good facilitator will be able to integrate all these levels and make the proper decisions on his/her actions, choice of methods and process steps (see Figure 6.4).

An effective workshop producing clear outputs needs to be well-planned and prepared. A facilitator alone without a solid feedback structure is likely to err, as many of the dynamics of the issues and groups are not known and are not immediately visible. A process steering group which consists of the organisers and other participants is important in defining what should come out of it before the workshop and to give feedback every evening. In community development, involving representatives of communities is critical to ensure their desired outcomes

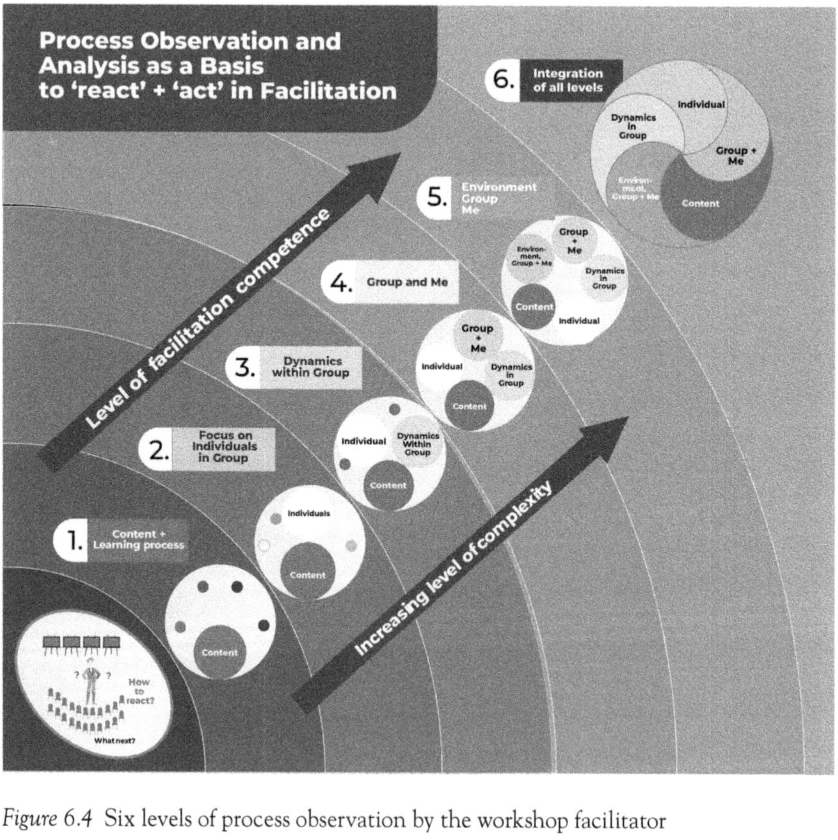

Figure 6.4 Six levels of process observation by the workshop facilitator

are built into the workshop design. Members of the process steering group become the bridge to the participants as they are co-creators and ambassadors for mobilising their constituency. Meetings of the steering group also give the facilitator the opportunity to anticipate reactions of the larger group and get prepared to react optimally to those.

The sequence of steps in a workshop is built around a rather universal logic in solution-oriented processes:

1. **Opening up/warm up:** familiarise people with each other, create an open and inclusive atmosphere, develop personal bonds between people, make everybody comfortable to speak, reduce hierarchies, match expectations.
2. **Exploring the topic**: overview, deepen analysis of issues, let people identify their own socially differentiated issues to address, validate with reality, create ownership of problems, challenge them in their way of thinking, provoke and deepen their analysis, get to the depth of things, brainstorm on the ideal situation.
3. **Identify ways to deal with issues in a creative way**: let them think unconventionally, consider new ways, challenge them about their usual ways.
4. **Converge around key issues**: bring it all together so that participants become aware of the systemic nature of problems and solutions, settle into commitments for action and next steps.
5. **Reflect on the process and outcomes**

Generally, the longer the exploration and analysis phase can be kept interesting, the easier it is to find common ground in the solutions (Figure 6.5). It's the deep thinking, the challenging through hard questions by the facilitator, and the open atmosphere which bring out the real issues which normally remain hidden, and which bring convergence towards solutions. The deeper you go in the analysis, the fewer solutions there are, which made us realise that:

> Consensus lies in depth!

The discussion challenges the communities' perceptions and behaviour patterns (e.g., a dependency syndrome, victim culture and belief in money as the solution) and breaks down the conventional communication patterns and barriers through different and changing seating orders and non-hierarchical communication. Often, this is a revelation and a relief; and high energy and participation emerge when the real issues and the truth have been found together and spelt out for all to examine.

F4C as a process has many facets and requires a high level of skill and, most importantly, the right attitude by the facilitator.

Facilitation for change in practice

In this section, specific aspects of facilitation processes will be elaborated based on practical experience.

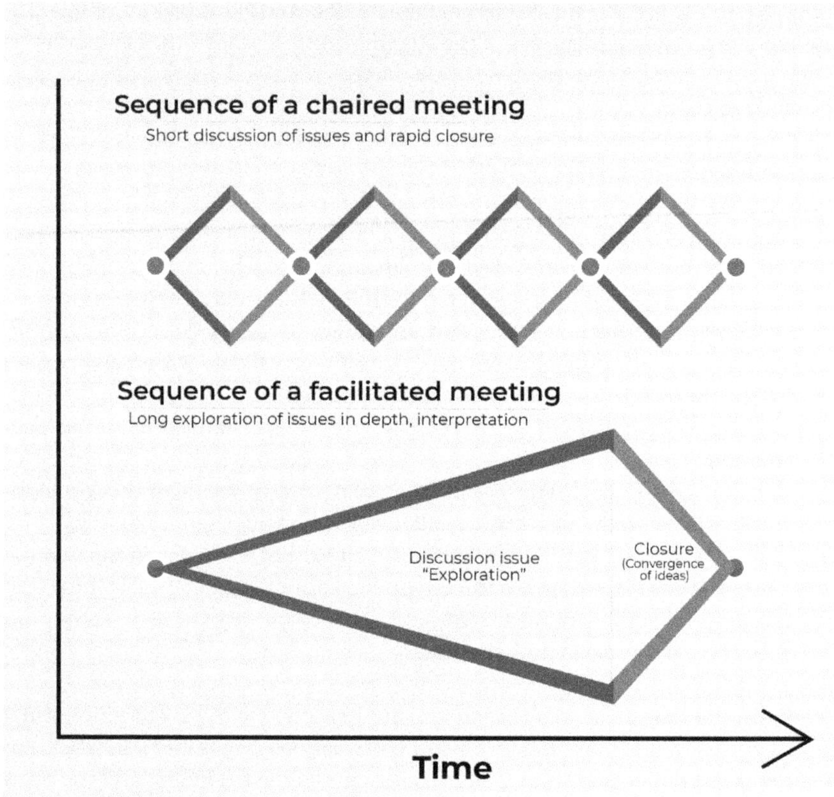

Figure 6.5 Long 'exploration' phase in facilitated meetings vs. chaired meetings

Getting excited – creating social energy and commitment to the process

In any community-based change process like ACM or community development, the first question for facilitators is always: why would people be excited to engage actively and take ownership? What would motivate them and bring out their energy? Why would they accept and trust outsiders starting social processes?

Outsiders/facilitators work normally on many differing assumptions about the motivation of local communities. To alleviate their fears of rejection, outsiders often try to obtain a smooth entry by promising the so-called 'trust building measures'. They may give the local people some direct immediate material benefits, which are envisaged to bridge the gap and reduce uncertainty. But once such benefits are given, be they a sitting allowance[1] or promises of development benefits, the perspective of entitlement will long persist and the foundation of the relationship is set: that of beneficiaries and benefactors. It is hard to transform such beginnings into self-propelled, truly community-owned processes, as expectations have been raised in a specific, dependent way.

Substantive trust is built through being genuine and by sharing the same values and expectations rather than by gifts. It requires consistency in explaining and demonstrating what the true intentions are over a longer period. Trust in such interventions needs to be earned by outsiders (and as Liswanti et al. and Fisher et al., both this volume, show, such trust can endure).

A community process which aims to become a self-driven and community-owned process – as ACM aspires to – requires more than a meeting for the start. Some basic principles and steps based on our experience include:

- **Addressing the community as an organisation**: First, the community needs to be seen and addressed as an organisational system, not a set of individuals. If one would do a project with a formal organisation, one would respect the different levels of management and authority and discuss with the respective levels to create commitment and agreements to start off. Similarly in a community, one needs to identify the different levels of leadership (often based on both traditional and more formal, government-initiated organisational structures), engage them in discussions, understand their perspectives, help them understand what the intentions are and get permission to do some scoping and enquiry by interviewing a range of different people from different social strata. This will provide a first basic understanding to the facilitators/outsiders; and the insights inform the facilitator in the development of an appropriate process design. This step builds relationships, clarifies expectations to some degree and reduces the likelihood of conflict.

  With these insights, the facilitators can return later to the leadership and inform them in a small meeting about some critical points, share ideas on what a future process could look like and agree on the way forward. Ultimately, this will result in organising a big community meeting as a next step to engage the broader community.
- **Opening up, breaking entrenched communication patterns and developing critical consciousness**: In the community workshop, the role of the facilitator is crucial. In most societies, a more hierarchical setting in meetings is the norm: the powerful talk, others listen, leaders stand in front of the group, giving long speeches, etc. This is what many people are used to, they expect it and at the same time they are often bored by these hierarchical communication patterns. They are thus not excited to come to meetings; the sitting allowances often become the biggest incentive to attend.

It is of great importance to start in a very different way: breaking these communication patterns by getting everybody to talk right from the beginning, contributing already in the introduction stage in small groups, responding to interesting questions. The goal is that each individual feels he or she has made a contribution, been recognised and is important, a source of knowledge.

Managing expectations is central at that stage, not promising outside solutions, but throwing the ball back to their own thinking. Challenging the status quo and through critical questions bringing out key factors which have led them to the status quo. This process of developing critical consciousness and analysis is supported through Paolo Freire's 'Pedagogy of the Oppressed' (Freire 1970, 1973). This thinking led to 'Training for Transformation' (Hope and Timmel 1984), which provides facilitation tools to open people up, make them challenge their situations, creating ownership of their problems and their own solutions. 'Codes' (see Figure 6.6) depicting certain common situations in stories, pictures and wisdom have been very effective in facilitation of this opening process as they raise real-life situations in a coded way, so people can relate and talk openly about similar issues they face. Codes can be in the form of pictures, videos, role plays and also learning tools which make biophysical/environmental process easy to understand and relate to (see Hagmann and Chuma 2002; Ramaru, Chuma and Hagmann 2014). Often social differentiation when dealing with issues is important so that different strata of society can bring out their perspectives on issues clearly and then negotiate a common perspective.

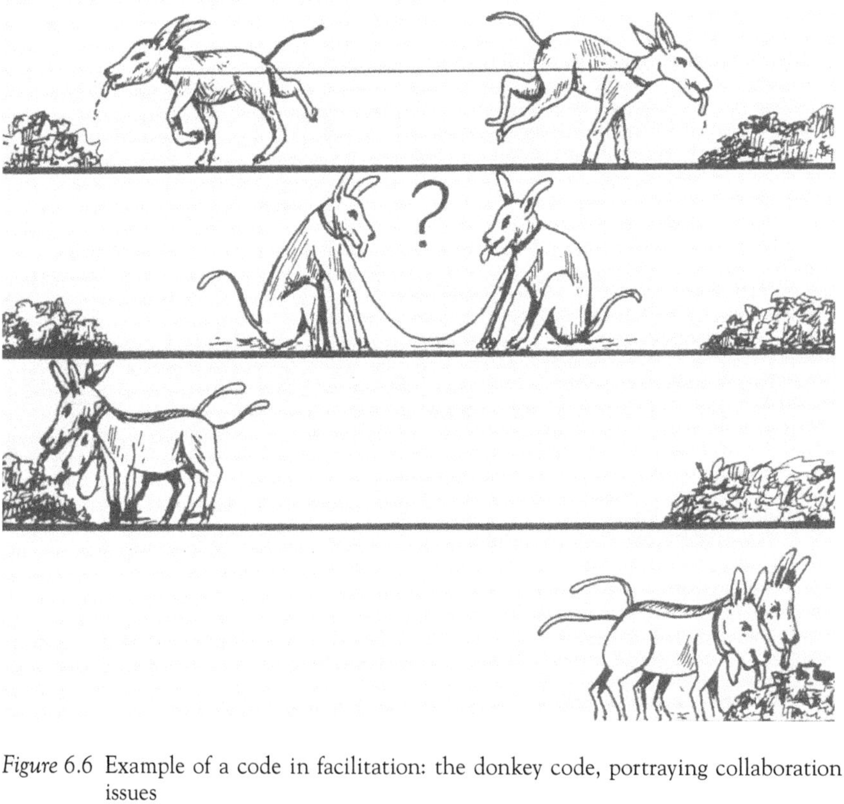

Figure 6.6 Example of a code in facilitation: the donkey code, portraying collaboration issues

This intensive, concentrated process has always been very energising (social energy) as it excites and empowers people to embrace their own values and issues, talk about the real issues and needs, and realise that they have the power to change the situation – this creates commitment to the process. It is important in such beginnings, not to dismiss the leadership, but equally challenge them in a positive and supportive way without any loss of face. They need to be supportive, while the facilitation process itself is reducing the hierarchy and power distance in local communication patterns and is socially inclusive.

Managing inclusiveness (gender, age, wealth, ethnicity, power)

Communities as organisations are typically extremely heterogenous in their composition. In most contexts, some groups are notoriously excluded or at least not adequately recognised in their needs and aspirations. Inclusivity through 'managing inclusiveness' is a central concept in F4C, in order to mobilise the whole community for positive action and sharing of benefits for everyone. It is a combination of social differentiation and negotiation to arrive at inclusive goals, actions and benefits (Ngwenya and Hagmann 2009).

The first step to manage inclusiveness is to know the diversity of groups. In the initial interviews, some groups/interests might reveal themselves but often more analysis and enquiry are needed. The fundamental social differentiation criteria are gender, age, wealth, power, ethnicity, but when going deeper it is important to understand the different interest groups as well (e.g., farmers, foresters and forest owners/users), also groups formed by development interventions and government. Once these are clear through an institutional analysis and other scoping measures, the facilitator can bring out the different groups' interests and expectations. Small working groups of the subgroups are an effective tool to bring out their interests, expectations and solutions, inviting them to present back to the plenary/community. This makes the differences transparent and allows then for proper negotiation in terms of solutions and benefits. In the end, what matters is the negotiation for agreed ways forward where the different interests are recognised and taken into account. Again, it is fundamental for the facilitator to change the patterns of communication for effective inclusion and give voice to the different groups in a subtle and socially acceptable way.

Managing complexity – first things first

In community processes, almost always a long list of problems/challenges emerges. Most are rather 'wants' than 'needs' and often inspired by what the outsider is perceived to be able to provide rather than what really would enable the community to act for themselves. How can we deal with such diverse and complex issues and interests – where to start?

The first principle is: don't start too quickly with just anything which seems reasonable. A lot of facilitated negotiation is required to come to the really pressing issues that hinder the community to move to the next level. For example, in Tanzania, we once had a case where after all the nice things people wanted to embark on, the women were assertive and insisted on priority No. 1 to extinguish the illegal brewing of a toxic alcohol which had devastating effects, particularly on the men. Initially not even on the list of possible actions, it turned out to be the biggest block for development of their area, emerging only after long discussions and several meetings. As we did not provide any resources except for facilitation, we made it a condition that any action must be accomplished by the community themselves. The women mobilised and organised the whole community, made a plan and within a month, to our surprise, all illegal brewers were stopped. After this positive experience of their power and dedication, they embarked on the next big things with great confidence (see similar examples in Johnson and Pokorny 2022).

What this example shows is that the real priorities are felt and need to be brought out. 'Drowning in complexity' and a myriad of things which all seem important simultaneously and are all totally interconnected… is paralysing rather than empowering. It is important to get rather quickly to the 'first things first' attitude, start with small things which can be done, create the feeling of success and empowerment, and move on to bigger things. This confidence building process develops pride in the community which breeds more excitement and social energy, readiness and stick-to-it-iveness to tackle other issues that come along.

Building a functioning community organisation

A community is a very diverse and heterogenous form of an organisational setting. Lots of formal and informal institutions and groups exist in parallel, often with similar functions. In addition, in many contexts, there is a whole graveyard of institutions formed by development interventions which have died once the incentives by the development agency were stopped. Officially, these groups and institutions remain, dysfunctional, but never dissolved – often without members or action. Indeed, this is part of the reality that spurred the development of approaches like ACM.

The first step in developing stronger organisations in communities is an institutional/organisational analysis which identifies what kind of local institutions/organisations are around, which ones are really functioning for what purpose, their strengths and weaknesses, and what the interactions and relationships between these entities are.

This analysis is ideally done by the outside facilitator team with a mandate from the community. The insights and results will be brought back to a community meeting where they will be discussed, validated (or corrected) and implications for the community's development brought out. Often, the perspectives on challenges and opportunities differ, but once the different perspectives of people's challenges in their communities and organisations are presented, it is a reality which cannot be wiped out and ignored. For example, in one case in Zimbabwe, people identified the fact that some individuals were monopolising leadership opportunities as a core problem. Surely that leader did not like it, but the public and intensive factual debates were powerful and accepted by the community. Ultimately, the community formed a new by-law which no longer allowed multiple leadership positions by one person. This opened up the community organisation drastically and many blockages disappeared. Important here is that it is not the facilitator who judges the findings; it is the community members who interpret, judge and make decisions. The facilitator only facilitates the discussions, making sure people understand that this is their own problem and not the outsiders' problem, that solutions lie in their own hands, rather than outsiders'.

Local organisational development is a core concept in F4C (Hagmann and Schwedersky 2000). There are many ways and forms for developing stronger organisations for different purposes. One model which was very successful in South Africa (Ficarelli et al. 2003), for example, is the 'umbrella organisation' which is composed of representatives of the institutions that really function and have a role in the community. This umbrella supports its members, making sure that knowledge is shared across these groups and institutions and dealing with broader community development aspects. It ensured considerable inclusiveness in a diverse setting. Before leadership was nominated, intensive debates on qualities of leaders, roles of leaders, dos and don'ts of leaders and members, how to remove a leader, etc., were discussed in small groups, and re-discussed several times until there was agreement within the large group. These were consolidating the values and principles of leadership. As these rules and norms came from the community, each new leader knew what he/she was supposed to do and not do. And the members felt empowered to claim accountability as this was agreed before he/she took the position. The facilitators' role was to guide the discussions, ask the right questions in a neutral way and share ideas of other places he/she had seen as an inspiration. All decisions were made by the community itself.

Managing conflicts

There are different types of conflicts in communities, which must be dealt with in very different ways: open conflicts vs. hidden conflicts, personal conflicts vs. interest-based conflicts. What they have in common is that they are sucking energy and are a blockage in the development process. In cultures where social harmony is a highly desired state, conflict has a very negative connotation and people can be ready to hide it and give in, rather than deal with it. As a facilitator, often it is even difficult to recognise a conflict as it is not spoken out, and only a good process observation indicates that something is not flowing, which might be an underlying conflict.

So, the first step is to identify conflicts and understand what type of conflict it is. For example, a conflict between different types of land use by different groups reveals itself relatively clearly (e.g., in an ACM process). Like in any conflict resolution process, one would go deep into the different interests of the parties (away from positions). Critical analysis hopefully would uncover some common ground and possibly rules, by laws and new technologies which could enable a better benefit and win-win for the different parties through cooperation.

When it comes to hidden conflicts, it is more complicated. We had cases where two families and individuals in a community had issues going back a generation; they could not work together. Each tried to sabotage whatever the other was doing. There is no way an outsider can know this, but all community members may understand. Often, these issues only come out in informal discussions in trustful relationships. Traditional societies also have their own ways to solve issues, which often are very different from a Western way, much more informal, using trusted third parties and mediating/negotiating behind closed doors. A facilitator for change has to have a feel for what conflicts are above or below ground and navigate accordingly. Only the local people can help to understand and often to solve it. A facilitator must allow local people to come into the process, support/help and create forums, like a process steering group, where process issues can be discussed rather openly.

Managing knowledge, innovation and creativity

Knowledge is key in facilitated processes aimed at emancipation and self-empowerment of communities. There are several challenges to manage:

- In oral cultures – mainly speaking of rural Africa – 'traditional' knowledge is mainly stored in stories, wisdom and experiences in an adaptive way and is often considered (by educated people) as inferior to the 'modern', science-based knowledge. The latter in rural communities has been pushed by state extension services, originating in colonial and often missionary systems and approaches. So, we are dealing with knowledge systems that are granted unequal prestige and value. The challenge is to combine traditional and modern in a positive way to find solutions to problems rather than classifying the knowledge.
- Access to external and alien knowledge and innovation has been rather limited in rural communities until recently through the internet. Technical services in forestry and agriculture, for example, are weak in many countries and often out of date with their technologies and approaches.
- Scientific knowledge generally comes in a rather academic form, often not grounded in the local context and experiences, which slows down adoption and adaptation.

Facilitated processes like ACM have to integrate technical know-how with indigenous knowledge and into social processes effectively. In our experience, an effective way to integrate different knowledge types and processes is experimentation by the land users. Once the real problems and interests are clear, people are encouraged to experiment with old and new ideas, combine them and analyse the results. This solution-finding process enables several benefits:

- First, communities and people become experimenters and unleash their creativity. No technology can be pre-assessed as superior, what counts is the result in their context and any idea is useful.
- Second, in this process peoples' analytical capacity by comparing and analysing different solutions is enhanced and results in deeper understanding of underlying issues.

- Third, the integration of scientific and traditional knowledge is happening naturally and adaptively and strengthens people's confidence in their own capacity.

The role of the facilitator is to link people to many sources of innovation, be it from research, innovative land users/communities, experts or academia. Exposure to new knowledge is critical to enhance creativity and expand experimentation. The facilitator team also needs to actively bring technical services and experts into the process.

The second role of the facilitator is to encourage experimentation at larger scales and sharing across people and communities. A range of mechanisms can be effective. For example, in Zimbabwe, South Africa and the Dominican Republic, we used competitions for the best ideas, where every community member could participate. The communities with the highest number of good ideas won. And the individuals with the best ideas won in each community, which then created a powerful incentive for everyone to experiment. In our experiences, this created enormous energy to be creative and think in solutions rather than problems, a very important perspective in ACM processes (Yuliani et al., this volume). And even the poorest could participate and be recognised for their great ideas. It lowered the barriers between the rich and the poor as both could make major contributions in their own right and both were recognised. It strengthened inclusiveness.

Knowledge management in ACM-type processes is critical. Active sharing of knowledge and experiences and the ideas coming out of experimentation and documenting the knowledge are major elements of that.

Facilitation at multiple levels (integration)

The integration of different levels of change was briefly mentioned above. The aim is to enable the development of a system which can perpetuate itself. For example, in ACM, the first level of change was initially in communities, and in some cases organisational levels (Colfer, Prabhu and Larson 2022; Komarudin et al. 2012). If this is successful, it is obvious that for large-scale implementation, the technical forestry services need to adapt, change the capacity of their field officers, change the way they provide services, their approaches, their internal working arrangements, etc. This change process in the technical forestry service does not come on its own, it needs to be facilitated. Once the technical services are on board, they will say that we need to change the policies and regulations at the national level, which are almost always initially antagonistic to such bottom-up approaches. Again, such changes don't come on their own; they require facilitation of change across the levels, encouraging interactions among the different levels, identifying the changes required to make things a success, considering new modalities on how to operate, working arrangements, etc. In the end, it is a multi-level change process of facilitation, all triggered and driven by the change needs of the primary delivery level, the community. Such multi-level processes become very complex very quickly, as Prabhu, Larson and Colfer (2022) also experienced. These processes need to be anticipated strategically at an early stage and designed smartly. The facilitation competence at different levels also differs as political levels operate differently than service providers and communities. In many cases, innovation platforms and multiple stakeholder platforms are required to get the different relevant actors together to make a system work better (Ngwenya and Hagmann, 2011; see also Fisher et al. and Kusumanto et al., both this volume) (Figure 6.7).

Managing the facilitator's biases

Facilitators in a change process are never really neutral, even if they don't have clear vested interests. Knowing one's biases is important in order to deal with them and be clear about what is happening. In the processes described above, the facilitator's main agenda is to make people empower themselves, be more creative, become better organised, more solution-oriented, better negotiators and enhance communication within and across people and communities. These in themselves are clear agendas with clear values and mind models one needs to be aware of. The facilitation methodologies may well also emanate from different knowledge systems than those of the communities.

It is desirable to bring in new ideas and ways of dealing with issues in systems – without innovation, there will be no development. The key is that the values

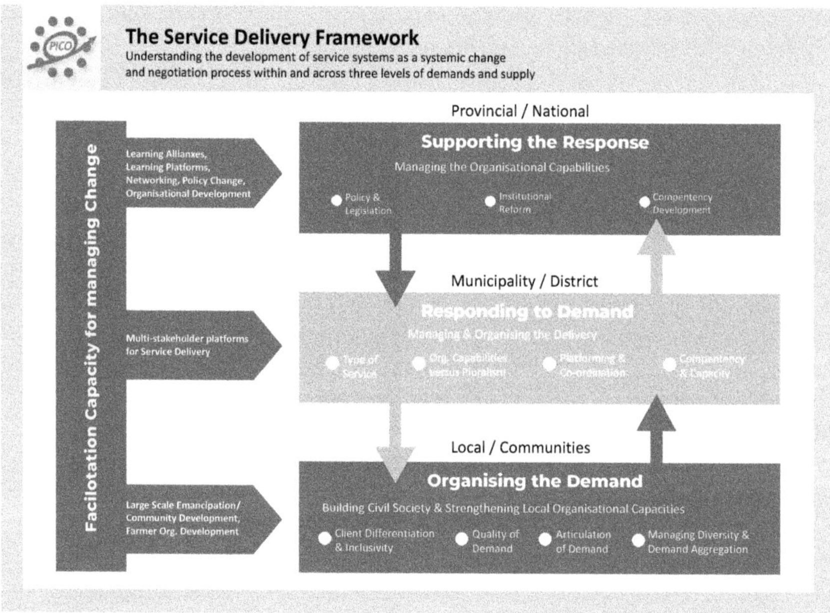

Figure 6.7 Example of a multi-level facilitation process to foster change in service delivery (Hagmann et al. 2002a)

underlying these ideas are shared, neither imposed by outsiders nor creating or reinforcing notions of 'inferior' and 'superior' ideas.

Fundamentally in our experience, the attitude of a facilitator is most crucial, respecting the way people handle their issues and creating space for them to do it in their own way, respecting different cultures and values. Once this attitude is achieved, and the values in the facilitation are open and transparent, the facilitator's biases will not lead to imposition. People have the choice of making their own judgements and decisions.

What this means for the facilitator's personality is that he/she should be highly self-aware and able to have a critical distance from his or her own work, allowing critical voices to be heard and engaging with diverse perspectives in a positive and humble way.

F4C is a demanding concept and practice. We have developed our own capacity over the years and were privileged to have ample opportunities to practise in a range of processes. We have systematised and conceptualised our learning and experience in frameworks, guides, etc. However, ultimately facilitation is exciting because it requires constantly 'thinking on your feet'; it is full of surprises which don't fit into any frameworks and depend heavily on one's personality and style. There is a lot of uncertainty in these processes and people have different ways to deal with uncertainty. Ultimately, a facilitator needs to feel secure in his/her own insecurity. Authenticity is a factor which makes it genuine and true and trustful and has a major effect on groups. Therefore, it requires more than just multiplying the facilitation methods in a standard training way with standard procedures. Developing facilitation capacity is never bringing out the same styles; we are all different as individuals. What is important then is the development of one's own style within the mindset and values of facilitation.

Capacity for facilitation of social learning processes

The sections above have clearly revealed that the process facilitators'/managers' capacity is one of the most crucial determinants for a successful ACM-type intervention. The development of this capacity is another substantive challenge in any intervention. These are capability profiles which cannot easily be picked from the labour market, so in the design of a programme such competences need to be built in, as a well-planned component. Here, we deepen the understanding of such capacity development processes.

Key skills and competences of a facilitator for social learning at multiple levels

The majority of capabilities required are in the domain of soft capacities and soft skills. In our experience, it is often a process of personal transformation, which has impacts on one's personality way beyond the professional arena, particularly when it comes to social and communication skills, emotional intelligence, creativity, curiosity (Goleman 1988) and the confidence to take and shape one's life on one's own initiative.

In our programmes, we found the following five competence domains critical for facilitators of community-based interventions.

Vision and values

Facilitators need to have a clear vision of what 'development' is supposed to be, with an emphasis on human development in communities. A process-oriented development process might be a different vision from that of the mainstream and needs to be nurtured. For example, if a forestry officer who has been working within the conventional paradigm and vision of most forest bureaus is supposed to become an ACM champion, he/she will clearly have to transform his/her vision/imagination for development.

The other dimension is human values. As elaborated before, sharing the same values in interventions is critical. Values are rarely obvious and explicit, but inherent. They show up in the form of behaviours. In our experience, it required substantive work to become self-aware of one's own values and other people's values and how to deal with value differences. This included the human values of interaction and its manifestation in communication (e.g., how do you communicate verbally and non-verbally in such a way that you respect other people, independent of status and hierarchy? What is important for people to feel they have co-created and contributed? etc.). It's a self-experienced 'learning as one goes'. Role plays and analysis have always been good tools to discover such aspects of oneself.

Conceptual understanding

Mental models are critical in guiding our actions. That is why a conceptual understanding of action learning and process approaches, as well as ACM itself, is crucial. Often, one shies away from discussing conceptual work with field officers as such ideas appear too difficult and intellectual. However, they are fundamental as a guiding framework for action. If a concept is too difficult to understand, it might be more a communication problem than the concept itself. It just requires effort to explain complex issues in simple words. Good analogies from real-world experience have always helped to illustrate complex issues and simplify (bringing the point 'home' – in other words use examples of daily life to explain).

Another conceptual dimension which a facilitator needs to understand is the functioning of social systems (also rural livelihood systems) and their behaviours and needs, including how self-organisation works, how technology contributes to development, and relevant methods and approaches.

Personal development, emotional intelligence and soft skills

The most difficult skill set for a facilitator is around personal development. It is not something which one can learn in a few sessions; it requires active engagement and working on self-development, which we all know is difficult.

The whole concept and skill set around emotional intelligence has been very helpful to enhance practitioners' empathy and self-awareness. Some other important skills and behaviours are as follows:

- Attitude – taking people seriously independent of status, education and power,
- Empathy, patience and authenticity,
- Creativity, innovation and curiosity,
- Flexibility and solution orientation,
- The ability to deal with uncertainty and insecurity,
- Humbleness – understanding that it is not about you as a facilitator – you are just a catalyst,
- The understanding that you can't force people to do things; you can only influence their decisions to change behaviour and thinking.

Most of those are reflected in the deep values and beliefs in facilitation: participatory engagement, recognising any knowledge in the system and appreciating the resources/the value of people's knowledge.

This personal development is often a challenge for technically trained specialists who may feel that they are more educated and have a higher status than the so-called uneducated villagers. Naturally, they cling to their educated knowledge which gives them superiority. It is a major transformation for them to accept local knowledge as equal and sit in the boat of rural communities and earn a different form of respect and recognition. But it is necessary.

Facilitation skills

The most fundamental skills a facilitator should have are:

Foundation facilitation techniques based on group dynamics and principles of adult education. These are a must for facilitators. Group dynamics provides a good skill set to manage groups, while adult learning provides deep insights and tools on how to engage adults through exploration of their experience and learning based on that experience.

Team development techniques are essential in facilitation to understand how teams function, their dynamics and how they can be developed.

Questioning techniques are the backbone of facilitation. Asking the right question at the right time is what triggers lively debates, solution searching and challenging of the status quo and people's own behaviours. It is a difficult skill, linked to one's own vision of development and understanding of human beings and their behaviour in organisations. Some concepts like 'strategic questioning' (Peavey 1990) and elements of provocative therapy (Farelly and Brandsma 1981) and organisational development provide good stimulation. Often, good questions originate in the facilitator's imagination and understanding of the issues, raising issues in the form of questions instead of comments, etc.

Visualisation is the visual language of facilitation. It's a skill which is very important for effective communication with the audience, for memorising and building on points agreed and discussed, and preventing 'going round and round'. Even more important with illiterate audiences, it does not need sophisticated ways of visualising; a simple flipchart visualisation can be effective. Creativity helps!

Storytelling, codes, analogies and a good sense of humour are powerful ways to bring out issues in an indirect way. They are great skills for a facilitator to make sessions lively and avoid loss of face for participants by being too direct.

Managerial and planning skills

A facilitator always has a process management task. This requires some basic management skills like action planning and different planning approaches as well as reporting skills. Process documentation is extremely important for both the audience and the facilitator. Without good process documentation, issues can get lost, making it hard to manage a good reflection process and build on the previous interaction. This does not have to be sophisticated – in areas without electricity it will be done by simple note taking.

Developing systemic facilitation competence

From a conventional training perspective, the skills and competences described are probably overwhelming and one may envision hundreds of training courses to develop them. Most likely, all these courses would not do the job in developing a rounded competent facilitator. All too often, there is no direct linkage between training inputs and challenges faced on the job. Consequently, most professionals do not apply what is learnt on training courses; training remains as mere information which, if not applied, will be forgotten over time. Knowledge does not develop through participating in a one-off training session: this requires well-designed learning processes.

Through our own learning over our first decade (Hagmann et al. 2003), we found an alternative to conventional training to support people in learning these skills and competences in a more iterative way, based on real-life practice. The learning programme is not about 'training' and then 'doing'; it is an integrated process of learning as we go, in practice and real life. The programme does not have training modules per se, but a set of core competencies which are developed, simultaneously guided by the field process and its challenges emerging.

The basic structure of such a learning programme called 'systemic competence development' (Hagmann et al. 2009) is:

1 At least four to five **learning workshops** with a group of 25–30 participants over 12–18 months.
2 **Peer learning teams** to support implementation of field practice.
3 Several months of **field practice** between the workshops where key steps of the process are managed by small teams.
4 **Coaching and mentoring in the field** by the learning facilitators, guiding the field practice and reflection.

In the first workshop, basic concepts are discussed, the overall process of F4C is elaborated and the broader context understood. Ideally, a field visit to sites where this work has been implemented helps participants to better imagine such processes. Then at the end of the first workshop, peer learning teams are formed as small groups of people who will be assigned to a community in which they will practise F4C in community development for the next two years. They will plan together as a team the first few steps in entering the community. These stages are role played to better understand and coach the teams to come up with a detailed workplan for the next two to three months in the field. The learning facilitators will coach these groups several times in the field during these first two to three months, so that they feel secure enough to do the work. They face a lot of challenges, things don't work out as planned and it can be messy. If not supported and coached, the field workers will prefer to return to their comfort zone and act as they always did – authoritarian and instructive instead of being consultative and facilitatory, experiences which Cronkleton, Evans and Larson (2022) faced in Latin America too.

In the second and subsequent learning workshops, the teams first process their experiences in depth, what happened, what worked and what did not.... and share with other teams. The workshop focuses on the big challenges they faced, identifies solutions, shows new methods and ways to deal with the issues they have experienced, organises some role plays, etc. Then, concepts are deepened, new ones introduced and the next few steps in the process are discussed in depth and another detailed workplan is developed. The teams go out and practise for another three to four months. The sequence continues until workshop 5, after which they have practised the whole process in the communities. Specifically, after the third workshop, a leap in understanding has often been observed. The participants break out of their linearity and start thinking more systemically, able to bring the complexity into one frame and deal with it (Figure 6.8).

As openness and self-development are such critical competences in participatory action research, the teams really practise a feedback culture and one can literally see how they grow in their confidence over time. The appreciation they receive in the communities also shapes a positive attitude to local people; they identify with the communities and sit in the same boat.

After 18 months to 2 years, the group has reached maturity – they can handle the process as competent facilitators by themselves and take on new ones. On average, a third of the participants have developed excellent facilitation skills (they become trainers/learning facilitators later on), another third is good in practice and about a third continues to struggle. As a follow-up and further development, learning networks and communities of practice are created where the different cohorts share their experiences and new ideas.

The systemic competence development process is intensive and requires a very serious commitment in terms of plans and finances. Often, we have been asked if this cannot be reduced to two training courses. It can't! It is an investment in human capacity which can turn an intervention into a great success within two years. But it needs a longer-term perspective.

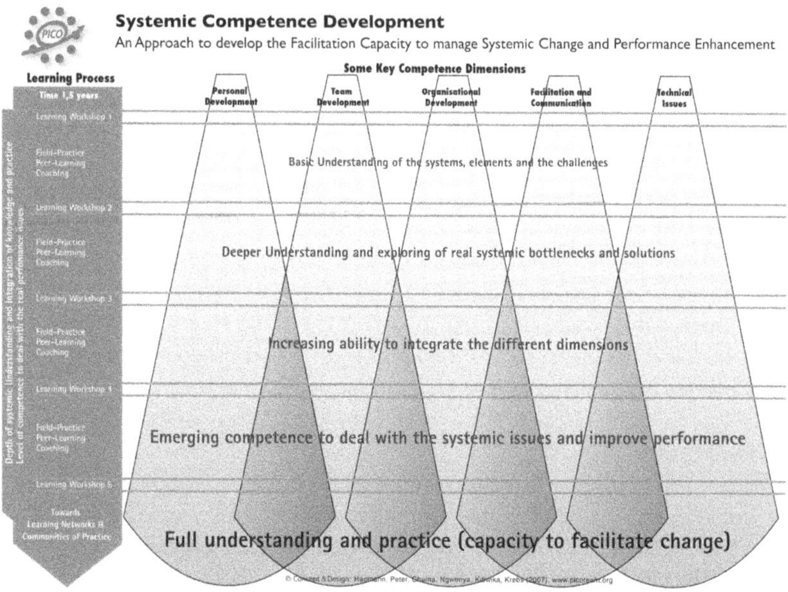

Figure 6.8 The process of systemic competence development (Hagmann et al. 2009)

In South Africa, we have developed more than 350 community facilitators through this process, in Zimbabwe also more than 100 at a time and in the Dominican Republic, some 50. Many of those well-skilled facilitators are highly marketable and often left their organisation for a better job within two years, as particularly the public sector could not provide incentives for them to stay. In all cases, we have rapidly developed second groups of learners and made the best facilitators their learning facilitators, so that the rapid turnover of good people could be buffered and did not undermine the future of good programmes.

In Uganda, we have developed this 'innovation' capacity within the university – with lecturers and professors, and the process was slightly differently focused but equally highly successful (Hagmann, Kibwika and Ekwamu 2009). The systemic competence development process works very well, but it is intensive and expensive. Looking at many existing development interventions, it might still be the most cost-effective way to invest. Deloitte (2019) comes to similar conclusions in their paper on the future of work where they conclude that capabilities are the fundamentals – not skills.

In future, with the rapid development of social platforms, many of the more technical skills involved can be learnt through blended learning with an increasing smartphone penetration in different contexts. Whether this will work well for the behavioural change aspects in becoming a facilitator for social learning, is yet to be seen.

Conclusion

There are five main conclusions of this paper:

1 The role of facilitation in ACM-type collective learning processes has often been unrecognised and underestimated. Few initiatives have experienced high-quality facilitation to appreciate the depth and quality needed. It looks like an easy skill which can be learnt quickly and so many programmes look for a weeklong facilitation skill course, send their field officers out in the communities and become terribly disappointed by facilitation, as the impact is very limited.
2 Facilitation in ACM-type social learning process interventions needs to be considered and planned as a multi-level change process driven by the experiences in the communities. When it is not planned like that in the beginning, resources are not available to address other levels, and programmes get stuck.
3 Investment is needed in the development of adequate facilitation and process management capacity at different levels, right from the beginning of initiatives. It is intensive and can be expensive but forms the foundation of longer-term success. Therefore, it needs to be in the programme design and plans.
4 For change to succeed, the incentives of the different players need to be considered, from villagers to bureaucrats. Social energy can mobilise good commitment and needs to be supplemented by rules that incentivise self-perpetuation of new approaches.
5 Facilitation is changing rapidly in the digitalised era through social platforms. The question remains how optimally the facilitation of social learning in rural/forest communities can be complemented by social platforms in a blended way, without losing the depth and quality required. Depending on the context and the increasing availability of smartphones, this might be a great future opportunity.

Note

1 A 'sitting allowance' is an amount of money paid to individuals for their participation in meetings, a common practice in many developing countries.

References

Bollinger, E. and T. Zellweger. 2007. *Facilitation: The Art of Making your Meetings and Workshops Purposeful and Time-Efficient*. Lindau, Switzerland: AGRIDEA.

Chambers, R. 2002. *Participatory Workshops: A Sourcebook of 21 Sets of Ideas and Activities*. London: Earthscan.

Checkland, P. 1999. *Systems Thinking, Systems Practice*. Chichester: Wiley and Sons.

Colfer, C.J.P., R. Prabhu and A. Larson, eds. 2022. *Adaptive Collaborative Management in Forest Landscapes: Villagers, Bureaucrats and Civil Society*. London: Earthscan and Routledge.

Cronkleton, P., K. Evans and A. Larson. 2022. "Capacity Building for ACM: Lessons Learned from Training in Distinct Contexts." In *Adaptive Collaborative Management of Forest Landscapes: Villagers, Bureaucrats and Civil Society*, edited by C.J.P. Colfer, R. Prabhu and A. Larson, 134–157. London: Earthscan and Routledge.

Deloitte. 2019. "Skills Change but Capabilities Endure: Why Fostering Human Capabilities First Might be More Important than Reskilling in the Future of Work." www.deloitte.com.

Farelli, F. and J. Brandsma, J. 1981. *Provocative Therapy*. Ann Arbor, MI: Shields Publishing. New edition.

Ficarelli, P., E. Chuma, J. Ramaru, K. Murwira and J. Hagmann. 2003. "Strengthening Local Organizations for Conservation Agriculture – Some Experiences from South Africa and Zimbabwe." Proceedings, Second World Congress on Conservation Agriculture, Iguassu Falls, Parana – Brazil, 11–15th August 2003. https://www.researchgate.net/profile/Juergen-Hagmann-2/publication/302469235_Strengthening_local_organizations_for_conservation_agriculture_-_some_experiences_from_South_Africa_and_Zimbabwe/links/575fe4bb08ae9a9c95619807/Strengthening-local-organizations-for-conservation-agriculture-some-experiences-from-South-Africa-and-Zimbabwe.pdf Freire, P. 1970. *Pedagogy of the Oppressed*. New York: Herder and Herder.

Freire, P. 1973. *Education for Critical Consciousness*. New York: Continuum

Goleman, D. 1998. *What Makes a Good Leader? Harvard Business Review*, November and December 1998.

Groot, A. and M. Maarleveld. 2000. "Demystifying Facilitation in Participatory Development." *Gatekeeper Series* 89: 1–23. http://www.iied.org/pubs/pdfs/X188IIED.pdf.

Hagmann, J. and E. Chuma. 2002. "Enhancing the Adaptive Capacity of the Resource Users in Natural Resource Management." *Agricultural Systems* 73, no. 1: 23–39.

Hagmann, J., E. Chuma, K. Murwira, and M. Connolly. 1998. "Learning Together through Participatory Extension: A Guide to an Approach Developed in Zimbabwe." Harare. Department of Agriculture, Technical and Extension Services, GTZ and ITDG. http://www.gtz.de/agriservice/areas/concepts/concepts.html#7.

Hagmann, J., E. Chuma, K. Murwira and M. Connolly. 1999. "Putting Processes into Practice: Operationalising Participatory Extension." *Agricultural Research and Extension Network Paper* 94: 24. http://cms.picoteam.org/files/PICOTEAM%202014/Publications/Operationalising%20Participatory%20Extension-1999.pdf.

Hagmann, J., E. Moyo, E. Chuma, K. Murwira, J. Ramaru, and P. Ficarelli. 2003. "Learning about Developing Competence to Facilitate Rural Extension Processes." In *Participatory Technology Development: Case Studies on Integration into Agricultural Research, Extension and Education*, edited by C. Wettasinha, L. van Veldhuizen, and A. Waters-Bayer, 21–38, 257. Silang, Cavite Philippines: IIRR, ETC Ecoculture, and CTA.

Hagmann, J., H. Peter, E. Chuma, H.J. Ngwenya, P. Kibwika and K. Krebs. 2009. *Systemic Competence Development – An Approach to Develop the Facilitation Capacity to Manage Systemic Change and Performance Enhancement*. www.picoteam.org.

Hagmann, J., K. Murwira, P. Ficarelli, J. Ramaru and H. Peter. 2002a. *PEA: Learning Together for Renewal in Community Development: Community Emancipation through Fostering Innovation and Local Organisational Capacity*. www.picoteam.org.

Hagmann, J., M. Connolly, P. Ficarelli, and J. Ramaru. 2002b. "The Service Delivery Framework: Understanding the Development of Service Systems as a Systemic Change and Negotiation Process within and across Three Levels of Demand and Supply." www.picoteam.org, www.researchgate.com.

Hagmann, J., P. Kibwika and A. Ekwamu. 2009. "Learning to Make Change – Developing Innovation and Change Competence in African Universities." In *Farmer First Revisited: Innovation for Agricultural Research and Development*, edited by I. Scoones, J. Thompson and R. Chambers, 238–247. Rugby, Warwickshire, UK: Practical Action Publishing.

Hagmann, J. and T. Schwedersky. 2000. "Lokale Organisationsentwicklung." In *GIZ/BfN:* Naturschutz in Entwicklungsländern – Neue Ansätze für den Erhalt der Biologischen Vielfalt, 241–248. Heidelberg, Germany: Kasparek Verlag.

Hope, A, and S. Timmel. 1984. *Training for Transformation. Handbook for Community Workers* (Books 1–3). Gweru, Zimbabwe: Mambo Press.

Kaner, S. 2007. *Facilitator's Guide to Participatory Decision Making.* San Francisco, CA: John Wiley and Sons.

Komarudin, H., Y.L. Siagian, C.J.P. Colfer, with Neldysavrino, Yentirizal, Syamsuddin, Irawan, and Deddy. 2012. "The Role of Collective Action in Securing Property Rights for the Poor: A Case Study in Jambi Province, Indonesia." In *Collective Action and Property Rights for Poverty Reduction*, edited by E. Mwangi, H. Markelova and R. Meinzen-Dick, 235–269. Philadelphia: University of Pennsylvania Press.

Mukasa, C., A. Tibazalika, A. Banana, E. Mwangi and T. Mutimukuru-Maravanyika. 2022. "Strengthening Women's Tenure Rights and Participation in Community Forestry in Central Uganda." In *Adaptive Collaborative Management of Forest Landscapes: Villagers, Bureaucrats and Civil Society*, edited by C.J.P. Colfer, R. Prabhu and A. Larson, 106–132. London: Earthscan/Routledge.

Neuland, M. 1998. *The World of Moderation. The Neuland Approach: Philosophy and Techniques for Extraordinary Group Performance and Active Learning.* Künzell, Germany: Neuland Verlag für lebendiges lernen.

Ngwenya, H.J. and J. Hagmann. 2009. "Facilitation for Change: Triggering Emancipation and Innovation in Rural Communities in South Africa." In *Farmer First Revisited: Innovation for Agricultural Research and Development*, edited by I. Scoones, J. Thompson and R. Chambers, 220–237. Rugby, Warwickshire, UK: Practical Action Publishing.

Ngwenya, H.J. and J. Hagmann. 2011. "Making Innovation Systems Work in Practice: Experiences in Integrating Innovation, Social Learning and Knowledge in Innovation Platforms." *Knowledge Management for Development* 7, no. 1: 109–124. http://dx.doi.org/10.1080/19474199.2011.593867.

Peavey, F. 1990. *Strategic Questioning Manual.* http://pnodn.pbworks.com/f/Peavey+strategic+questioning+manual.pdf.

Johnson, J. and B. Pokorny. 2022. "Local People's Perspectives on Action Learning: Impressions from the Amazon". In *Adaptive Collaborative Management in Forest Landscapes: Villagers, Bureaucrats and Civil Society*, edited by C.J.P. Colfer, R. Prabhu and A.M. Larson, 31–60. London: Earthscan/Routledge.

Ramaru, J., E. Chuma and J. Hagmann. 2014. "Facilitating Community Engagement and Water User Associations (WUAs): 'Toolkit' for Project Implementation and Project Design." IMAWESA Project Report. IWMI/IMAWESA/PICOTEAM. www.picoteam.org.

Introduction to Chapter 7

We begin our journey through Africa – as we did with Indonesia – with a revisit to an ACM site from the early 2000s. These authors, all participants in the earlier work, bring us up to date on what has happened in a country beset by chaos ever since, Zimbabwe. After briefly describing the original work with its significant successes, the authors outline the disasters that have plagued the country (land invasions, political upheaval, drastic currency devaluation, HIV/AIDS and COVID-19 epidemics, and more).

Piece by piece, Kozanayi and his colleagues unpack both the community-level successes that remain and the unravelling of the once-incipient process of institutionalizing ACM in the Forestry Commission and the Forestry College. They trace the losses of donor funding and facilitators, of key personnel and advocates. Some key actors were re-assigned, some died of disease, some emigrated due to political or economic necessity. In short, the critical mass that seemed to be growing in the mid-2000s dissipated before the necessary institutionalization could be firmly established.

Like the Indonesia case (Chapter 2), this case shows how ACM action at the local level, valuable and enduring as much of it has been, is 'necessary but not sufficient' insofar as we want to effect change on a broader scale. Where in Jambi, an explicit policy of the central government favoring oil palm over forest changed the landscape that surrounded the community's carefully protected forest, in Gokwe the broader scale effects were more multi-strand. Governmental encouragement of land invasions by onetime freedom fighters and the landless led to takeover of parts of Mafungautsi Forest. Nationally, the effects on previous landowners (mainly white) raised the ire of the international community and sanctions were imposed; the economy and its currency went into freefall. Whatever prosperity the country had had dissipated exacerbated further by raging HIV/AIDS in early days, and most recently COVID-19.

Products the communities had begun to sustainably manage disappeared as forest areas were taken over and turned into agricultural and residential plots by newcomers. Products were bartered rather than sold, as inflation rendered the value of money lower and lower. Conflicts arose between settlers and oldtimers

DOI: 10.4324/9781003325932-12

and between proponents of one political party or the other – all complicating the systematic management processes co-developed under ACM.

This case provides dramatic evidence of the need to work at multiple, critical scales simultaneously. We will need to be able to embed the seeds that allow us to scale up and out from the beginning, in our work with communities. As we tackle climate change, forest restoration, land tenure and other environmental issues, we will have to build on what we have learned here. That will mean adapting the approach to address *sets of nested and interacting circles*, as suggested in Chapters 4 and 11 that can reinforce, echo and generate movement capable of transformative change.

7 Sustaining adaptive collaborative management processes

Challenges and opportunities from Mafungautsi State Forest, Gokwe, Zimbabwe

Witness Kozanayi, Richard Nyirenda, Tendayi Mutimukuru-Maravanyika, Frank Matose, Mkhululi Ngwenya and Lizwelabo Sibanda

Introduction

Following the failure of top-down resource management initiatives, participatory resource management initiatives rose steadily from the 1990s to the early 2000s (Chambers, Pacey and Thrupp 1993). Such initiatives range from passive, where local people's participation is merely lip-service, to genuine participation that is empowering and transforming for forest-dependent communities. Zimbabwe pioneered state-community participation in the forest sector through a resource-sharing project (RSP) initiated in Mafungautsi State Forest in 1994 (Mandondo, Prabhu and Matose 2008; Matose 2002; Mutimukuru-Maravanyika 2010). The project aimed to resolve tenurial conflict between the state and local resource users, with the envisaged collaborative harmony failing to take hold in part because collaboration was conceived in a top-down manner (Mandondo and Mapedza 2002), with participation tokenist rather than substantive (Matose 1994; Sithole and Kozanayi 2002). Broadly, the resource-sharing approach had ambitious objectives of:

- Changing the mindset of forest rangers from criminalization of local people to collaboration
- Opening the Forestry Commission (FC) itself to alternative models of managing forests beyond physical 'fences' and criminalization
- Beginning a transformative journey to redefine the relationship between resource users and managers, and
- Using this pilot to demonstrate to government that new models of coexistence between people and nature were needed and possible.

While the launch of the resource-sharing initiative was itself a bold move at that time and laid the foundation to build on these early lessons and evolve new partnerships and approaches, we suspected that the approach could not achieve its objectives in the limited time frame (five years).

DOI: 10.4324/9781003325932-13

Thus, by 2000, under the RSP, there remained numerous conflicts between the two key stakeholders who were supposed to collaborate; and the status of the forest had not improved (Matose 2002). Hence, the Center for International Forestry Research (CIFOR) joined hands with the FC and other stakeholders to facilitate adaptive collaborative management (ACM) in order to more effectively bridge the gap (Prabhu et al. 2009).

The ACM research project was implemented from 1999 to 2006, with an aim to gain more understanding of the usefulness of such people-centered approaches in community-based natural resource management initiatives. The main thrust of the research was to facilitate reform of management practices, institutional arrangements and policies to promote sustainable forest management systems and practices for both human and ecological benefits. To that end, the project aimed to facilitate a shift away from blanket prescriptions for solving forest-related problems. Rather, the ACM approach was anchored in locally based management that was adaptive to rapidly changing circumstances in a manner that was in accordance with sustainable forest management principles. The specific purpose was to facilitate widespread use of self-improving and equitable forest resource management systems that built on local capacity, 'vertical' and 'horizontal' stakeholder interactions, and responded positively to external pressures. Broadly, ACM in Mafungautsi sought to achieve the following: (a) create the conditions necessary for collaboration among stakeholders, (b) develop suitable approaches that would enable the diverse stakeholders to participate meaningfully in sustainable use and management of the forest, and (c) analyze the impact of the improved collaboration on the livelihoods of local people and forest resources.

The outcomes of the Mafungautsi ACM project and their determinants are fairly well documented (Matose and Mutimukuru-Maravanyika 2009; Mutimukuru-Maravanyika 2010; Mutimukuru-Maravanyika, Kozanayi and Nyirenda 2006, 2007; Mutimukuru-Maravanyika et al. 2008; Nyirenda and Kozanayi 2007; Sithole 2005). Key outcomes include (1) resolution of deep-rooted and complex tenurial challenges that had characterized Mafungautsi Forest since the time it was reserved in 1954 (Matose 2002); and (2) resultant improvements in resource status arising from collaborative management by local communities and the FC. Intermediate outcomes comprising factors that appeared to determine the final outcomes included (1) improved transparency and downward accountability by local resource management committees (RMCs, an institution that was put in place to represent local communities in the management of the forest resource); and (2) improved participation in decision making by local communities in the management of Mafungautsi Forest. Factors that led to these improvements are discussed in detail in the citations above.

As the above outcomes gained traction, Zimbabwe embarked on radical land reforms that came with steep economic decline and widespread agrarian instability. Consequent upheavals reverberated across most sectors at all levels, including around state forests that became subject to new land claims by radicalized land-hungry groups. The mostly subaltern claimants in Mafungautsi over the past decade included those displaced when the forest was reserved in 1954, those retrenched from mostly urban jobs, and former commercial farm workers evicted from acquired white-owned farms (Hanlow, Manjengwa and Smart 2013; Scoones et al. 2010). To date, the state has used various strategies to resolve conflicts between local forest edge communities and the FC, including coopting the new claimants into collaborative natural resource management initiatives or evicting them through instruments of state coercion. These pathways to resolve state-people conflicts around state forests have remained ineffective as they have been accompanied by resultant resource deterioration and compromised livelihoods for local people.

In light of the political and economic changes that have taken place in the country, in this chapter, we revisit the Mafungautsi ACM project 18 years after its initiation to track if any of the processes initiated have been sustained, in an attempt to build on work done by other scholars (Mutimukuru-Maravanyika 2010). The aim of the chapter is to critically reflect on the Mafungautsi case and assess factors underlying the endurance of the facilitated ACM processes operating within the resource-sharing groups,[1] after CIFOR withdrew from the study site.

Specific questions underlying this aim are as follows:

1 What socio-political processes have taken place in Zimbabwe and Mafungautsi since the ACM project?
2 What ACM processes are still operational in Mafungautsi?
3 What lessons can we learn from Mafungautsi on how ACM processes can be sustained?
4 How does an ACM approach feature in current forest management discourse in Zimbabwe?
5 What inferences can we draw about the search for local, equitable solutions within a context dominated by larger scale drivers of change/problems?

In addition to the country level challenges highlighted, there were other global challenges that have had a confounding impact on the local people, including our efforts to implement the ACM program. Such challenges included climate change, the COVID-19 pandemic and a drive to use nature-based solutions to address local level environmental challenges.

The next section provides background to the study sites and Zimbabwe's ACM project. This is followed by a description of the research methods and a literature review which serves as the bed rock against which the results of the study are analyzed. Evidence from the field is presented next and is followed by a discussion of these results. The chapter ends by concluding on key findings, drawing on the discussion of the results.

Background of the study site and the ACM project in Zimbabwe

The advent of fortress conservation

Zimbabwe has about 800,000 hectares (ha) of reserved forest land. The forests were gazetted between 1926 and 1960 in order to control the wanton harvesting of commercial indigenous timber species used to produce mine props, railway sleepers, flooring parquets and furniture (McGregor 1991). These forests are managed by the FC's Conservation and Extension Division. The forests are commonly referred to as 'Kalahari Sand Forests' and are important for watershed and soil protection (catchment area protection), biodiversity conservation, wildlife habitats and as a source of commercial timber and nontimber forest products such as honey, mushrooms, edible insects and indigenous fruits (Matose 2002). The FC occasionally awards concessions to companies to harvest hardwood timber, while wildlife is touted for tourism purposes. Local people can harvest a range of nontimber forest products such as broom and thatch grass, insects and mushrooms upon payment of a harvesting permit, and in accordance to harvesting conditions set by the FC.

Mafungautsi Forest: a historical overview

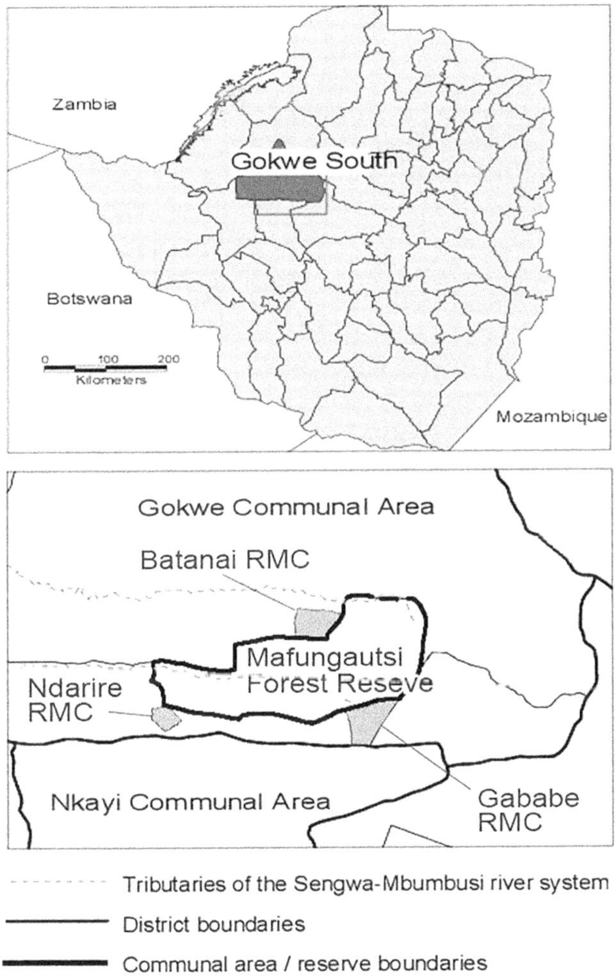

Figure 7.1 Location of Mafungautsi Forest

From fortress conservation to co-management

Mafungautsi Forest, one of Zimbabwe's state forests, is a protected forest located in west central Zimbabwe (see Figure 7.1). It was reserved in 1954 principally for ecological reasons, especially conserving the watershed and headwaters of tributaries of the Zambezi River. The Sengwa, Ngondoma, Mbumbusi and Lutope are the four main rivers that drain from the Gokwe-Charama Plateau of the Mafungautsi area. The Zambezi River is of strategic economic importance because it feeds into the Kariba Dam from which Zimbabwe and Zambia generate hydropower. The Mafungautsi State Forest is 82,000 ha in size. It was 101,000 ha on its gazettement but the government ceded 20,000 ha to restless and landless local people in the 1980s. The FC managed the forest through exclusion or the fortress conservation model from 1954 to 1992 (Matzke 1993). Endless conflicts between the FC and indigenous people who had been displaced to make way for reservation characterized this exclusion management phase (Matose 2002). Tenurial conflicts mostly manifested through establishment of settlements, cultivation and poaching of nontimber forest products by communities bordering the forest. Cases of arson were also reported as local people retaliated against perceived FC excesses in enforcing exclusion (Matose 2002). These problems date back to the colonial period's inequities and an entitlement failure, lack of access to land and economic opportunities. Thus, conflicts around the forest were, and are an indicator of the absence of a comprehensive national solution to forest use challenges.

To stem the conflicts, the Canadian International Development Agency broached a co-management program that the FC adopted and initiated from 1993. The co-management model provided was meant to create win-win arrangements to reconcile the concerns and interests of both the FC and local communities. In the co-management scheme, villages surrounding the forest were organized into the above-mentioned RMCs to represent local people in the management of the forest (Sithole 2005). Altogether, 14 RMCs were set up representing different villages and communities living around Mafungautsi Forest. Each RMC was allocated a resource area in Mafungautsi to manage and from which members of the RMC could harvest a designated set of forest resources such as thatch and broom grass, mushrooms, mopane worms, firewood and fruits. The co-management program allowed RMC members to harvest these resources following specified guidelines (Matose 2002). Members of these RMCs were elected through a 'democratic' process. The co-management scheme was initiated as a pilot scheme that would generate lessons for the scaling up of such schemes to other protected forests in Zimbabwe (Büscher and Mutimukuru-Maravanyika 2007).

Overall, the co-management initiative fell short of meaningfully resolving the endemic conflict, mostly because the FC ended up patronizing the communities that still lacked co-equal partnership status at the negotiating table (Vermeulen 1997). Although most of the available literature applauds this earlier effort to involve local people in forest management, it also raises several reservations on the pilot initiative. For example, Mandondo and Mapedza (2002) argue that the initiative was top-down rather than bottom-up, while Matose (1994) and

Sithole and Kozanayi (2002) question the extent of community involvement and the nature of the relationship between the FC and communities. Matose and Sithole and Kozanayi argue that community involvement was cosmetic and the relationship between the FC and the community was filled with mistrust and conflicts. Sithole and Kozanayi (2002) opine that the FC had retained its top-down management approach, ignoring local people and other stakeholders. They go on to say that local respondents described the FC as a 'leopard that has not changed its spots'. The FC continued to enjoy broad discretionary powers on the shape and form of privileges to be bestowed upon the local people (Mapedza 2006). Conflicts between the FC and people living around Mafungautsi continued unabated. As aptly noted by Hill and Katerere (2002), "despite the Forestry Commission's legal rights, its power to enforce its rights has been severely weakened by political intervention and lack of political will by the government to enforce the law" (p. 262).

From co-management to ACM implementation around Mafungautsi Forest

The ACM program was implemented in Zimbabwe in two phases from 1999 to 2005, of which this study covers the second. Phase I of the project was implemented between September 1999 and December 2002. Phase I focused on facilitating adaptive management processes, first by understanding the context within which the project was being implemented, then building capacity of the various stakeholders through several approaches that included training for transformation (TFT), creation of platforms for social learning and group experimentation and conflict resolution techniques. CIFOR was at the forefront of facilitating all the processes in Phase I. Phase 2 (2003–2005), however, aimed at handing over such facilitation to the main collaboration partners, i.e., the FC and the local resource user groups. In the second phase, CIFOR's role was peripheral, only advising on collaborative monitoring approaches and their role in the iterative learning cycle.

Three RMCs, namely Batanai, Ndarire and Gababe, were selected as pilot sites for the ACM program. However, over time more RMCs were involved, and as the initiative gained momentum, all stakeholders gained confidence. Beyond Gokwe, six other districts adopted the ACM approach (Nyirenda and Kozanayi 2007), providing diverse settings from which comparative evaluations of the approach could be done (*ACM Newsletter* 2006).

The ACM program was implemented by an interdisciplinary team which comprised stakeholders from a diverse set of organizations, including local communities, forestry officials, extension staff, government departments and CIFOR. A steering committee constituted at the national level gave a strategic direction to the process. A principal investigator led an interdisciplinary team which comprised experts in Participatory Action Research (PAR), sociology, resource economics, extension, forestry and ecology in implementing the project. This was over and above consultants/experts occasionally hired to provide expert advice as needed. Community researchers conducted community mobilization at the local level. Relevant government departments supported the program, with banks later coming on board to provide financial services to the RMCs.

The stakeholders facilitated ACM through several pathways from 1999 to 2005, namely joint planning and monitoring, communication, learning processes and platforms.

Joint planning and monitoring

Joint planning and monitoring provided the platform on which stakeholders institutionalized their unique mandates and interests through ACM. For instance, the then Ministry of Youth, Development, Gender and Employment Creation had a special interest in community projects funded by the revenue from the RMC permit system and would audit RMC accounts; the FC and Agricultural and Technical Extension Services jointly had an interest in conducting training courses in beekeeping; the banks' mandates lay in providing financial services to the RMCs; and the mandate of the District Administrator involved providing policy guidance to the whole process. Every month, the FC convened and chaired meetings about ACM processes around Mafungautsi. All the stakeholders participated in these meetings. Stakeholders also conducted joint field monitoring tours. Findings of such monitoring missions were reflected upon in reflection platforms to inform next steps.

Social learning processes and platforms

The stakeholders created several platforms to enhance social learning. These included 'look and learn' tours within and beyond the district; launching a biannual Zimbabwean ACM *Newsletter* for dissemination and sharing of ACM issues; and commissioning of PAR sessions between resource harvesters and the FC. The stakeholders also initiated TFT to empower and build capacities of communities, transforming them from passive subjects to active agents in the ACM arrangement. TFT is a Paulo Freirean philosophy on adult learning aimed at changing the nature, form and function of actors in order to enhance their capability to act or do something about their situation (Hope and Timmel 1995).

Communication

Seminars provided platforms for stakeholders to generate and disseminate policy-relevant discussions around ACM. The seminars, at district and national levels, brought together many forest-sector stakeholders, including government, civil society and academia. After Wollenberg, Edmunds and Buck (2000), visioning and scenario-building workshops facilitated at district and national levels also enabled participants to reflect on their current activities and approaches as a step toward redefining their future development pathways, and in the process providing a rallying point for all the stakeholders involved. The FC provided a home to consolidate and retain institutional memory of the visions, with CIFOR underwriting the costs of the visioning and scenario-building processes.

Efforts to institutionalize ACM in forestry syllabi

By 2005, FC management had started mulling over the idea of mainstreaming ACM within the FC establishment on the basis of the promise the approach presented. The FC envisaged scaling up the approach through two pathways. The first involved extending ACM to six other districts. The second approach targeted incorporating ACM into the syllabus for the forest diploma offered by the FC's Forestry Training College. The college produces a steady pool of graduates from which the FC recruits forest extension officers. CIFOR funding for the ACM project, however, drew to a close in 2006 before the curricular changes could be finalized.

Research methods

Our study used qualitative methods to gather data used in this chapter. Our primary sources included participant observation and key informant interviews, supplemented by a review of reports and communication on Mafungautsi by the FC. The story of Mafungautsi has attracted both private and public print media; reports from such media were also used to track its story.

Two of this study's authors continued to work on donor-funded agricultural projects in Gokwe District from the inception of the project up to 2018 enabling them to observe unfolding processes in the district in general and Mafungautsi Forest in particular. These authors also sat in Rural District Council meetings enabling them to capture official perspectives on the Mafungautsi Forest whenever they came up in council deliberations. One of the authors resides in Gokwe District and thus kept abreast of quotidian developments there.

Key informant interviews covered a wide array of respondents, including local resource users such as thatch and broom grass harvesters and loggers, local FC managers, RMC members, the Forest Protection Unit (FPU) and civil servants at district level. Interviews with these key informants provided more nuanced data to issues reported in media as well as those deliberated in council meetings.

ACM: a literature review

A huge body of literature has addressed the conceptual basis of the ACM approach (Colfer 2005a, 2005b; Colfer et al. 2011; Diaw, Aseh and Prabhu 2009; German et al. 2010; Mandondo, Prabhu and Matose 2008; Ojha, Hall and Sulaiman 2013; Prabhu et al. 2009, among others). Broadly, ACM enables stakeholders to interact, negotiate a vision for their resource and consciously undergo shared-learning in developing and implementing their plans (Colfer 2005a, 2005b; Fisher, Prabhu and McDougall 2007). Implementation of these plans is jointly monitored, and outcomes are observed and reflected upon to generate lessons for subsequent activities. When using the ACM approach, management processes are influenced by conscious and deliberate lessons generated by stakeholders involved in joint learning processes (Fisher, Prabhu and McDougall 2007; Mandondo, Prabhu and

Matose 2008). The approach is characterized by conscious efforts among stakeholders to communicate, collaborate and seek opportunities for joint learning about the impacts of their actions (Colfer 2005a, 2005b). ACM is expected to lead to self-improving systems of resource management based on improved flows of information; decision making that follows experimentation; communication and negotiation among stakeholders; and learning that results in changes in management systems among resource users (Mandondo, Prabhu and Matose 2008). Thus, the ACM approach is seen as enhancing and fostering genuine partnerships through proper identification of stakeholders and deliberate efforts to strengthen their engagement based on trust, common interests and objectives; providing platforms for different stakeholders to negotiate, resolve conflicts and learn together from experience; and facilitating and enhancing social development of communities (Hartanto et al. 2003). Reconciling and satisfying stakeholders' interests are key to ACM processes (Colfer 2005a, 2005b). In sum, an ACM approach should result in the development of adaptive, self-improving systems of forest management based on improved flows of information (feedback loops), decision making that follows rules of 'experimentation', communication and negotiation among stakeholders, more responsible representation among stakeholder groups and social learning about forces of change in forest systems (Prabhu 2003).

ACM processes are driven through PAR cycles. PAR is a process through which members of a group or community identify a problem, collect and analyze information, and act upon the problem in order to find solutions and promote social and political transformation (Selener 1992). PAR combines four principal activities: research, education, learning and action. These steps can stimulate social learning by bringing different groups together through a conscious and deliberate cycle of inquiring, observing and monitoring, reflecting, planning and acting (see Figure 2.1) to improve the current situation in a way that is appropriate for collaborators.

While learning and action are supposed to take place in a conscious manner after each reflection session, the way stakeholders organize themselves and design solutions in dynamic systems has been found to be complex. Cleaver (2012) uses institutional *bricolage* (French for 'making do' with whatever is available) scholarship to illustrate how stakeholders design or adapt to changes in complex situations such as in Mafungautsi. Drawing on 'post-institutionalist' perspectives, Cleaver (2012) rejects over-formalized managerial approaches, preferring instead to embrace a variety of partial and contingent solutions that are more reflective of the ever-changing evolution of institutions. Actors (regulators and harvesters alike) are constantly interacting and reconfiguring themselves in ways that result in different desired outcomes, so much so that neat fit, institutional design is elusive (Kozanayi 2018). Institutional components from different origins are continuously reused, reworked or refashioned to perform new functions. Adapted configurations of rules, practices, norms and relationships are attributed meaning and authority (Cleaver and de Koning 2015, p. 4). While ACM offers a plausible option to deal with complex situations, considering the degree of sociopolitical and economic changes such as hyperinflation (Jones 2010) witnessed in

Zimbabwe in the past two decades, ACM's suitability, if implemented in its ideal form, is questioned. Our review offers an opportunity to unpack this question, by assessing if there are some elements of ACM still operational in Mafungautsi.

Findings

Ever since our last conscious facilitation of ACM processes in Mafungautsi in 2006, Zimbabwe has witnessed striking changes, including land reform, economic meltdown, socio-political challenges and removal from power of long-term ruler Robert Mugabe. In the sections that follow, we present the different processes that have taken place from 2006 to 2020 and how such processes affected or were affected by ACM processes facilitated in three RMCs around Mafungautsi State Forest. We next consider these macro-level changes in relation to ACM processes emphasizing how both shaped each other from the FC level to the local RMC level.

Overview of the national macro-level socio-political and economic processes

The past two decades have seen unprecedented decline in the socio-economic and political landscape in Zimbabwe. In fact, such changes started in the late 1990s when the ruling Zimbabwe African National Union – Patriotic Front (ZANU PF) – embarked on populist policies to stem political disgruntlement. Prominent among these policies was a violent and somewhat chaotic land reform (Scoones et al. 2010) resulting in an acute decline in agricultural production, drastically collapsing the country's agro-based economy (Jones 2010). The violence associated with agrarian reforms also spawned a culture of impunity that characterized both state-sanctioned as well as opportunistic land grabs across almost all land tenure categories, including protected forests such as Mafungautsi. Over time, the FC appeared to suffer what Murphree and Cumming (1991) term a double expropriation: (a) loss of considerable swathes of protected forestland, and (b) the hemming in of the FC's technical managerial function by other politically radicalized government organs. This resulted in considerable loss of control over the management of the forest, including via the ACM approach.

Invasion of state forests

In 2000, Zimbabwe unleashed a colossal land reform program dubbed Fast Track Land Reform Program (FTLRP) ostensibly to address historical land imbalances in the country (Hanlow, Manjengwa and Smart 2013; Scoones et al. 2010). Initially, the land reform program targeted annexing white-owned farms on prime land and redistributing it to landless black people (Matondi 2010; Sadomba 2008). Over time, FTLRP assumed a violent and chaotic character, with state forests being invaded by land-hungry people.[2] Early invasions coincided with the ACM project, with project personnel joining hands with other stakeholders to diffuse encroachment onto the forest. This was done on the basis that the FTLRP exempted all demarcated indigenous forests from land invasions as provided for in the Forest Act Chapter 19:05. Efforts at stemming land invasions were not entirely successful, in part because the invasions had become politicized, with some state actors tacitly aiding and abetting them. Hence, Mafungautsi Forest has suffered waves of invasions since 2000,[3] with 700 families illegally settling there between 2010 and 2011. By 2014, the number of settlers in Mafungautsi had increased exponentially (Table 7.1).

Other invasions also occurred beyond the first decade of land reforms, with new settlers in Mafungautsi further increasing by ~1,960 families near the end of

Table 7.1 State of forest settlement in Mafungautsi, September, 2014

Settlement No.	*Location description*	*Estimated land area affected (ha)*	*Estimated no. of households*	*Year settlement started*	*Comments*
1. Zanda	Northwestern part of the forest, extending eastward toward FC's Lutope camp	2,701	880	2000	The FC attempted to evict the first 85 households, but this was indefinitely postponed on the advice of the FC Board
2. Ngondoma	Eastern part of the forest with homes concentrated on the verges of Ngondoma vlei – marshy ground, covered with water most of the time.	2,835	520 and increasing	2005	Political considerations incessantly interfered with FC efforts to remove the illegal settlers.
3. Nyamazana	Northeastern part of the forest, with most of the occupants coming from the neighboring Nyamazana in Chief Njelele's area	548	90	2005	This is the Minister of State for Provincial Affairs' targeted area. Currently, it is estimated there are 90 households, up from 48 in 2013
Sub-totals		**6,084**	**1,490**		
Proposed expansion	Extension of Nyamazana westward to join Gokwe-Nkayi road	5,145	Nil	Proposed	This is currently an intact forest, the source of Sengwa River and its tributaries Mzola and Chifura
Totals		11,229	1,490		

Source: FC report (2014).

the second decade. The new settlers included local people and other land claimants from afar, including civil servants. In some cases, indigenous people invaded state land to farm or harvest natural products, or reclaim ancestral land lost during the colonial era (Mubvami 2004). The new claimants settled in the *vlei* (seasonally flooded, marshy areas) with the tacit support of the local chief, who did not consult the FC. Infrastructural investments in invaded areas, by some arms of the state and some members of the development community (non-governmental organizations, NGOs), appeared to secure the invasions. NGOs like CARE-International helped build makeshift schools, in addition to sheltering and feeding the new settlers. Polling booths were also put in the forest during the 2018 harmonized elections. In sum, the land reforms have had far-reaching impacts on the governance of natural resources, especially state forests.

FC's management and control over protected forestlands politically hemmed in

At the peak of the land reform program, the invasion of Mafungautsi Forest began to be treated as a land rather than a forestry issue. Land transfers through invasions were now handled through the District and Provincial Land Committee and not by the FC as a legally designated custodian of protected forestlands. Membership of the land committees conflated technocrats with security forces and political appointees, resulting in the FCs technical managerial functions being hemmed in, with ACM receding backstage as the FC became marginalized. All issues pertaining to Mafungautsi were now under the purview of the Land Committee.

In March 2012, the then Resident Minister and Provincial Governor for Midlands convened 'settlement regularization' meetings in the Ngondoma and Zanda areas reportedly 'to officially hand-over the settled land to the illegal settlers'. Key line ministries and traditional leaders (four chiefs and one headman) among others attended the meeting. This made official the increase in the total population in Mafungautsi from the previously estimated 3,500 people to an unknown figure as the newcomers were promised a 'trouble-free' stay in Mafungautsi in perpetuity. A total of 11,229 ha of forest was requested by the governor for these folks.

A second high-profile meeting to discuss the request by the Minister of State for Provincial Affairs to settle peasants in Mafungautsi was convened on 27 August 2014. In attendance were the Minister of Environment, Water and Climate, the Minister of State for Provincial Affairs – Midlands, the Deputy Minister of Environment, Water and Climate, the Director of Environment and Natural Resources, the General Manager – FC, the FC Acting Deputy General Manager, the Provincial Administrator, Midlands and the Provincial Lands Officer.

The FC sought, in vain, to counter the double expropriation by invoking laws mandating themselves as custodians of protected forestlands and regulators of forest resources. The FC protects reserved forests through the Forest Act (1954) and regulates the utilization of forest resources beyond protected areas through the Communal Lands Forest Produce Act (1987), both of which comprehensively provide it with legal backing (Hill and Katerere 2002). Drawing on this legal support, the FC especially relied on Sections 35 and 36 of the Forest Act (Chapter 19:05) to reverse the parceling out of protected forestlands to invaders (FC 2013). Both sections provide for the gazetting and de-gazetting of any piece of land in Zimbabwe, a process that should legally be sanctioned by Zimbabwe's President, the Minister of Environment, Water and Climate, the FC and Parliament.[4]

Moreover, the subtraction of any area of land from a gazetted forest [paragraph (d), subsection (2) of Section 35] is effected through a notice made by the President on the recommendation of the FC, and with the subsequent approval of Parliament. Subsection (5) of Section 35 states: "*No notice may be made in terms of paragraph (d) or (e) of subsection (2) unless the proposal to make such notice has been approved by Parliament*". The FC argued that no such legal process had been followed, implying that the presence of both the previous and recent settlers in Mafungautsi Forest remained illegal. The FC further invoked the terms of subsection (2) of Section 34 of the Forest Act that the total extent of any gazetted forest land shall not be reduced by more than 1% of the total extent of forest land. The settlement of people on the 11,229 ha of forest land then proposed in 2014 would result in the reduction of the gazetted forest by a staggering 13.7%, which far exceeded the legal benchmark.

Though Cabinet approved the eviction of the 'illegal settlers', its overture was overwhelmed by the groundswell of populist pressure engendered by FTLRP, resulting in lack of political will to enforce the evictions (FC manager, personal communication, 09/03/2020). The political will did not materialize even as the FC appeared to ease its stance from total exclusion[5] to regulated inclusion entailing the re-introduction of plains game for ecotourism and safaris in collaboration with surrounding communities (an incentive for voluntary relocation from the reserve). But there was relentless pressure from the politicians to cede part of the forest to the settlers resulting in some senior FC managers now reportedly considering the more politically pragmatic position of ceding 10,000 ha (12.2%) of the forest. This was seen as an avenue for formally settling the people already in the forest. However, there was fear that this would shrink the forests further (a total of over 20,000 ha would have been ceded to the communities since gazettement of the forest) and jeopardize their protection purposes, e.g., protection of the headwaters of the four rivers in Mafungautsi. However logical the argument sounded, the politicians pushing the settlement of people in the forest did not relent on their demands.

The major lesson to be drawn from the saga appears to be that appealing to the letter and spirit of the law in an environment characterized by a breakdown in the rule of law is futile. Pragmatically, the FC could have tried resolving the conflicts through approaches like ACM. ACM's iterative and interactive approaches to reconciling interests offer a greater scope for crafting common visions among haggling stakeholders. Indeed, the FC management would later, in 2014, consider using ACM to resolve Mafungautsi Forest's endemic conflicts. Such recourse to ACM though proved to be too little too late, as a combination of other intervening developments – such as high staff turnovers, further economic meltdown and political polarization – may have already conspired to deny ACM processes formerly underway much needed momentum.

Combination of factors that enfeebled ACM in Mafungautsi

High FC staff and stakeholder turnover[6]

Instability in the forest may further have interacted with other factors, including high turnovers of ACM-experienced staff within and outside Gokwe District and Mafungautsi Forest to stall ACM processes. Staff members from key line ministries and departments with which the ACM team had worked in the early 2000s either resigned, changed jobs or moved to other districts. Table 7.2 presents changes in person-power that occurred at the district level.

Table 7.2 Staff turnover affecting ACM since the early 2000s

Government Department/Agency & private sector	*Role in ACM*	*Current situation*
District Administrator	Policy guidance and scaling up of the project to the provincial level	District Administrators have changed more than twice and the effects of ACM were lost through time. Focus is on managing illegal settlers.
Forestry Commission	Management of the forest, supporting RMCs' ACM activities around the forest, e.g., through creating dialogue platforms, joint planning and management of the forest	All officers at district and provincial levels who had been driving the ACM processes left the FC. Replacements were coming from districts that had not been exposed to the ACM philosophy.
Ministry of Women's Affairs and Small and Medium Enterprises[a]	Support local level projects specifically targeting women and youth	All officers at district level transferred to other provinces. No longer auditing books of accounts- not much money being generated and resource constraints – no vehicle.
Environmental Management Agency (EMA)	Implementation of EMA, the supreme law on all issues related to forest and natural resources	Both trained officers left the Agency long ago to join academia and new officer is not familiar with ACM.
Chiefs and traditional authorities	Governance of natural resources in communal areas – outside Mafungautsi forest, and also mobilizing residents to participate in community projects	While most of these have not changed, some became actively involved in encouraging people to settle in the forest at the height of the land reform program.
Banks – e.g., Agribank	In the early 2000s at the peak of the ACM project, RMCs were approached by banks so that they could open savings account with the banks	No longer actively involved in providing this service – RMCs only making small savings and deposits – if at all. Culture of saving money in banks eroded due to lost savings due to hyperinflation up to 2008!
Ward Councilors	Ward councilors as elected local leaders were instrumental in spearheading ACM at the local level and also articulating it in meetings at the district	New crop of ward councilors who were not actively involved in ACM processes are now in office.

[a]Formerly known as the Ministry of Youth, Development, Gender, Employment and Community Development.

Interviews with the new staff in the entities listed in Table 7.2 showed that in the notes provided at handover, no clear mention of ACM activities was found. [Source: Key staff stakeholders at the district members in the district.] Literature on ACM was also not readily available, other than a few publications at the FC; no one else had literature on the ACM experience in the District.

Within the FC, there was massive staff turnover at the district, provincial and national levels. Other notable changes were at the Forestry College in Mutare where all Forestry Extension officers are trained. Staff members who had been exposed to the ACM approach and were incorporating it in the local curriculum left the college; similarly at the ministerial level, the Permanent secretary from the parent ministry of the FC had also embraced the ACM approach,[7] but left.

Economic downturn and its domino effects

Zimbabwe's economic meltdown characterized by a wide repertoire of 'making do' activities dubbed the *kukiya kiya* economy, resulted in widespread impoverishment and the collapse of the middle class, resulting in a greater pool of citizens scrounging for a living (Jones 2010). Families falling from the middle class joined the pool of subalterns harvesting natural resources for own use or for sale to augment their ever-diminishing earnings and livelihoods (Kozanayi 2018). New patterns of forest dependence in Gokwe South included the harvest of timber from the reserved forest for firewood. An ox-drawn cart of firewood cost USD 5–8. The firewood was gathered from areas that were being cleared for crop fields by the new land occupants, collected as dead wood, or by felling trees. The new modes of harvesting disregarded resource-sharing arrangements that were in place. Demand for wood also increased in response to persistent power cuts due to low power generation in the country. Gokwe center itself has been expanding since the 1990s, having been accorded town status in 2016, resulting in an ever-spiking demand for new farmland and biomass energy – both locally perceived as abundant within Mafungautsi Forest. Further, the hyperinflation triggered by the economic meltdown concomitantly collapsed the formal economy, accentuating forest dependence and *kukiya kiya* activities, with some (mostly civil servants) joining the diaspora in search of better opportunities (Jones 2010). The majority of the authors of this chapter were included in this group – some of us are still in the diaspora, while others are now back in Zimbabwe.

Austerity measures put in place by the government to stem economic collapse additionally entailed drastic cuts in government financial support to the FC that was (as a parastatal) henceforth expected to generate adequate revenues to fend for itself. Prospects for securing much needed extra support for the Commission from donors turned bleak, in part because of the country's poor human rights record. Western donors had been a major source of soft funding before the onset of FTLRP. As it consequently struggled to fund its operations, the FC imposed cutbacks in financial support to its non-income-generating operations, including resource-sharing programs and social forestry in general. The FC directed much of its funds to its corporate operations.

Polarization of societies

Gokwe has been one of Zimbabwe's political hot beds since the emergence of the opposition Movement for Democratic Party (MDC) in 2000. The opposition has made some inroads in the Midlands province in which Gokwe District falls. Though the province had supported the ruling ZANU- PF party, the MDC won a significant number of its parliamentary and council seats in 2000. ZANU PF used violence against perceived supporters of MDC to regain its influence.[8] Political polarization stalled ACM by undermining the goodwill required for communities to collaborate in the planning, reviewing and redesigning of resource use strategies. In fact, some of the ACM movers perceived to be sympathetic to the opposition party were hounded out of the district. For instance, one of the authors of this chapter is now based in South Africa after being targeted for harassment for his perceived inclination to the opposition. Political tensions added to existing latent tensions, especially those centered around competing claims between communities settled in and outside the forest. Forest dwellers, whose prime motive was farming, ended up on a collision course with communities living just outside the forest whose motives have been to extract NTFPs from the forest. Most of these latter were more easily inclined to participate in ACM by forming RMCs through which they obtained permits to generate revenue.

Global problems impact local processes

The period under review witnessed the outbreak of the COVID-19 pandemic and climate change-induced droughts in Gokwe District. The hard lockdowns that were imposed by the government to contain the spread of COVID-19 resulted in local people who had depended on selling agricultural products in urban centers resorting to harvesting wild products for sale such as firewood. This exerted more management pressure on the local RMCs that remained functional. We discuss their functioning and the challenges they faced in more detail below. Furthermore, group meetings, which are key pillars of ACM for purposes of reflection and dialogue, were banned for more than a year in 2020. With some farmers not having mobile phones for communication, and a high cost of data for those residents with mobile phones, communication among stakeholders involved in ACM was drastically reduced. Recurrent droughts caused residents to harvest forest resources such as firewood and fruits from Mafungautsi. Some residents settled in *vleis* where they could grow crops from the residual moisture in those areas.

In sum, a combination of factors has conspired to undermine the extent and scope for ACM within Mafungautsi. These include intra-state contestations over control of the forestlands that leave the FC politically hamstrung and side-lined, high staff turnover, economic decline on a grand scale and political polarization. So, what is ACM's current status and what factors and processes account for the outcomes?

ACM's current status, and factors and processes shaping it

This study assessed the current status of ACM processes in Mafungautsi in relation to the existence and vibrancy of RMCs, changes in attitudes of traditional leaders toward ACM, scope for evolving an institutional memory on ACM, and the vibrancy of collaboration and social learning and communication.

Existence, vibrancy and adaptiveness of RMCs

Of the three RMCs where ACM was actively promoted, two, Batanai and Gababe, were still actively using the ACM concepts in 2020, notably for joint planning, monitoring and social learning. For Batanai, the village head is at the forefront of facilitating these processes at his village court. He oversees the granting of harvesting permits to resource harvesters as well. Two other RMCs, around Mafungautsi, are reportedly still operational and using elements of the ACM approach which they had copied from CIFOR's three original ACM sites.[9] These four RMCs still convened regular planning and reflection meetings, had committees in place and were still using the resource access permit system introduced in 1994.

On its own part, the FC is still holding refresher courses at the local level for these RMCs, albeit with FC officers exhibiting fatigue and disgruntlement over "people from far off who have come and settled in the forest, while residents from these obedient RMCs are being kept outside".[10] Fortuitously, there has been no or insignificant settlement in the *vleis* where the two former ACM RMCs harvest resources such as broom grass and thatch grass. The third RMC in Ndarire has since become dormant though its once vibrant timber user group still remains intact. The group still meets and discusses timber harvesting as well as strategies to use. The community reported no active engagement with the FC. This is probably due to the location of the RMC, which lies over 75 km on bad earth road from the FC office, whose officers lack resources for mobility, resulting in this minimal contact. The FC and the timber group of the Ndarire RMC had a hostile relationship before the RMC project. By 2006, these two groups were closely working together. Fourteen years later, though the RMC is now almost defunct, the timber user group is still intact and keen to engage with the FC. The group's resolve to work with the FC might have been necessitated by some Chinese who were harvesting commercial hardwood timber in the Chemagora Area and were keen to expand their extraction into Mafungautsi. By working with the FC, the timber user group would possibly be able to keep their timber harvesting territory.

RMCs have adapted to changes in various ways. First, courtesy of technology, virtual meetings are now possible as people in an RMC are now constantly communicating using phones or social platforms like WhatsApp. With a cyber-penetration of over 90% in Zimbabwe, a further increase in the use of mobile technology is envisaged.[11] Increased use of mobile technology should lessen the burden via virtual digital communication. RMCs have also adapted by embracing new stakeholders such as politicians and settlers, thus avoiding the disruptive disharmony that could come with antagonizing them. For instance, the Gababe

RMC chairperson had the following words of wisdom: "We have not been hostile to the settlers. We try to cohabit with them. We engage and tell them not to settle in the areas where our RMC harvests thatch and broom grass. And they understand" (20/08/2019).

Lastly, RMCs have reviewed and changed the payment and distribution of proceeds from the permit system and incentives to RMC committee members, with both adaptations accounting for the longevity and vibrancy of the RMCs. Perks for RMC committee members have been increased and made disbursable in non-monetary form to motivate them. For example, in Gababe, committee members are allocated areas where they harvest thatch and broom grass for free. In Batanai, such perks have included free beehives. The use of non-monetary benefits has been a more meaningful motivational pathway than monetary rewards that did not provide an effective hedge against the country's then spiraling inflation that stood at 500%. Additionally, harvesting permits for broom and thatch grass are now paid per season and pegged at Z$100, freeing resource users from the onerous burden of regular and fragmented payments.[12] If harvesters do not have the money to pay for a thatch grass permit, they can share their harvest with the RMC. For firewood, no permit is required but harvesting is regulated such that it is only permitted on the first and last day of the month. Such adaptive adjustments have been key in strengthening and invigorating RMCs.

What has also helped to sustain ACM processes in Gababe and Batanai is the institutional memory of current leaders. The chairperson and village head of Gababe RMC and Batanai village, respectively, were part of the RMC during the ACM phases, and they were actively involved in ACM processes. They have been able to sustain ACM processes in their areas, using lessons and skills learned during the ACM period.[13] In most democratic processes, leaders are not encouraged to be in office for long, but this example illustrates that in some instances, to deal with processes that have long temporal scale, longevity in supportive leadership might be needed to drive processes.

The Janus-faced role of traditional leaders

Some traditional leaders have been actively involved in resettling people in the forest on the one hand, while on the other, they have played a central role in driving ACM processes. For instance, the village head for Batanai appears not opposed to new occupants falling under his jurisdiction; yet, he also claimed that he summons and fines anyone breaking the rules regarding harvesting of resources from the Batanai RMC resource area at his community court. Usually, the fine is a chicken. Though small and tokenistic, the fine is a deterrent because the spirit behind the court is to name and shame, which is more costly morally and socially. And in a society where social fabric and cohesion are key for survival, being named and shamed carries a huge stigma and social cost. The issue of traditional leaders assuming a more assertive role in the governance of Mafungautsi resonates with Mutimukuru-Maravanyika's (2010) finding that traditional authorities appeared to quickly move in to claim the institutional space left by the FC due to its incapacitation. Traditional leaders are inclined to tacitly support new settlers in areas of forest they perceive as lying within their jurisdiction in order to extend their influence by increasing the pool of residents under their control. Hence, future ACM interventions have to be wary about where and when they can count on traditional leaders as potential allies in securing conservation through ACM.

ACM *collaboration and factors undermining its vibrancy*

The FC did not draw on the support of its partners within the government establishment[14] to counter its being outflanked by populist government functionaries through ACM collaboration. Going solo appears to have deprived the FC of the institutional critical mass that could have proved crucial in buffering it against two adverse pressures against its tenurial and institutional-administrative forest protection functions. The first being the Resident Minister's patently unlawful demands for protected forestlands to be occupied by illegal settlers, with the second being the loss of control over reserved land to Provincial Land and District Land Committees. The dissipation of collaboration among the respective inter-government agencies may have been further accentuated by high staff turnovers within the agencies. Uncertainties in the tenure status of the illegal settlements additionally rendered donors reluctant to be associated with collaborative endeavors within these settlements. Hence, as collaboration receded, the FC's forest management approach appears to have slid back toward a less permissive regime allowing communities to harvest only minor forest products and not timber, even in ACM areas.

The FC appears to have abandoned leadership of ACM collaboration not only at the district level but also at the national level when the National-level ACM steering committee became defunct. Loss of strategic guidance from forestry sector experts from government, civil society and academia blunted the ACM approach's lobbying edge. Among the collaborative partners, CIFOR used to provide funding and advisory support to the committee, with the committee failing to outlive the end of such support. The ephemeral nature of the ACM processes in general, and the National Steering Committee in particular, suggests that longer term 'hand-holding' than was achieved under this project would have been desirable. This recommendation nevertheless appears out of sync with the short-cycle nature of most donor-funded projects.

Social learning and communication

Evidence of internalization and endurance of some aspects of the ACM process nevertheless remain evident, with emerging lessons providing scope for its quick resuscitation and strengthening through surviving media of social learning and communication. Admittedly, many such media and platforms – including ACM's biannual newsletter, 'look and learn' tours and monthly review meetings at the RMC, district and national levels – have become defunct since CIFOR stopped underwriting the costs of their operation. But some have survived, providing reason for optimism. For instance, RMCs are still active at the local level where the internalization and sustenance of ACM have most materially mattered. By and large, RMCs remain active at the local level where the Committees still organize review and planning meetings with occasional support from FC, suggesting some commendable modicum of the internalization of the ACM process among local actors. Though most of the RMCs no longer pursue ACM through a more formal and ordered PAR cycle, there is evidence that a few still engage in these more rigorous if onerous learning cycles. For instance, the Gababe RMC chairperson indicated that they still use results of foregoing trials to convince people not to harvest broom grass through the ecologically more damaging method of uprooting the grass. PAR may have ended up more internalized and better sustained in Gababe because participatory action broom grass research was piloted through its RMC (Mutimukuru-Maravanyika, Kozanayi and Nyirenda 2007 and Nyirenda and Kozanayi 2007). Longevity of some ACM processes thus appears to be a function of duration of exposure to learning and internalization.

Position of the FC on ACM

Although the FC has generally been appreciative of ACM of forests, currently there is very little activity on the ground for various reasons. The main reason being that most of the state-controlled forests have been invaded by 'illegal settlers' from elsewhere and the FC is battling to prevent these settlers from totally destroying the forests. Second, the FC is reverting to the default management style of using force to prevent access to the forest by communities surrounding state forests. According to the general manager, the FC is now accused of using a 'Fortress Management' style, trying to keep out all real and imaginary enemies. Third, there has been no planned or deliberate efforts to institutionalize the major concepts of participatory forest management initiatives like ACM; though in a moment of crisis (see below), the ACM concept was considered again by the FC as a management option for Mafungautsi. Any activities happening now are residual efforts driven by communities themselves: the FC reports that in Mafungautsi there are at least four RMCs that are still operational. This section looks at the current position of the FC on ACM, its institutionalization within FC structures, including (a) research on ACM or participatory forest management initiatives, (b) evident effects of ACM within FC structures and on the ground, and (c) FC plans on management of indigenous forests going forward.

Institutionalization of ACM by FC

The ACM team attempted to institutionalize the ACM approach with government through incorporating it into the Diploma course on forestry practice and science. That would have meant that every graduate from the Forestry College, who would normally end up as forestry extension officers employed by the FC, would have been exposed to the ACM approach and would be confident to use the approach later, on the job. More importantly, we hoped that the ACM approach would be included in any forestry policy in the country as a way to sustain it. However, we have no evidence of recent, deliberate efforts to institutionalize ACM by the FC. As noted above, in Table 7.2, most of the institutional memory remained with the particular officers who were involved in the projects in the early 2000s, most of whom no longer remain with the FC.

An explanation given by one of the FC managers was, "The tendency is that externally driven processes are abandoned once the project or funding ends" (personal communication 2020). Although one of the rationales for developing ACM in the first place was to avoid just such a scenario – by placing so much responsibility and action at the community level – the broader scale, institutional elements of ACM seem to have suffered this same fate within the structures of FC.

Some elements of ACM though have been incorporated into three participatory forest management projects that the FC has been involved in:

- Hwange-Sanyathi Biological Corridor project funded by the World Bank through WWF (ended 2019);
- Community Forest Management Project in Hurungwe funded through WWF (three-year project, ended in 2018);
- SADC Participatory Forests Management Project funded by JICA (ended 2020).

In 2014, in response to immense resistance from the settlers and politicians, the FC toned down its rhetoric for use of force (FC 2014), "The Forestry Commission welcomes the exploration of further opportunities for participatory forest management models to complement the Resource Sharing and Adaptive Collaborative Management approaches to enhance the continued existence of the forest and its utility to its neighbouring communities" (p. 11). The ACM concept at that time was considered at the highest level of the FC as a solution to the myriad challenges in Mafungautsi Forest. However, in all the arguments made by the FC, curiously, no mention was made of local livelihoods. Livelihoods of the local people have been at center stage in the conflictual FC/local people interface.

Discussion

The ACM concept had been introduced at different geographical and institutional levels, and residual elements of ACM can be found at several institutional scales – i.e., RMC user groups and District and Forestry HQ levels. It may be that when conditions on the ground are conducive, ACM as a concept and approach can take center stage in the governance of Mafungautsi Forest. The new government that came into office in 2018 after a coup that dislodged the longstanding president of the country has promised to address the Mafungautsi Forest issue by resettling the forest dwellers elsewhere. Though no deliberate mention of the ACM approach is made in the plans to resolve this impasse, perhaps if this happens RMCs will be able to reclaim their niche resource harvesting areas, which are currently occupied by the settlers, and revive ACM processes. That the FC senior management is considering the ACM approach as an appropriate model to use around Mafungautsi makes this idea more credible. Further, the problems affecting Mafungautsi are endemic around most state forests and without any promising plausible options/models to deal with them, ACM – perhaps complemented with other approaches such as conservation concessions or ecosystem restoration concessions (Robertson and Wunder 2005)[15] – stands a good chance of being adopted. Attempts by the FC to protect itself and assert its authority over state forests using the law alone have failed dismally.

More than a decade and half later, ACM processes in Mafungautsi have been driven by several factors, notably institutional memory (and in some case the lack thereof), lack of plausible and better models to deal with the complex issues involving different stakeholders with varying and competing interests, and strong social capital between local people and those acquainted with the ACM processes. To sustain the ACM approach, in Mafungautsi, process movers in the form of donors, civil society, researchers, and enthusiastic and energetic people will be required. Funding is also needed to drive all the ACM components such as social learning, communication and PAR. CIFOR used to underwrite these costs. Without external (donor) support and limited support from the national treasury due to the decade-long economic meltdown, the FC struggles to facilitate ACM processes at the local and national levels.[16]

Among the many factors that affected ACM's implementation, the land reform and economic upheaval since the early 2000s were central. They confounded and complicated the ACM processes that were unfolding in Mafungautsi Forest. The colossal land reform and occupation of state forests by land-hungry peasants added another layer of complication to the governance of Mafungautsi Forest. The issue changed from being a forestry to a land issue, and it also assumed a national dimension as the invasion of the forest was not an isolated case. At its core, ACM is characterized by multiple stakeholders with many and sometimes different interests in the resource, collaboration, negotiation, observation, planning, executing and adjusting plans and strategies along the way in response to lessons learned. As stakeholders engage iteratively to reconcile their differences, find solutions to problems and unlock potential, they steer toward a common goal for all involved and for the good of the environment and their own livelihoods. ACM thrives on some level of stability in terms of access to resources and the rules, norms and institutions governing resources access and the relationships between the different stakeholders.[17] The haphazard nature of the land reform program and the economic crises of the early 2000s disrupted that much needed underlying condition for ACM.

There were also major changes in human capital within the FC at district, provincial and national levels. All the influential staff at these levels who were supportive of the ACM approach left the FC for greener pastures. Changes in staff coincided with a radical paradigm shift in government's priority, which was now on wholesale land reform and economic policies that centered on empowerment of the indigenous people.[18] At the core of these policies is (unfettered) utilization of natural resources for economic empowerment. The drive toward forest-based economic development posed the challenge of balancing the need for economic development versus the need to sustainably use nature, the foundation for development itself. As well, local institutions which had been at the core of resource governance snapped under immense pressure from demand for forest and land resources.

Enthusiastic leaders and technocrats were side-lined by politics. Further, the diversity of stakeholders involved in Mafungautsi continues to increase (traditional chiefs, land seekers, politicians, schools, the army and police). This makes the reconciliation of divergent interests more complicated, requiring robust social learning platforms and facilitators who can cope with the changing environment (as was the case with CIFOR). The FC, because of its deeply entrenched interests in Mafungautsi Forest, might be an inappropriate actor to drive the ACM process. Resource management in Mafungautsi Forest is no longer a legal issue but a political one and the FC needs to recognize this in order to explore new pathways to resolving the problem.

Finally, current issues around Mafungautsi which the ACM approach is struggling to address are as old as the history of the country. Mudekwe (2007) and Wily (2000) noted that, in Zimbabwe, the processes of demarcating and gazetting state forests alienated the indigenous peoples who had lived in the protected forests and subsisted on the forest products. The indigenous peoples' rights to customary use of the protected forests and forest resources were abolished through the application of the forest legislation. Use was only possible through permits or licenses in order to achieve set forest conservation and management objectives. Local communities living around protected forests have been contesting ownership and control of the forests and forest products for a long time (Matose and Clarke 1993). The permit system dates back to the colonial era (e.g., 1893, amending the 1886 Act). This Game Law Amendment Ordinance banned locals from selling, hunting or hawking game meat without a license. The people were required to acquire a license to kill, catch, capture, hunt or shoot game and each license cost three pounds, an amount that was way beyond the reach of most Africans. By and large, the permit system has had little success in Zimbabwe before or after independence.

There is a diversity of contingent plans, much like the *kukiya kiya* economy (Jones 2010) and bricolage scholarship (Cleaver 2012) at the local level. These include traditional leaders assuming a more prominent role in driving the ACM and RMC processes and barter trade to curb the erosion of money generated as permit fees. Some of these arrangements borrow elements of ACM, such as those requiring collaboration (alliances) and adaptive responses to changes. Livelihoods have not been at the core of the Mafungautsi discourse. The FC is obsessed with returning Mafungautsi Forest to its pristine status, principally for ecological purposes, while settlers in the forest and politicians want a land use model that recognizes human occupancy in the forest. The menu of resources that local people can harvest from the forest also needs revisiting. High-value resources such as timber should be considered. Already a permit mechanism that allows for a flexible win-win situation for the resource users and the state are in place, albeit for low-value resources such as thatch and broom grass.

More importantly, as discussed in the next section, institutionalization of the ACM approach in forestry policy would have helped insulate the ACM approach (and the advantages that flow therefrom) from all these dramatic changes.

Whither ACM in Mafungautsi?

Institutionalization of ACM through enactment of supportive policies, as happened to some degree with other successful conservation approaches in Africa (Kamoto, Missanjo and Djenontin, this volume), helps to sustain the ACM approach long after the implementing institutions have left the scene. When the ACM approach is recognized through policy, resources (financial and otherwise) are allocated toward its implementations, and institutions working in the forestry sector will be required to report on progress made in terms of ACM. In the process, ACM becomes institutionalized within the government system and has more chances of success in the long term. The approach that was most likely to successfully institutionalize ACM was when CIFOR attempted to facilitate incorporation of the approach in the syllabus of the Forestry College course. To institutionalize ACM within the FC will call for a paradigm shift by the officers, supervisors and top management on externally/donor-funded projects. Though some residues of ACM can be found in and around Mafungautsi Forest, perpetuation of the practice has faced many challenges, viz:

The legacy of inequitable land and resource distribution remains unresolved and has fueled resource depletion and associated human insecurity, mainly in the communal areas (Hill and Katerere 2002). The state's indecision about settlers in Mafungautsi and the paralysis that followed resulted in the land issue in the forest becoming more complex, with more people and actors getting involved (e.g., civil servants and people from far-off places coming to settle in the forest, chiefs encouraging their landless subjects to settle there).

Economic meltdown – the *kukiya kiya* economy. These 'making do' activities included opening up crop fields and harvesting resources such as firewood, thatch grass, timber and game from Mafungautsi for sale by both local communities and the middle class whose savings were eroded by hyperinflation. An additional compounding factor included long hours of power outages which created demand for firewood as an alternative source of energy for residents in Gokwe town. Further compounding the impacts of the economic meltdown was the fact that the RMCs were only allowed to harvest low-value products from Mafungautsi (broom grass and mushroom); and this had two main drawbacks: first, because of the low value, harvesters had to supplement their income through harvesting other (prohibited) forest products such as firewood for sale; and second, crop production in the forest offered a better opportunity cost. Thus, the cost of being ACM-compliant – social learning, joint planning and review meetings – could not be offset by the income from the harvest and sale of the low-value NTFPs permitted under the provisions of the co-management program.

Polarization also arose between settlers in the forest and communities from RMCs who used to benefit from resources in the forest. For example, settlers established crop fields in areas where some RMCs had harvested thatch grass and firewood. By privatizing such 'commons', the settlers disrupted local livelihoods which formed the basis of local ACM projects.

Considering the diversity of people settling in the forest, an approach that ensures dialogue and monitoring resource use is imperative. ACM seems such an approach. However, beyond the latent processes of engagement and adaptation that are hallmarks of the ACM approach, other subtle power and interest configurations and design of adaptive strategies need to be embraced. These may escape the normal analytical eye and fall within the rubric of Cleaver's (2012) institutional bricolage. In Mafungautsi, there was overwhelming evidence of this in RMCs where the ACM concept was still being used.

Finally, ACM as an approach has been altered on the ground, in response to the prevailing situation. For example, in the face of restricted movement of people, courtesy of mobile technology, RMC members have been able to interact virtually. The RMCs that are still active have also had to break 'rules' of the state and engage illegal settlers in the forest as key stakeholders in resolving the resource use challenges in the area under the jurisdiction of the RMC. That institutional change, which shows elements of institutional bricolage, needs to be embraced as part of an evolution of the ACM approach. It makes the ACM approach more responsive and therefore appropriate to local needs. Last, but not least, external support is needed to underwrite some of the costs associated with ACM processes such as social learning platforms.

Conclusion

ACM works well at a local level because the social learning platforms are effective at that level but it needs to be institutionalized and such platforms replicated at higher levels in order to transform forest management. For ACM to deliver change at a national level, ACM processes must establish links with advocacy initiatives. Advocacy initiatives can use the evidence from local level ACM processes to lobby for lasting policy and structural reforms that can ensure more sustainable ACM processes at different scales.

The ACM interventions in Mafungautsi have been superseded by land rights over which the polity of Zimbabwe was fought in the 1970s leading to its independence in 1980. These unresolved issues are being played out around the state forest which was not included in the land reform program of 2000. The state has not included restitution of lost rights around protected areas under its jurisdiction, focusing only on privately held land; yet as the invasions around the forest are illuminating, state forest land needs to be part of the restoration of lost rights through facilitating dialogue and vision building among stakeholders. ACM will need to engage with these broader political economy issues around the forest in order to get to the heart of vexing issues that have manifested themselves over the last decade. Furthermore, with population growth on a steady increase, demand for land will inexorably rise,[19] putting pressure on forests like Mafungautsi. To remove pressure on state forests in marginal areas like Gokwe, there has to be parallel investments in the communal areas to address the wider entitlements failure that Zimbabwe inherited at independence.

The broader political economic context and the natural disasters ushered in by climate change and the COVID-19 pandemic in Zimbabwe and around Mafungautsi have created very deep challenges for ACM as an approach. These challenges have implications for broader rural livelihoods and overall forest management in order for the approach to be sustainable and scaled up. One of the key elements of ACM is its adaptable nature which is driven by social learning. When the conditions for secure access to resources by stakeholders are under threat and when stakeholders lack the space and facilitation of the social learning and deliberative process, ACM struggles to survive. The Mafungautsi case shows that if ACM is not embedded in the broader policy processes, its sustainability remains in jeopardy. While there is very little evidence of ACM institutionalization within the Zimbabwe FC, this cannot be blamed on the FC but rather on the broader political economic context of the country. At the same time, ACM needs to tap into the opportunities offered by climate change initiatives that include REDD+. The Mafungautsi Forest and other protected state forests in Zimbabwe and elsewhere have huge potential for nature-based, climate change mitigation and adaptation. The Mafungautsi Forest was demarcated primarily for watershed management and as such there is a scope of enhancing its management using ACM as an approach to deliver a payment for environmental-services scheme. Such initiatives would provide revenues and other incentives that can be used to sustain the needs of communities and deter further clearing of forest land for agriculture.

Notes

1 These were groups that were created during the era of the resource-sharing program for purposes of facilitating equitable and sustainable resource-sharing by communities living around Mafungautsi forest.
2 Some of the settlers who were first to invade the forest were autochthones who were displaced from their ancestral lands when Mafungautsi was gazetted in 1954. At the onset of the land reform program, they took advantage of it to reclaim these ancestral lands (Matose 2002). But later, opportunists also joined the invasion of Mafungautsi.
3 Even global efforts to diffuse the tense situation on the ground had been futile, e.g., the Abuja Conference of 1998.
4 The **Third Schedule** under the Forest Act lists all the demarcated forests in Zimbabwe and describes in detail the boundaries of each forest.
5 For instance, an FC report notes: "We recommend to the authorities to remove them all (illegal settlers) as a matter of urgency" (FC Report 2014, p. 8).
6 Cronkleton, Evans and Larson (2022) show how vital experience and training in an ACM approach is to its successful conduct.
7 CIFOR annual report, 2004.
8 *Newsday* 26 October 2011 ("Chokuda: The man who fought his own battle") and Sachikonye (2011).
9 During the time we were implementing the ACM project, cross site/RMC learning was encouraged and promoted; hence, some non-ACM sites came to learn about and adopt ACM as a management strategy.

10 FPU officer, personal communication, 11 August 2019, Gokwe South.
11 A negative outcome of this was that mobile technology enabled forest invaders across the country to coordinate their activities. Using mobile technology, land occupiers in all state forests came to know of what was happening in each state forest and this encouraged the occupiers to be relentless in the occupation of the forests.
12 USD1.2 at the current (2022) interbank rate.
13 Such skills include communication, giving and receiving feedback, conflict resolution, democratic processes, etc. (see Mutimukuru-Maravanyika 2010, for a detailed account).
14 Such as the Environmental Management Agency, Ministry of Youth, Development, Gender and Employment Creation, banks, Agricultural and Technical Extension Services, the District Administrator and Rural District Council.
15 An emergent approach whereby national authorities or local resource users agree to protect spatially well-defined ecosystems in exchange for a stream of structured compensations from conservationists or other environmental-service users. Local stakeholders are compensated for loss of access to certain ecosystem services benefit streams which are lost when a landscape is conserved usually for the provision of ecological services which are enjoyed as public goods (Robertson and Wunder 2005).
16 After the government weaned the FC and cut off its budgetary support, the FC reportedly went for several months without paying its employees (FC staff, personal communication, 13 March 2019).
17 It is worth noting that *some* level of conflict also seems necessary to stimulate sufficient interest within communities for them to take the trouble to gather, plan, monitor and revise (Colfer 2005a).
18 'Land is the economy, and the economy is land', a mantra of the ruling party and the Indigenous Act and ZIMASSET program.
19 One element in this equation, population rise, is amenable to stabilization with the voluntary cooperation of the people, with significant positive potential, particularly for enhancing women's lives (see, e.g., Colfer, Dudley and Gardner 2008).

References

ACM *Newsletter*. 2006. "Reflecting on the ACM Experiment: A Synthesis of the Experiences from the Learning Centres in Zimbabwe." ACM *Newsletter (Zimbabwe)* 3. no.1 (February). Harare, Zimbabwe: CIFOR Regional Office for Eastern and Southern Africa.

Büscher, B.E., and T. Mutimukuru-Maravanyika. 2007. "Buzzing too Far? The Ideological Echo of Global Governance Concepts on the Local Level: The Case of the Mafungautsi Forest in Zimbabwe." *Development Southern Africa* 24, no. 5: 649–664. https://doi.org/10.1080/03768350701650512.

Chambers, R., A. Pacey, and A.L. Thrupp, eds. 1993. *Farmer First: Farmer Innovation and Agricultural Research*. London: Intermediate Technology Press.

Cleaver, F. 2012. *Development through Bricolage: Rethinking Institutions for Natural Resources Management*. New York: Routledge.

Cleaver, F., and J. de Koning. 2015. "Furthering Critical Institutionalism." *International Journal of the Commons* 9, no. 1: 1–18.

Colfer, C.J.P. 2005a. *The Complex Forest: Communities, Uncertainty, and Adaptive Collaborative Management*. Washington, DC: Resources for the Future. https://doi.org/10.4324/9781936331727.

Colfer, C.J.P. 2005b. *The Equitable Forest: Diversity, Community, and Resource Management*. Washington, DC: Resources for the Future. https://doi.org/10.4324/9781936331673.

Colfer, C.J.P., E. Andriamampandry, S. Asaha, E. Lyimo, E. Martini, J.L. Pfund, and J. Watts. 2011. "Participatory Action Research for Catalysing Adaptive Management: Analysis of a 'Fits and Starts' Process." *Journal of Environmental Science and Engineering* 5, no. 1: 28–43.

Colfer, C.J.P., R.G. Dudley, and R. Gardner. 2008. "Forest Women, Health and Childbearing." In *Human Health and Forests: A Global Overview of Issues, Practice and Policy*, edited by C.J.P. Colfer, 113–133. London: Earthscan and Routledge.

Cronkleton, P., K. Evans, and A.M. Larson. 2022. "Capacity Building for ACM: Lessons Learned from Training in Distinct Contexts." In *Adaptive Collaborative Management of Forest Landscapes: Villagers, Bureaucrats and Civil Society*, edited by C.J.P. Colfer, R. Prabhu, and A. Larson, 134–157. London: Earthscan and Routledge.

Diaw, M.C., T. Aseh, and R. Prabhu. 2009. *In Search of Common Ground: Adaptive Collaborative Management in Cameroon*. Bogor, Indonesia: Center for International Forestry Research. https://doi.org/10.17528/cifor/002633.

FC (Forestry Commission). 2013. "Position Paper on all Illegal Settlements in State Forests." 13 October 2013, Harare, Zimbabwe.

FC (Forestry Commission). 2014. "Report on the Visit to Mafungautsi Forest to Assess Land Requested by the Minister of State for Provincial Affairs (Midlands) to Settle More People." September, Harare, Zimbabwe.

Fisher, R., R. Prabhu, and C. McDougall. 2007. *Adaptive Collaborative Management of Community Forests in Asia: Experiences from Nepal, Indonesia and the Philippines*. Bogor, Indonesia: CIFOR.

German, L.A., A.-M. Tiani, A. Daoudi, T. Mutimukuru-Maravanyika, E. Chuma, N. Beaulieu, H. Lo, C. Jum, N. Nemarundwe, E. Ontita, G. Yitamben, and V. Orindi. 2010. *The Application of Participatory Action Research to Climate Change Adaptation: A Reference Guide*. Bogor, Indonesia: IDRC and CIFOR.

Hanlow, J., J. Manjengwa, and T. Smart. 2013. *Zimbabwe Takes Back its Land*. Sterling, VA: Kumarian Press.

Hartanto, H.M., C. Lorenzo, C. Valmores, L. Arda-Minas, L. Burton, and R. Prabhu. 2003. *Learning Together: Responding to Change and Complexity to Improve Community Forests in the Philippines*. Bogor, Indonesia: CIFOR.

Hill, R., and Y. Katerere. 2002. "Colonialism and Inequity in Zimbabwe." In *Conserving the Peace: Resources, Livelihoods and Security*, edited by R. Matthew, M. Halle, and J. Switzer, 247–272. Winnepeg, Canada: International Institute for Sustainable Development and IUCN.

Hope, A., and S. Timmel. 1995. *Training for Transformation. A Handbook for Community Workers* (rev. ed.), Books I–III. Gweru, Zimbabwe: Mambo Press.

Jones, L.J. 2010. "Nothing is Straight in Zimbabwe: The Rise of the *Kukiya-kiya* Economy 2000–2008." *Journal of Southern African Studies* 36, no. 2: 285–299.

Kozanayi, W. 2018. "Influences of Customary and Statutory Governance on Sustainable Use and Livelihoods: The Case of Baobab, Chimanimani District, Zimbabwe." PhD Thesis, University of Cape Town, Cape Town, South Africa.

Mandondo, A., and E. Mapedza. 2002. "Allocation of Governmental Authority and Responsibility in Tiered Governance Systems: The Case of Environment-Related Laws in Zimbabwe." *Environmental Governance in Africa Working Papers*. Washington, DC: World Resources Institute.

Mandondo, A., R. Prabhu, and F. Matose, eds. 2008. *Coping Amidst Chaos: Studies on Adaptive Collaborative Management from Zimbabwe*. Bogor, Indonesia: CIFOR. https://doi.org/10.17528/cifor/002558.

Mapedza, E. 2006. "Compromised Co-management, Compromised Outcomes: Experiences from a Zimbabwean Forest." *Africa Development* 31, no. 2: 123–146.

Matondi, P., ed. 2010. *Inside the Political Economy of Redistributive Land and Agrarian Reforms in Mazowe, Shamva and Mangwa Districts in Zimbabwe.* Harare, Zimbabwe: Ruzivo Trust.

Matose, F.M. 1994. "Local People's Uses and Perceptions of Forest Resources: An Analysis of State Property Regimes in Zimbabwe." MSc Thesis, University of Alberta, Edmonton, Canada.

Matose, F.M. 2002. "Local People and Reserved Forests in Zimbabwe: What Prospects for Co-Management?" PhD diss., University of Sussex, Falmer, Brighton. http://hdl.handle.net/10625/31221.

Matose, F.M., and J. Clarke. 1993. "Who is the Guardian of the Indigenous Forests?" In *The Ecology and Management of Indigenous Forests in Southern Africa*, edited by G.D. Piearce and D.J. Gumbo, 62–70. Proceedings of an International Symposium, 27–29 July 1992, Victoria Falls, Zimbabwe: Zimbabwe Forestry Commission and SAREC.

Matose, F.M., and T. Mutimukuru-Maravanyika. 2009. "'Squatting' as a Means of Establishing Authority over Forest Land in Zimbabwe: A Missing Dimension to Land Reform." Working Group on the Politics of Authority, Land, and Natural Resources at the Workshop on the Workshop 4, Bloomington, Indiana University, 3–6 June 2009.

Matzke, G. 1993. "Resource Sharing Schemes for State Owned Land in Zimbabwe: A Discussion of Conceptual Issues Needing Consideration in the Development and Planning of Co-Management Regime." CASS *Occasional Paper* 54/93. Harare, Zimbabwe: University of Zimbabwe.

McGregor, J. 1991. "Ecology, Policy and Ideology: An Historical Study of Woodland Use and Change in Zimbabwe's Communal Areas." PhD Thesis, Loughborough University of Technology, London.

Mubvami, T. 2004. "Impact of Land Redistribution on the Environment." In *Post Independence Land Reform in Zimbabwe: Controversies and Impact on the Economy*, edited by M. Masiiwa, 71–89. Harare, Zimbabwe: Friedrich Ebert Stiftung and Institute of Development Studies, University of Zimbabwe.

Mudekwe, J. 2007. "The Impacts of Subsistence Use of Forest Products and the Dynamics of Harvested Woody Species Populations in a Protected Forest Reserve in Western Zimbabwe." Unpublished PhD Thesis, Stellenbosch University, Cape Town, South Africa.

Murphree, M.W., and D.H.M. Cumming. 1991. "Savanna Land Use Policy and Practice in Zimbabwe." Paper Presented at the UNESCO/IUBS on Savanna Land Use, Nairobi, Kenya, January.

Mutimukuru-Maravanyika, T. 2010. "Can We Learn Our Way to Sustainable Management?" PhD Thesis, Wageningen University and Research Centre, Wageningen, The Netherlands.

Mutimukuru-Maravanyika, T., R. Prabhu, F. Matose, R. Nyirenda, and W. Kozanayi. 2008. "Facilitating Adaptive Collaborative Management in Forested Landscapes: The Mafungautsi Case Study." In *Coping Amidst Chao:. Studies on Adaptive Collaborative Management from Zimbabwe*, edited by A. Mandondo, R. Prabhu, and F. Matose, 15–64. Bogor, Indonesia: CIFOR.

Mutimukuru-Maravanyika, T., W. Kozanayi, and R. Nyirenda. 2006. "Catalyzing Collaborative Monitoring Processes in Joint Forest Management Situations: The Mafungautsi Forest Case, Zimbabwe." *Society and Natural Resources* 19, no. 3: 209–224. https://doi.org/10.1080/08941920500460674.

Mutimukuru-Maravanyika, T., W. Kozanayi, and R. Nyirenda. 2007. "Initiating a Dynamic Process for Monitoring in Mafungautsi State Forest, Zimbabwe." In *Negotiating Learning: Collaborative Monitoring in Forest Resource Management*, edited by I. Guijt, 114–123. Washington, DC: Resources for the Future.

Newsday Newspaper. 2011. "Chokuda: The Man who Fought his Own Battle." 26 October. https://www.newsday.co.zw/2011/10/2011-10-26-chokuda-the-man-who-fought-his-own-battle/ (accessed 11 October 2021).

Nyirenda, R., and W. Kozanayi. 2007. "Tracking Broom Grass Resources for Equity in Zimbabwe." In *Negotiated Learning: Collaborative Monitoring in Forest Resource Management*, edited by I. Guijt, 66–72. Washington, DC: Resources for the Future.

Ojha, H., A. Hall, and R. Sulaiman. 2013. *Adaptive Collaborative Approaches in Natural Resource Governance: An Introduction*. London: Earthscan.

Prabhu, R. 2003. "Developing Collaborative Monitoring for Adaptive Co-Management of Tropical African Forests (Contract Number: B7–6201/99-05/FOR): Final Report for the Period: January 1, 2000–December 31, 2002." Harare, Zimbabwe: CIFOR.

Prabhu, R., C.J.P. Colfer, C. Diaw, C. McDougall, and R. Fisher. 2009. "Action Research with Local Forest Users and Managers: Lessons from CIFOR's Research on Adaptive Collaborative Management." In *Farmer First Revisited*, edited by I. Scoones and J. Thompson, 66–70. Warwick, UK: Practical Action Publishing.

Robertson, N., and S. Wunder. 2005. *Fresh Tracks in the Forest: Assessing Incipient Payments for Environmental Services Initiatives in Bolivia*. Bogor, Indonesia: CIFOR.

Sachikonye, L. 2011. *When a State Turns on Its Citizens. Institutionalised Violence and Political Culture*. Auckland Park, South Africa: Jacana Media.

Sadomba, W.Z. 2008. "War Veterans in Zimbabwe's Land Occupations: Complexities of a Liberation Movement in an African Post-Colonial Settler Society." PhD Thesis, Wageningen University, The Netherlands.

Scoones, I., N. Marongwe, B. Mavedzenge, J. Mahenehene, F. Murimbarimba, and C. Sukume. 2010. *Zimbabwe's Land Reform: Myths and Realities*. Oxford, UK: James Currey. Weaver Press.

Selener, J.D. 1992. "Participatory Action Research and Social Change: Approaches and Critique." PhD Thesis, Cornell University, Ithaca, NY.

Sithole, B. 2005. "Becoming Men in our Dresses! Women's Involvement in a Joint Forestry Management Project in Mafungautsi, Zimbabwe." In *The Equitable Forest: Diversity, Community and Resource Management*, edited by C.J.P. Colfer, 171–185. Bogor, Indonesia: Resources for the Future and CIFOR.

Sithole, B., and W. Kozanayi. 2002. *Development of the Framework in Zimbabwe: Case Material from Mafungautsi Joint Forest Management Project and Simudzirayi Micro-Catchment Scheme*. Bogor, Indonesia: CIFOR.

Vermeulen, S. 1997. "Sharing of State-Owned Resources with Local Residents: The Case of Wood in Mafungautsi State Forest and Gokwe Communal Area." *IES Working Paper* 4. Harare, Zimbabwe: University of Zimbabwe.

Wily, A.Z. 2000. "Land Tenure Reform and the Balance of Power in Eastern and Southern Africa." *Natural Resource Perspectives* 58 (June). London: ODI. https://cdn.odi.org/media/documents/2839.pdf (accessed 21 November 2022).

Wollenberg, E., D. Edmunds, and L. Buck. 2000. "Using Scenarios to Make Decisions about the Future: Anticipatory Learning for Adaptive Co-Management of Community Forests." *Landscape and Urban Planning* 47: 65–77.

Introduction to Chapter 8

In the previous chapter, on Zimbabwe, we focused on communities with tentacles tentatively reaching up in scale to the national level. We saw the impacts of broader scale chaos on our local level action. In Malawi, we start at the national level and make use of the Participatory Forest Management (PFM) program, with its rich repertoire of repeat assessments of its own implementation throughout the country over time.

The lead author, Judith Kamoto, led the ACM work in Malawi in the early 2000s, and the PFM program has had the benefit of her advice and influence ever since. This chapter provides us with an example of genuine institutionalization of many ACM-like principles at a national level. As these authors assert:

> Malawi's PFM has a similar iterative process of setting goals, like visioning in ACM planning, then institution building, and then implementing practical actions, followed by performance monitoring and learning.

Key contributions from this chapter include

- the wealth of information available from the numerous studies evaluating the program,
- the diverse aspects that have been assessed, from human well-being to environmental impacts, including information on community participation in management, and
- the straightforward description of one example of how an ACM-like process can be partially institutionalized.

This chapter is an excellent segue also to the following chapter, which discusses a similar approach in Uganda (Collaborative Forest Management), but one which did not have routine access to the long-term, ACM-style input from someone like Kamoto.

DOI: 10.4324/9781003325932-14

8 An assessment of Participatory Forest Management inspired by adaptive collaborative management in Malawi

Judith F.M. Kamoto, Edward Missanjo and Ida N.S. Djenontin

Introduction

Participatory Forest Management (PFM) refers to processes and mechanisms that enable people who have a direct stake in forest resources to be part of decision-making in all aspects of forest management, from managing resources to formulating and implementing institutional regulatory frameworks (Klooster 2000). Since the famous but controversial article, 'The Tragedy of the Commons', was published (Hardin 1968), management approaches on shared natural resources – common pool resources – have been widely discussed. In the context of sustainable forest resource management, people-centered participatory approaches have been given more attention in developing countries. Numerous studies have shown that PFM approaches have been instrumental not only in the recovery and maintenance of forest conditions, but also to ensure sustainable use by local communities who rely on the resources to meet their livelihoods (Banana et al. 2012; Dhakal and Masuda 2008; Gobeze et al. 2009; Hajjar et al. 2021; Islam et al. 2014; Kamoto et al. 2013; Kibria, Jashimuddin and Makoto 2014; Matiku, Caleb and Callistus 2013). Indeed, these approaches are central to organizing people, making a community-based institution and implementing forest management activities based on collective interests at the community level (Iversen et al. 2006). Above all, PFM plays a key role for livelihood improvement of communities living near the forests (Hajjar et al. 2021; Mbuvi et al. 2009).

Malawi has a unique PFM experience built upon initial endeavors that piloted the adaptive collaborative management (ACM) concept. A number of studies have been conducted to assess whether PFM initiatives in Malawi have achieved their objectives as expected, and to draw lessons for future applications/replications to other sites. Taken collectively, they contribute to providing an unusual degree of monitoring and outcome assessment for the PFM experience in Malawi. We review and discuss a number of these studies that were conducted to assess PFM processes, performance and outcomes in Malawi. We draw lessons learned and best practices for sustainable forest management (SFM), especially as the country transitions to embrace forest landscape restoration (FLR) policy. The implementation of FLR highly relies on the country's experience of PFM. Our analysis also offers reflection on what (and how) to improve PFM approaches and implementation, including improving Malawi's Standards and Guidelines for PFM.

DOI: 10.4324/9781003325932-15

Forests and participatory governance in Malawi

In 1975 and 2010, Malawi's forest cover was about 47% and 20% of its land surface area, respectively (AAS 2012; Mauambeta et al. 2010). This illustrates the severe degradation of forest resources and the considerable change in forest cover that the country experienced. Scholars of forest management in Malawi recognize that social factors such as biomass-based energy needs, livelihood needs and poverty status affect forest degradation. Therefore, combating poverty is a prerequisite for sustainable forest resource management. The Government of Malawi has set utmost priority on people-oriented forest management approaches since the 1990s. PFM is seen as a way to accomplish these goals and is stipulated in the National Forest Policies of 1996 and 2016 (GoM 1996, 2016) and operationalized by the National Forest Act of 1997 (GoM 1997).

Malawi's 1965 Land Act (GoM 1965) and 2002 Land Policy (GoM 2002) recognize three types of land: public, customary and private land. Public forested lands are managed by the Department of National Parks and Wildlife and the Department of Forestry (DoF). Public lands are held in trust and managed by the government or traditional authorities (TAs) and openly used or accessible to the public at large. This category of land includes land gazetted for use as national parks, recreation areas, forest reserves (FRs), conservation areas, and historic and cultural sites. Customary land is all land held, occupied or used by communities under customary law and is under the jurisdiction of TAs. Private land is all land that is exclusively owned, held or occupied under (a) **freehold tenure** and (b) **customary land** allocated exclusively to a clearly defined community, corporation, institution, clan, family or individual. Such exclusive allocations of customary land will henceforth be known formally as a **'customary estate'** (GoM 2002, 28). It is important to note that a **leasehold estate** can be created out of government land or any private land, including customary estates, so long as the term of the lease is less than that of the owner (GoM 2002, 13).

The law recognizes two main types of PFM, namely Co-management and Community Based Forest Management (CBFM). Co-management, in contrast to CBFM, is based on a management agreement between local communities and government authorities regarding the management of state government FRs or plantations. With Co-management, land ownership remains with the government, while local communities are duty bearers and, in turn, get user rights and access to some forest products and services (GoM 2005). CBFM, however, takes place in forest on village lands and villagers take full ownership of village forest areas (VFAs; GoM 2003, 2005). In brief, PFM takes place on customary land through the management of VFA by communities, or in state forest reserves and plantations through co-management between communities and the DoF (Box 8.1; USAID-Malawi 2015; Zulu 2013). In 1999, there was only one FR under co-management; by 2010, the number had grown to 12 and to date there are 14 FRs under co-management (USAID-Malawi 2021). In 1996, establishment of VFAs was sanctioned by GoM Forest Policy. Each village had to establish a VFA in customary-degraded lands. Kamoto (2007) reported a total of 1,100 VFAs in

Lilongwe district, of which 300 were VFAs with indigenous trees. In 2012, Lilongwe recorded 438 indigenous VFAs (GoM 2017). This number has grown since that time (District Forest officer Lilongwe, personal communication). Malawi has committed 4.5 million hectares to restoration under the AFR100 Bonn Challenge. This means that communities will be encouraged to establish more VFAs as various projects roll out to support this initiative.

BOX 8.1 Overview of Participatory Forest Management in Malawi

PFM activities in Malawi are carried out primarily through establishment of VFAs or co-management agreements in forest reserves. These programs are summarized below.

Village Forest Areas: VFAs enable forest communities to establish formalized rights to manage customary forest lands. In order to establish a VFA, a Forest Management Agreement is developed and signed with the District Forest Office (DFO). Plans are developed by communities in conjunction with local extension agents on the basis of a Participatory Forest Resource Assessment. At the community level, VFAs are managed by Village Natural Resource Management Committees (VNRMCs).

Forest Reserve co-management: Co-management of forest reserves aims to distribute the costs and benefits of managing FRs between DFOs and village communities living within the buffer zone of the reserves. Co-management plans are developed by the DFO and communities, in line with the strategic plan for the FR. These plans define roles and responsibilities as well as set out objectives and rules for resource management within the reserve. FRs are divided into blocks, which are managed by a Block Management Committee (BMC) composed of representatives from member villages. In addition to BMCs, Local Forest Management Boards (LFMBs) are also established around FRs to serve as a multi-stakeholder entity for convening community representatives, TAs, civil society groups and government officials. Benefits derived from income-generating activities in the blocks are divided between the communities (60%), DFO (30%) and LFMB (10%).

It is important to note that, in the attempt to improve co-management operations, some districts have registered some amendments to the institutional arrangements related to the BMCs and LFMBs. The amendments include either overhauling BMCs and replacing them with existing VNRMCs or up/downgrading the scale of operation of the BMCs. LFMBs, which in most cases have become obsolete, have been replaced with newly established institutions integrated in the local governance structure following the decentralization policy (see Djenontin and Zulu 2021, for more details). However, in many other districts, where the initial arrangements still apply, BMCs and LFMBs might have become obsolete and not be operating as intended.

Malawi has established Standards and Guidelines for PFM (GoM 2005), which outline the basic framework for implementation of PFM activities as well as guidelines to support best practices (Figure 8.1). The Standards and Guidelines for PFM in Malawi were developed based on lessons learnt from implementation of co-management and CBFM initiatives in Malawi. Of importance are the initial co-management pilot project (1992–1999), the ACM project (2000–2005) at Chimaliro Forest Reserve and a Social Forestry Project promoting CBFM (1997–2004; GoM 2005). Using the experience gained from implementing the ACM project at Chimaliro Forest Reserve, the first author of this chapter was involved in developing Malawi's Standards and Guidelines for PFM. Specifically, ACM was implemented in a pilot collaborative management in Chimaliro Forest Reserve, which had been a 'no go' zone for local communities. However, following the Earth summit in 1992, Chimaliro opened up for community involvement, first as a co-management site under a World Bank project with the Research Institute of Malawi. Challenges experienced in the 'new' paradigm shift were embraced by the ACM project from 2000 to 2005, and lessons were drawn for future co-management experiences. With the experience gained from the pilot ACM project, the DoF developed what were considered 'best practice' standards for promoting improved and sustainable forest management in support of rural livelihoods and sustainable development in Malawi.

The Service Standards for Participatory Forestry (Figure 8.1) were built on both theoretical and practical perspectives in forest governance. First, the service standards drew heavily on the 'worm approach' (shown in Figure 2.1) that the ACM project used at Chimaliro Forest Reserve and Ntonya Hill for its Participatory Action Research (PAR) for forest management. PAR drove the ACM process using the worm (with its four stages of observation, action, monitoring and reflection) as it proceeded in a systematic, iterative mode, allowing 'a process within the local community in which people…can jointly plan improvements in local conditions…gain power and skills in dealing with others and develop a self-monitoring system to enhance sustainability' (Colfer 2005, 5). The four main areas of the Service Standards for Participatory Forestry (setting strategic goals and roles; institutional building, strengthening and prioritizing actions; implementing practical actions for sustainable forestry and livelihoods; and finally, performance monitoring and learning) mirror the four stages of the worm. This indicates the overlaps in the stages between ACM, PAR and the Service Standards for Participatory Forestry, as the latter largely draws from the worm of PAR in the ACM approach. Second, Ostrom's design principles for managing the commons under governance of common pool resources (Ostrom 1990) also informed some specific service standards, including service standards 1, 4, 5, 6 and 16.

Several PFM programs were initiated to apply these Standards and Guidelines, with support from different donor agencies. The most significant was the Improved Forest Management for Sustainable Livelihood Programme (IFMSLP) supported by the European Union. The program was implemented in two phases, Phase I (2005–2010) and Phase II (2011–2014), and in 12 districts out of the total 28 districts in Malawi. Carrying over the same activities from Phase I, Phase II

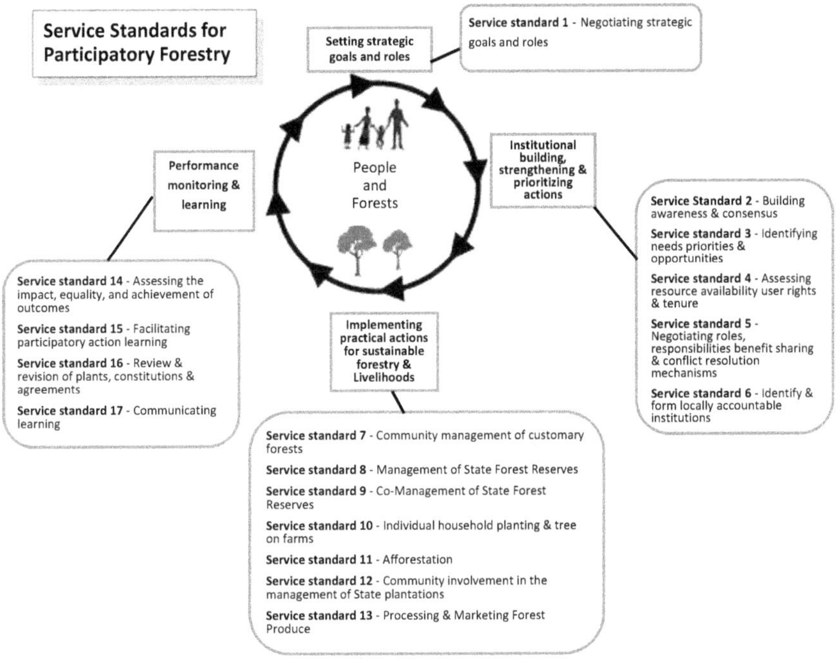

Figure 8.1 Standards and Guidelines for PFM in Malawi

was deemed important to sustain the momentum instigated and to allow long-term processes to solidify outcomes. The main aim of IFMSLP was to improve the livelihoods of forest dependent communities through a combination of three strategies: (1) PFM in FRs – co-management, (2) PFM in VFAs – CBFM and (3) forest-based enterprises (Olivier and Mwase 2012; Remme et al. 2015; Zulu 2013). With co-management, groups of villages were organized as BMCs that featured elected members from villages associated with delineated forest blocks. In addition, an LFMB, which comprised district officers for forestry, agriculture, fisheries, water, community services, the chief executive of district assembly, TAs and non-state actors, was created to coordinate management across the blocks of an FR. Under CBFM, VNRMCs managed the designated VFAs on communal land areas (Zulu 2013). The introduction of PFM in the program sites was expected to improve forest conditions, socio-economic status of the local community and sustainable management of the forest resources.

In addition, the Protecting Ecosystems and Restoring Forests in Malawi (PERFORM) project was implemented from 2014 to 2019 to consolidate and improve the legacy of the IFMSLP. PERFORM worked in 3 of the 12 IFMSLP districts. However, given the scant success of the LFMBs as an umbrella body of local governance structures, the project made some noteworthy changes in the co-management institutional arrangements, which are yet to be legally endorsed in the forest policy. The changes involved diverse scale rearrangements and changes in the functional roles of the community-led forest governance structures (Djenontin and Zulu 2021).

Overall, PFM has changed forms and scales from the IFMSLP and PERFORM eras to today and its evolution and dynamics mirror the kind of adaptive and collaborative processes encouraged within ACM. ACM used the iterative 'worm' of PAR to capture this process (Colfer 2005; and more recently, see Mukasa et al. 2022, their Figure 5.1). The worm reflects the iterative process of observation, planning, action, monitoring and reflection. Malawi's PFM has a similar iterative process of setting goals, like visioning in ACM planning, then institution building, and then implementing practical actions, followed by performance monitoring and learning. There is a strong congruence between the Standards and Guidelines for PFM in Malawi (Figure 8.1) and the ACM PAR worm.

Literature assessing PFM performance

Our analysis and insights are based on a variety of studies that assessed the governance processes, institutional arrangements, institutional performances, socio-economic, institutional and biophysical impacts of PFM in Malawi. The studies covered all the 12 districts in which IFMSLP was implemented (Figure 8.2). We reinforce the analysis with our experiences of and research on the forest management context and dynamics in Malawi.

Kamoto and Milner (2003) facilitated ACM processes – using visioning, PAR, focus groups and content analysis – and examined its impacts on Chimaliro Forest Reserve in Kasungu district and Ntonya Hill in Zomba district. Kamoto (2007) further examines the impacts of ACM in Chimaliro Forest Reserve, where she focused on collaborative monitoring of bee-keeping activities in two co-management blocks.

In 2013, Kamoto et al. (2013) assessed the implications of neglecting local institutions in policy development for community-based natural resources management (service standard #6). These authors raised the issue of policies that do more harm than good in community-based forest management. Specifically, they highlighted elite capture, negative consequences of external incentives, worrying decision-making processes, and conflicts between traditional and imposed institutions.

Zulu's (2013) study supplemented qualitative analysis with descriptive statistics based on a household survey of 45 men-headed and 20 women-headed households between 2009 and 2012 in Ntchisi Forest Reserve. His study covered 36.1% of the 180 households in the study site. He also conducted focus groups, observed and compiled secondary data. He examined the challenges in implementing co-management and achieving SFM and improved livelihoods. The study, conducted in Ntchisi district after the fourth and sixth years of co-management, used mixed social science research methods.

When the IFMSLP was in its seventh year (2012), Chinangwa, Pullin and Hockley (2016) assessed the impact of forest co-management on community livelihoods and welfare in Zomba and Ntchisi districts. They interviewed 32% of the

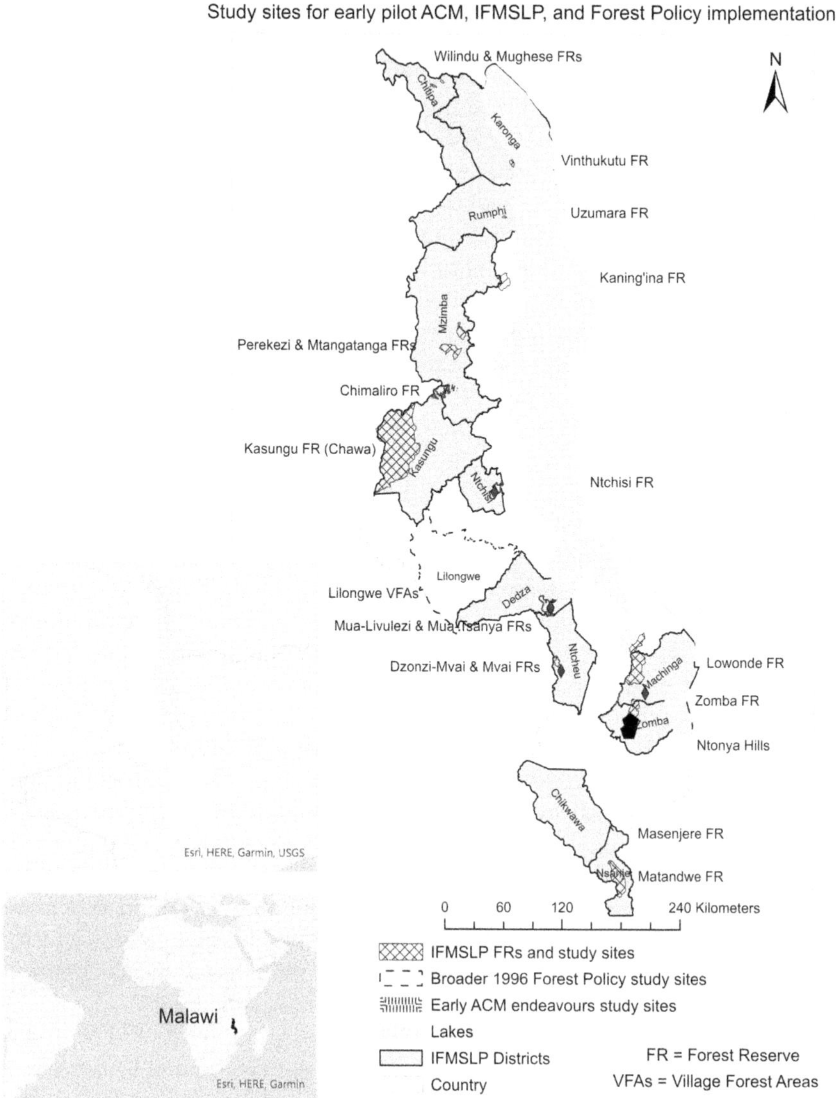

Figure 8.2 Districts and forest reserves with PFM under the IFMSLP

households (n = 213) in the study communities and used a sustainable livelihoods framework and a quantitative analysis. The authors used Probit and Tobit regression models, implemented in STATA, for statistical analysis of their survey.[1]

In 2013, Mtambo and Missanjo (2015) analyzed CBFM biophysical outcomes in terms of tree species richness and diversity. They compared four VFAs under the CBFM approach and four others outside CBFM in Kasungu district. These authors inventoried 160 plots (20 from each of eight VFAs), using systematic line transect sampling. Tree species stocking, for all woody species of all stages, was enumerated and their species names were also recorded. Tree species richness and diversity were determined by using a Rènyi diversity profile in Biodiversity R. Biodiversity R. is software that does all the biodiversity analyses, while Rènyi diversity profiles are curves that provide information on richness and evenness. The shape of the profile is an indication of the evenness. A horizontal profile indicates that all species have the same evenness. The starting position at the left-hand side of the profile is an indication of the species richness. A profile that starts at a higher level has higher richness. The major advantage of Rènyi diversity profiles is that sites can easily be ordered from high to low diversity. If the profile for one site is everywhere above the profile for another site, then this means that the site with the higher profile is the more diverse of the two.

Chingaipe et al. (2015) assessed the effectiveness of local co-management institutions in sustainable management of forest resources in Dedza district. The study interviewed approximately 35% of the population (n = 214) in the selected communities and adopted mostly a qualitative analysis (with Chi-square tests).

Banda, Senganimalunje and Missanjo (2015) examined community attitudes and perceptions toward management of Kaning'ina FR in Malawi, with much emphasis on (1) determining if people are aware of the environmental problems in the reserve, (2) identifying types of illegal activities being conducted in the reserve and (3) identifying possible strategies to combat those illegal activities. The study interviewed approximately 30.2% of the population (n = 42) in the selected communities and adopted a qualitative analysis with Chi-square tests.

Two other studies were conducted in Mua-Livulezi FR in Dedza district. Senganimalunje, Chirwa and Babalola (2015) examined the potential and effectiveness of local institutions and institutional arrangements associated with co-management. Senganimalunje et al. (2016) evaluated the effect of PFM on community organization, forest access, forest use, product availability and commercialization of forest products. Both studies used data collected with mixed social science research methods in about 30% of the villages that were part of the targeted co-management, covering about 49% of the population (n = 300) in the community.

In 2015, Remme et al. (2015) conducted a review of the IFMSLP. The review critically assessed PFM in Malawi and identified lessons learnt and best practices based on a mixed-method approach that involved both primary and secondary data. The exercise focused on obtaining evidence-based conclusions, which also contributed insights into the PERFORM project. Site visits were done in all the 12 districts that were implementing the IFMSLP. This included 25 co-management blocks, 16 VFAs and 40 forest-based enterprises. GIS and remote sensing techniques were used to assess changes in forest cover over time.

In addition, in 2018–2019, two other studies were conducted in Mua-Livulezi FR and Ntchisi FR in Dedza and Ntchisi districts, respectively. Djenontin, Zulu and Ligmann-Zielinska (2020) analyzed the nature of the decisions to engage in restoration at individual (farm-household) level and in collective actions (community-level). This study used a mixed-methods approach with qualitative data from 7 focus group discussions and role-playing games, and quantitative data from a household survey of 480 participants. Then, Djenontin and Zulu (2021) investigated the structure and functions of the current governance system supporting collective landscape-scale agro-forest resources restoration. The authors used a polycentric governance system lens, implemented through a novel theory of polycentric governance – the Ecology of Games Theory. This analysis was based on 35 focus group discussions with local level resource-governance bodies leading restoration efforts, 21 key informant interviews with district level officers and local TAs, and 16 such interviews with national level officers.

In 2019, Gondwe et al. (2019) assessed woodland/forest cover status through Land use/land cover (LULC) classification across Malawi and compared forest cover within and between forest governance strategies, including 11 co-management and 12 government-managed FRs between 1999 and 2018.

Together, the portfolio of studies examined cuts across the country, in north-south and east-west directions, with a variety of local communities exhibiting a diversity of ethnicities. Insights from these studies on PFM implementation and outcomes and subsequent PFM dynamics contributed to provide an overview of a contextualized ACM in Malawi.

PFM performance and outcomes

This section critically assesses the performance of PFM in Malawi in accordance with the requirements of good forest governance in achieving the intended outcomes as designed by Malawi's DoF for its relevant projects and initiatives. Mainly, IFMSLP had four results areas: (1) sustainable livelihood strategies promoted within impact areas; (2) equitable access to forest resources secured by increasing the area under SFM arrangements; (3) strengthened governance of key forest resources within the forest sector and (4) communication and advocacy enhanced among stakeholder groups along with administrative and technical support. By giving communities legal rights to access and use forests sustainably, PFM is thought to potentially provide opportunities and capabilities for accessing the different forms of capital that forest and forest systems provide and support. More broadly, these PFM goals are in fact integral to, and respond directly to, the recent COP26 Declaration on forests' importance in addressing climate change.

One outstanding feature of the results presented here is the variation in the conclusions from one study to another. This is not an uncommon problem in assessing ACM-style efforts, but it is one that interferes significantly with 'proving' the nature, extent and quality of ACM impacts. One example of this problem, highlighted, for example, in the section on livelihoods and welfare, revolves around definitions. What exactly is meant by livelihoods? For some research, only monetary income directly attributable to restoration may be considered; yet from an ACM livelihoods perspective, one needs to look more holistically at a family's access to the various sources of subsistence 'income'. This recurrent definitional problem is evident in our material below.

Performance of PFM with regard to improving livelihoods and welfare

Since the implementation of IFMSLP, a number of studies have been conducted to assess the impacts of PFM on forest communities' livelihoods and welfare. No or minimal positive impacts of PFM in terms of livelihood improvements of forest communities have been reported (Chinangwa, Pullin and Hockley 2016). Authors have reported varying perceptions of communities on PFM livelihood outcomes. For example, Chinangwa, Pullin and Hockley (2016) reported that approximately 57% of Zomba-Malosa communities and 71% of Ntchisi FR communities perceive that the IFMSLP has had no impact on their livelihoods. Similar findings were reported by other authors. Zulu (2013) and Senganimalunje, Chirwa and Babalola (2015) reported that overall project outcomes from all 12 FRs under the IFMSLP showed significant progress toward meeting some quantitative targets, but income generation was generally disappointing. Low and generally disappointing cash benefits burdened poor communities with conservation costs and created perverse incentives to overharvest forest resources. These findings imply that contrary to the IFMSLP plan of improving livelihoods and welfare of poor communities, communities in some instances were made worse off.

The measure used to assess livelihoods was very narrow, but there were other livelihoods benefits (Colfer 2005; Senganimalunje, Chirwa, and Babalola 2015). Firewood was the dominant forest enterprise for the FRs, many of which had wood extraction and marketing challenges. Money-generating activities in FRs, such as firewood sales and pottery that the co-management program initiated were of low value. The minimal positive impact on livelihoods creates uncertainties for the program's long-term success with regard to livelihoods (Chinangwa, Pullin and Hockley 2016).

Zulu (2013), for example, found that the formalized firewood group in Ntchisi only raised Malawi Kwacha (MK) 12,996 or ~US$95 net from firewood sales after paying MK9,000 in license fees to the BMC, without accounting for labor costs; and group members received only MK 1,000 (a measly US$ 7.32) for 15 weeks of work, two days weekly. Consequently, the firewood group abandoned commercial firewood production after one (2009) harvesting season.

However, IFMSLP has helped forest communities to attain new income sources, such as (1) wage labor during firebreak construction and maintenance, (2) income-generating activities, for example, sale of timber, firewood, pottery (clay pots), as well as bee-keeping and mushroom farming, and (3) indirect benefits in the form of dry season irrigated agriculture. These activities were not directly spelled out in IFMSLP plans as drivers of livelihoods improvement; instead, they provide evidence of the importance of the ACM emphasis on responsiveness to local conditions and opportunities; Emphasis was put instead on sale of forest resources such as firewood and other non-timber forest products (NTFPs).

Performance of PFM with regard to strengthening forest management institutions

Institutions are defined as systems of established, prevalent and social rules that structure social interactions. Local resource management institutions exist to control resource governance in order to ensure sustainability and reduce problems of access to the resources in and around the local communities (Chingaipe et al. 2015). In this case, formal local forest institutions are mandated to make decisions in a participatory manner, a role that development agencies have often assumed cannot be done by communities (Chingaipe et al. 2015). To participate effectively in forest management, these institutions need to develop SFM capacities.

Zulu (2013) reported some progress in capacity building of forest communities. For example, in Ntchisi, the IFMSLP invested in communities focusing on exchange visits for forest users, on-demand training for diverse skills, and grants of basic start-up forest management tools and enterprise. The program also invested in extension-staff training focusing on gaps in social facilitation skills, forest policy, participatory indigenous forest management and enterprise development. The series of capacity building activities significantly improved the organizational, institutional and technical capacity of communities and extension staff.

Such institutional capacitation is critical as many recognize the importance of strengthening local institutions in adapting to and even mitigating climate change – an additional environmental risk that both forest resources and local populations bear. Effective mobilization of and response to local communities' interests in addressing climate change will require the kinds of institutional strengthening that PFM has tried to promote.

In contrast, despite efforts by the IFMSLP to build the capacity of local institutions in SFM, local institutions did not attain adequate knowledge in fire management and management of regeneration, for instance (Chingaipe et al. 2015). Other studies have found similar results; there is a knowledge gap in forest communities due to the complexity of forest management and the PFM model, and inadequate training (USAID-Malawi 2015). Managing forests using the PFM model requires financial resources for both natural and human capital.

Despite having clear roles and responsibilities for co-management as defined by IFMSLP using the Standards and Guidelines for PFM, local forest institutions at the community level and DFO staff are challenged to carry them out in practice due to lack of capacity. Unsustainable forest management and unacceptable harvest levels in a few co-management blocks were observed (USAID-Malawi 2015).

Another issue that constrains efforts to strengthen forest management institutions is trust building and transparency. Participation of communities in co-management activities is highly influenced by communities' level of trust in co-management leadership with regard to financial accountability and transparency (Chinangwa, Pullin and Hockley 2016). Most participants in co-management programs in Malawi perceive benefit distribution as unfair and that only a few influential members of the community, for example, committee members and chiefs, obtain benefits (Chinangwa, Pullin and Hockley 2016). For example, both Zomba-Malosa (87%) and Ntchisi (72%) forest users who had indicated no willingness to participate in co-management attributed their decision to lack of benefits from the program and lack of trust in the leadership with regard to financial accountability and transparency.

Performance of PFM with regard to communities' participation in forest management

A policy goal under PFM is to empower rural communities to conserve and develop Malawi's forest resources for the economic and environmental benefit of the present and future generations. This requires transferring certain management responsibilities to the community level; this decentralized form of natural resource management is seen as a mechanism for sustainability as local communities are empowered to make decisions over natural resource use (Kamoto et al. 2013; Senganimalunje, Chirwa and Babalola 2015). High levels of social capital are seen to increase collective action and conformity to rules required for long-term sustainability.

Kamoto et al. (2013) and Remme et al. (2015) reported that in Malawi, forest management had been dominated by men and efforts have been made to encourage women's participation, particularly in PFM. A significant achievement was obtained when it was decided that in most community bylaws a quota of positions in the forest committees (VNRMC and BMC) would be for women. In some cases, 50% and in others 30% women's representation was stipulated. Remme et al. (2015) reported that women's participation in forestry activities under IFMSLP and other projects was generally high, especially with respect to forest produce and services that are important for household needs, including the collection of fuel wood, and NTFPs. Men appeared generally more interested in commercial activities such as timber production or securing water for irrigation. Despite the high level of women's participation in forestry activities, their representation in decision-making and leadership positions remained relatively low. Such a situation is generally due to cultural factors that place women as subordinate to men. In addition, women are time-poor and already overburdened by household management requirements. These authors also reported that the IFMSLP and other projects contributed to a high level of participation by women in PFM but many of the committees and leadership positions were dominated by men. They recommended that there was a need for further participation and a greater role of women in decision-making structures. Despite this finding, in some atypical areas, especially in Zomba, a great majority of women were found to participate, including as officials. For example, Mtogolo BMC had 67 members, of whom 64 were women, and the chairperson, treasurer and secretary were all women. The explanation given was that women are prime beneficiaries/users of forest products, while men were more concerned with earning an income and were also often the ones involved in illegal activities. Although women were mostly well represented, their decision-making power could be constrained by cultural conditions as the men were usually most vocal during meetings.

Literature indicates that communities participate more in PFM when they have a village forest (Banana et al. 2012; Chang and Andersson 2021; Mukasa et al. 2012; Pham et al. 2021; Riggs et al. 2021; Schweizer et al. 2021; Wilson and Cagalanan 2016; Zulu 2013). Zulu (2013) reported that having a VFA had significant synergistic effects on communities' participation in various co-management activities under IFMSLP. He found that communities that are well organized and have strong social capital and traditional leadership are more successfully implementing PFM.

Under IFMSLP, local ownership of FRs was high in Ntchisi (Nyanja) with 69% of community members considering communities to be the owners of the assigned block, while a few (29%) cited the government as the owner. As an indicator of communities' participation in forest management, one in three respondents had informed DoF staff or BMC members of rule-breaking by others. Approximately 88% of those who had participated in co-management (46.2%) were very satisfied with their participation and 91% of respondents were willing to continue or start participation in co-management (Zulu 2013).

Other authors have reported similar findings from other IFMSLP impact areas. Although a majority of communities perceive the program did not economically benefit them, approximately 83% (Zomba-Malosa) and 81% (Ntchisi) of respondents were willing to pay membership fees to participate in the forest co-management program. With approximately MK 1,000 (US$ 3.5) in Zomba-Malosa and MK 400 (US$ 1.4) in Ntchisi mean annual willingness to pay, it can be argued that the estimated willingness to pay is due to communities' optimism of future benefits that forest recovery could potentially provide.

However, other authors have observed different levels of communities' participation in co-managed FRs and VFAs with the latter enjoying more community participation. USAID-Malawi (2015) reported that community ownership is considered stronger on VFAs, where stricter community rules are put in place and enforced by traditional leaders. It can therefore be argued that communities participate more in VFAs and prefer to deplete forest resources in FRs' blocks rather than on VFAs. Having FRs further away from villages, and in some cases large forest blocks, negatively affects participation levels of communities. In the case of the VFA, the VNRMC performs control and patrolling, fire break construction, and weeding and slashing. In co-management blocks of Chimaliro Forest Reserve where PAR was facilitated, the BMCs intensified monitoring of resources through patrols and through the reflective and learning cycles of PAR. This collaborative monitoring had lasting impacts on improving honey production and communities' livelihoods. This was because the illegal thefts of honey were completely halted by the rigorous monitoring and learning sessions.

Despite IFMSLP efforts to include forest communities at all levels of forest management, some authors have reported exclusion or weak participation of communities at the forest management policy level. Communities are rarely consulted at the policy level, and when consulted their contributions are not taken on board (USAID-Malawi 2015). Lack of involvement of communities at policy levels has resulted in some activities, such as harvesting of forest resources, being implemented with little knowledge of communities and consequently not implemented according to co-management plans. Kamoto et al. (2013) had similar findings and argued that when voices of local authorities, communities and NGOs are not heard during policy formulation, policies are implemented with little knowledge of the institutions already in place in local communities. Such findings can be attributed to lack of effective facilitation of the reflective learning embedded in ACM/PAR and the standards and guidelines for PFM by the extension agents. Effective and efficient facilitation of the processes of ACM or the Standards and Guidelines for PFM should empower communities to develop strategies to overcome any challenge in forest management.

Incentives for communities' participation in sustainable forest management

Overemphasis on cash incentives, initially considered the primary motivation for forest communities to participate in co-management, overlooks locally significant non-cash motivations, inflates local expectations and creates perverse incentives that undermine socio-ecological goals of PFM.

Despite the project focus on income as the overriding incentive for communities' participation in co-management, non-cash motivations relating to the rainfall regulation role of forests (forests are seen to bring rain) or rights-based issues of equitable access to forest resources emerged as more important (Djenontin, Zulu and Ligmann-Zielinska 2020; Kamoto 2007; Kamoto and Milner 2003; Remme et al. 2015; Zulu 2013). For instance, Kamoto and Milner (2003) showed that, in a situation of multiple and overlapping claims on land, negotiations on land rights and learning changed the trajectory of forest management in Chimaliro Forest Reserve and Ntonya Hill, the two earliest ACM sites. In both sites, the government had allocated forest land previously under their domain to communities for co-management; however, the government did not follow ancestral land rights when allocating the land. This was a contentious issue among communities and proved problematic. However, after facilitation of PAR using the worm, the communities understood the government rationale and began again managing the allocated forest land effectively. This finding is supported by the second-generation commons theory, which states that humans have the need and ability to cooperate for broader social benefits or altruistic motivations under certain conditions. In the case above, the community benefits and motivation were their ability to access ancestral forest land with its resources, which had been denied for decades.

While incentives encouraged local interest and desire to be involved in co-management, expectations were raised and acted as an additional focus and opportunity for elite capture. Some authors have suggested that the use of incentives to encourage individuals and communities to implement policy and activities devised and brought in from 'outside' can be interpreted by local people as 'payment' for doing what a project or government wants (Kamoto et al. 2013; see also Chapter 9); and that this has become divisive and the source of unhealthy competition among both community members and NGOs (Djenontin and Zulu 2021).

Performance of PFM with regard to ecological health and other environmental outcomes

Since the implementation of IFMSLP, forests under PFM have enjoyed more abundant tree species than non-PFM forest. Restrictions imposed on access to forest products found in the VFA's under PFM resulted in higher tree species richness. It appears that co-management may be an effective method to manage/protect the reserve – certainly more so than bureaucracy-based management and community-only management, i.e., CBFM (Mtambo and Missanjo 2015; Zulu 2013).

The PFM approach has provided sufficient incentives and consensus to promote behavioral change reducing deforestation and forest degradation in forest areas. Approximately 32% and 24% of respondents in Zomba-Malosa and Ntchisi, respectively, attributed the reduction in access to forest resources to the strict laws and regulations being enforced under the co-management program (Chinangwa, Pullin and Hockley 2016). Illegal forest activities have declined, and this could indicate and foster forest resource restoration and better management of FRs in Malawi.

Such positive outcomes are directly relevant to the recent COP26 Declaration, which explicitly emphasizes resource restoration and reducing deforestation and forest degradation as key elements in addressing climate change, despite the short- and long-term trade-offs that need to be factored in (Miller et al. 2021).

However, other authors suggest that co-management has coincided with declining forest resources due to increased illegal use of the forest resources. Increased time used to collect desired forest resources compared to the past five years, as reported by communities, could indicate forest degradation, despite the co-management program (Senganimalunje, Chirwa and Babalola 2015). Other studies supported the argument by mentioning that despite forest co-management, forest resources have dwindled due to high illegal forest activities (charcoal production, timber sawing and encroachment for farming), which accounted for 53.5% of the illegal activities. Further, the empirical comparison of forest condition in selected co-managed and solely DoF-managed FRs between 1999 and 2018 by Gondwe et al. (2019) showed no advantage for co-management. Their study reveals that forest cover declined by 37% in co-managed FRs, with 9 of 11 FRs studied showing declines; and by 11.6% in DoF-management FRs, with 10 of 12 FRs declining in cover. USAID-Malawi (2015) also reported patches of deforestation in Machinga Forest Reserve.

The seemingly conflicting findings across studies point to common challenges in assessing environmental outcomes of forest management and governance with much certainty. This also highlights issues of temporal and spatial scales of assessment and of methodological choices/approaches of impact evaluations. Improving assessment and accuracy of environmental and ecological status of forest resources, using robust methods combining GIS, remote sensing and appropriate modeling approaches, is important to understand more fully the role of PFM experiences in Malawi, so as to build on the positive and reduce the negative. In addition, the conflicting findings can be explained by how well the facilitation of co-management/CBFM was done by the government to enhance social learning. The experience in facilitating ACM shows that the processes require investment in time, financial resources and human capacity (knowledge and skills). In the absence of any of these, despite having well-crafted standards and guidelines for PFM, the results may be varied.

Lessons learned and best practices

This section critically discusses the lessons learned and the best practices for improving the implementation of PFM for achieving SFM and FLR, including addressing climate change:

- **Continuity of support:** PFM is a complex process that requires proper introduction, facilitation and mentoring of communities over a longer period. Experience from IFMSLP shows that intensive capacity building, financial support and monitoring are required. A target-oriented and rushed approach, trying to reach too many areas within a short time, contributes to underperforming and weak local PFM structures. Discontinuity of support for some time can also have negative effects on people's motivation and be disruptive. An analysis of blocks and VFAs in the IFSLMP showed that the areas that were more frequently visited by the extension officers did better than those left on their own. It is important that extension officers be in frequent contact with the BMCs and VNRMCs and share responsibilities.
- **Forest management plans**: These plans are necessary tools for PFM. However, the development and approval process are time-consuming and have required enormous inputs from DoF staff at the cost of providing direct support, coaching and monitoring of the BMCs and VNRMCs. Experience from IFMSLP shows that the delays in approval of the forest management plans have been a cause of frustration for the involved communities who would get demoralized over the long wait time. However, learning from this experience, the DoF has devolved some of its functions of forest management to district councils and therefore no longer requires the Director of Forestry to sign the forest management plan; it can now be signed at the district level. This will expedite the process and reduce frustrations.
- **Impact area:** The selection of FRs for co-management requires an adequate needs assessment and prioritization based on clear criteria. The argument of some officials in the DoF that co-management should only be undertaken in FRs that face a lot of pressure and problems of encroachment might seem logical but the chances for success are much lower than for FRs that are still in better condition. Examples in the North, such as in Chitipa where the FRs are still intact, show that co-management works well. We argue that it is better to assess the conditions and prioritize based on a set of criteria that includes better returns to communities for managing the forest than applying a one-size-fits-all approach, based on levels of forest degradation.
- **Harmonization of extension approaches:** There is a strong need for harmonization of extension approaches. Many organizations provide handouts or

pay farmers for project activities, such as seedling production or tree planting. The use of incentives to encourage communities to implement activities can be interpreted by local people as 'payment' for doing what a project or government wants. This has created a dependency syndrome, a tendency to forget the people's own agency (see similar conclusions in Chapter 6, this volume).

- **Capacity building:** PFM is a long-term process and would benefit from ongoing capacity development and mentoring. Through IFMSLP, Malawian institutions (Malawi College of Forestry and Wildlife, Bunda College) were trained to build capacities which contribute to sustainability. However, there is a need for in-service capacity development programs within the DoF and other relevant institutions to ensure that all field staff are trained in PFM. In addition, as TAs, group village headmen and village heads (customary land authorities) are considered critical in the effective implementation of PFM, consequently ongoing capacity development is necessary for TAs and chief structures, to strengthen their leadership and organizational capacity to support PFM.
- **Multi-sector approach**: The IFMSLP experience shows that the use of collaborative stakeholder platforms in the district, such as the District Environmental Sub-Committee, is useful but also inadequate if these institutions are not actively involved. PFM must not only rely on the DoF but should involve a multi-sectoral approach, including state and non-state actors to ensure that adequate expertise is provided and resources are shared. This is especially important if a more holistic approach is followed that also considers alternative (non-forest-based) livelihood activities and that recognizes the important interlinkages among forest management, other livelihood resources, and addressing climate change.
- **Income-generating activities:** The focus of PFM should not be predominantly on income generation but should equally consider other non-cash-based and environmental objectives and benefits from forest management for the community, including sustained access to firewood and NTFPs and continued water supply for consumption and irrigation especially where access to these resources is denied in the absence of PFM. A narrow pursuit of cash-based benefits also creates incentives to overharvest and to focus predominantly on a few activities. The 'tangible' benefits that are often referred to as a necessary pre-condition for communities to enter into PFM do not have to be direct cash. Intangibles are also important to people.
- **Factors influencing PFM performance:** Experience from IFMSLP indicates that there are important conducive as well as disturbing factors that influence the performance and outcomes of PFM. The main factors are grouped into four clusters. The first cluster refers to the PFM support mechanisms, i.e., the approach followed, the quality of services delivery systems, and the available resources and institutional capacity to support the PFM process. The second cluster refers to the conducive environment to support PFM in terms of policies and legal framework. The third cluster refers to the community-level factors that influence PFM implementation; and the last cluster refers to the

local external conditions that facilitate or hamper PFM. The performance of ACM was related to the four clusters above in the sense that the early ACM had a support mechanism as a CIFOR-funded project, and enormous support mechanisms were put in place for all processes from human resources and site selection and other initial processes, including understanding the policy environment through background studies. ACM processes of visioning, building future scenarios and PAR using the worm in reflective cycles made the facilitators understand the community and external factors that influenced adaptive collaborative management of forest resources. The community and facilitators were all immersed in the process at regular intervals and therefore this enhanced the social learning. PFM, if implemented with the same rigor, might yield better sustained outcomes.

Conclusion

Despite the conflicting evidence in this chapter, we have identified a number of positive impacts from the PFM approach as implemented through IFMSLP in Malawi. A crucial one has been the transformation of the relationship between many communities and the DoF. In many places, the approach has provided incentives[2] and consensus to promote behavioral change, thereby reducing deforestation and forest degradation. ACM, which inspired PFM implementation in Malawi, aims to level the playing field, resolve conflicts, foster collaboration and negotiation, build skills and capacities and promote gender equality even among communities with diverse views. This is the lesson that was drawn and that inspired PFM in Malawi. The same ACM approach that was learnt and applied in PFM at a broader scale has opened up opportunities to improve local livelihoods and demonstrated gains to SFM, especially on the restoration of degraded forests in customary land forests and co-management of FRs. However, experience from IFMSLP indicates that despite the forward progress, many improvements in the performance and outcomes of PFM are possible. Therefore, we need to continuously apply the ACM concept which is aimed at catalyzing change while continuously monitoring performance and consciously learning from it. Policy makers and practitioners need to embrace the ACM-inspired standards and guidelines for PFM by more explicitly and consistently implementing the reflective cycles at all levels, so that the learning can more consistently inform the next steps.

Notes

1 These combinations of qualitative and quantitative approaches are replicated in the work of Mukasa et al. (2022) and Bomuhangi et al.'s (2022), qualitative and quantitative analyses of ACM impacts, respectively, in Uganda.

2 For example, Kamoto et al. (2013) reported that some village heads established VFAs because they were envious of the incentives from NGOs and development partners such as irrigation equipment, boreholes, training of VNRMCS and subsequent daily subsistence allowances that go with it, among other benefits that other villages received.

References

AAS (Air Asia Survey). 2012. "Forest Resource Mapping Project under the Japanese Grant for the Forest Preservation Programme to Republic of Malawi: Final Report for Implementation Phase." Zomba, Malawi: Department of Forestry.

Banana, A.Y., M. Bukenya, E. Arinaitwe, B. Birabwa, and S. Ssekindi. 2012. "Gender, Tenure and Community Forests in Uganda." *CIFOR Working Paper* 87. Bogor, Indonesia: CIFOR.

Banda, W., T. Senganimalunje, and E. Missanjo. 2015. "Community Attitudes and Perceptions Towards Management of Kaning'ina Forest Reserve in Malawi." *Journal of Basic Applied Research International* 8, no. 1: 34–40.

Bomuhangi, A., A.Y. Banana, J. Bushoborozi, C. Mukasa, A. Tibazalika, and E. Mwangi. 2022. "Gender and Adaptive Collaborative Management in Forested Ugandan Landscapes." In *Adaptive Collaborative Management in Forest Landscapes: Villagers, Bureaucrats and Civil Society*, edited by Carol J. Pierce Colfer, Ravi Prabhu, and Anne Larson, 87–104. London: Earthscan and Routledge.

Chang, K., and K. Andersson. 2021. "Contextual Factors that Enable Forest Users to Engage in Tree-Planting for Forest Restoration." *Land Use Policy* 104: 19.

Chinangwa, L., A.S. Pullin, and N. Hockley. 2016. "Livelihoods and Welfare Impacts of Forest Co-management." *International Journal of Forestry Research* 12. http://dx.doi.org/10.1155/2016/5847068.

Chingaipe, B., L. Mwabumba, E. Missanjo, and T.C. Senganimalunje. 2015. "Effectiveness of Local Institutions in Sustainable Management of Forests in Selected Communities in Dedza, Malawi." *Journal of Basic and Applied Research International* 12, no. 3: 181–189.

Colfer, C.J.P. 2005. *The Complex Forest: Communities, Uncertainty, and Adaptive Collaborative Management*. Washington, DC: Resources for the Future.

Dhakal, M., and M. Masuda. 2008. "Role of Participatory Forest Management in Rural Livelihood Improvement: A Case of Nawalparasi District, Nepal." *Tropics* 17, no. 2: 159–168.

Djenontin, I.N.S., and L.C. Zulu. 2021. "The Quest for Context-Relevant Governance of Agro-Forest Landscape Restoration in Central Malawi: Insights from Local Processes." *Forest Policy and Economics* 131: 1–24.

Djenontin, I.N.S., L.C. Zulu, and A. Ligmann-Zielinska. 2020. "Improving Representation of Decision Rules in LUCC-ABM: An Example with an Elicitation of Farmers' Decision Making for Landscape Restoration in Central Malawi." *Sustainability* 12: 5380. http://dx.doi.org/10.3390/su12135380.

Gobeze, T., M. Bekele, H. Lemenil, and H. Kassa. 2009. "Participatory Forest Management and its Impacts on Livelihoods and Forest Status: The Case of Bonga Forest in Ethiopia." *International Forestry Review* 11, no. 3: 346–358.

GoM (Government of Malawi). 1965. "Land Act." Zomba, Malawi: Government Press.

GoM (Government of Malawi). 2017. "Forest Landscape Restoration Opportunities Assessment for Malawi." Lilongwe, Malawi: Government Press.

GoM (Government of Malawi), Department of Forestry. 2005. "Standards and Guidelines for Participatory Forestry in Malawi." Zomba, Malawi: Government Press.

GoM (Government of Malawi), Ministry of Lands and Housing. 2002. "Malawi National Land Policy." Zomba, Malawi: Government Press.

GoM (Government of Malawi), Ministry of Mines, Natural Resources and Environmental Affairs. 1997. "Forestry Act." Zomba, Malawi: Government Press.

GoM (Government of Malawi), Ministry of Natural Resources. 1996. "National Forest Policy of Malawi." Zomba, Malawi: Government Press.

GoM (Government of Malawi), Ministry of Natural Resources. 2016. "National Forest Policy of Malawi." Zomba, Malawi: Government Press.

GoM (Government of Malawi), Ministry of Natural Resources and Environmental Affairs. 2003. "Community-based Forest Management: A Supplement to the National Forest Policy of Malawi." Zomba, Malawi: Government Press.

Gondwe, M.F., M.A. Cho, P.W. Chirwa, and C.J. Geldenhuys. 2019. "Land Use Land Cover Change and the Comparative Impact of Co-Management and Government-Management on the Forest Cover in Malawi (1999–2018)." *Journal of Land Use Science* 14, no. 4–6: 281–305.

Hajjar, R., J.A. Oldekop, P. Cronkleton, P. Newton, A.J. Russell, and W. Zhou. 2021. "A Global Analysis of the Social and Environmental Outcomes of Community Forests." *Nature Sustainability* 4, no. 3: 216–224.

Hardin, G. 1968. "The Tragedy of the Commons." *Science* 162, no. 3859: 1243–1248.

Islam, K.K., K. Hyakumura, M. Tani, M. Krott, and N. Sato. 2014. "Actors' Power, Livelihood Assets and Participatory Forestry in Bangladesh: Evidence from the Sal Forests Area." *Open Journal of Forestry* 4, no. 5: 1–9.

Iversen, V., B. Chhetry, P. Francis, M. Gurung, G. Kafle, A. Pain, and J. Seeley. 2006. "High Value Forests, Hidden Economies and Elite Capture: Evidence from Forest User Groups in Nepal's Terai." *Ecological Economics* 58, no. 1: 93–107.

Kamoto, J. 2007. "Improving Forest Beekeeping through Monitoring in Chimaliro, Malawi." In *Negotiated Learning: Collaborative Monitoring for Forest Resource Management*, edited by I. Guijt, 105–113. Washington, DC: Resources for the Future.

Kamoto, J., G. Clarkson, P. Dorward, and D. Shepherd. 2013. "Doing More Harm than Good? Community Based Natural Resource Management and Neglect of Local Institutions in Policy Development." *Land Use Policy* 35: 293–301.

Kamoto, J., and J. Milner. 2003. "Negotiating Multiple and Overlapping Claims on Land Rights: Experiences from Malawi." Paper Presented at the Hawaii International Conference on Social Sciences, Honolulu, Hawaii.

Kibria, A.S.M.G., M. Jashimuddin, and I. Makoto. 2014. "Effects of Participatory Forest Management on Livelihood Capitals of the Community in Cox's Bazar, Bangladesh." *Journal of Forest Research* 19, no. 1: 42–51.

Klooster, D. 2000. "Institutional Choice, Community, and Struggle: A Case Study of Forest Co-Management in Mexico." *World Development* 28, no. 1: 1–20.

Matiku, P., M. Caleb, and O. Callistus. 2013. "The Impact of Participatory Forest Management on Local Community Livelihoods in the Arabuko-Sokoke Forest, Kenya." *Conservation and Society* 11, no. 2: 112–129.

Mauambeta, D.C., D. Chitedze, R. Mumba, and S. Gama. 2010. "Status of Forests and Tree Management in Malawi: A Position Paper Prepared for the Coordination Union for Rehabilitation of the Environment (CURE)." Blantyre, Malawi.

Mbuvi, M.T.E., J.O. Maua, P.O. Ongugo, C.K. Koech, R.A. Othim, and J.K. Musyoki. 2009. "Status of Participatory Forest Management Impacts on Poverty for Buyangu Non-PFM area Adjacent Community: Kakamega Forest: Kakamega District. Status of Participatory Forest Management Impacts on Poverty for Buyangu Non-PFM Area." Nairobi, Kenya: KEFR.

Miller, D.C., S. Mansourian, M. Gabay, R. Hajjar, P. Jagger, J. Kamoto, P. Newton, J. Oldekop, O.H. Razafindratsima, P. Shyamsundar, T. Sunderland, and C. Wildburger. 2021. "Policy Implications of Current Knowledge on Forests, Trees, and Poverty Alleviation." *Forest Policy Economics* 131: 102566. https://doi.org/10.1016/j. forpol.2021.102566.

Mtambo, C., and E. Missanjo. 2015. "The Impact of Participatory Forest Management on Tree Species Abundance and Diversity in Selected Village Forest Areas in Kasungu, Malawi." *Journal of Ecology and Environmental Sciences* 3, no. 2: 15–20.

Mukasa, C., A. Tibazalika, A. Mango, and H.N. Muloki. 2012. "Gender and Forestry in Uganda: Policy, Legal and Institutional Frameworks." *CIFOR Working Paper* 89. Bogor, Indonesia.

Mukasa, C., A. Tibazalika, A.Y. Banana, E. Mwangi, and T. Mutimukuru-Maravanyika. 2022. "Strengthening Women's Tenure Rights and Participation in Community Forestry in Central Uganda." In *Adaptive Collaborative Management in Forest Landscapes: Villagers, Bureaucrats and Civil Society*, edited by C.J.P. Colfer, R. Prabhu, and A. Larson, 106–132. London: Routledge and Earthscan.

Olivier, R., and W. Mwase (2012). *Final Evaluation of Improved Forest Management for Sustainable Livelihoods Programme in Malawi*. Zomba, Malawi: Department of Forestry.

Ostrom, E. (1990). *Governing the Commons: The Evolution of Institutions for Collective Action*. New York: Cambridge University Press.

Pham, T.T., F. Kengoum, M. Moeliono, and B. Dwisatrio. 2021. "Forest Governance in DRC: An Analysis from Actors' Participation in REDD+ Policy Development." *International Forestry Review* 23, no. 1: 79–89.

Remme, H., F. Muyambi, J. Kamoto, and E. Dengu. 2015. "A Technical Review of Community-Based Forest Management on Both Customary Land and Forest Reserves (Participatory Forest Management)." Zomba, Malawi: Department of Forestry, Malawi.

Riggs, R.A., R. Achdiawan, A. Adiwinata, A.K. Boedhihartono, A. Kastanya, J.D. Langston, H. Priyadi, M. Ruiz-Pérez, J. Sayer, and A. Tjiu. 2021. "Governing the Landscape: Potential and Challenges of Integrated Approaches to Landscape Sustainability in Indonesia." *Landscape Ecology* 36: 2409–2426.

Schweizer, D., P. Meli, P.H.S. Brancalion, and M.R. Guariguata. 2021. "Implementing Forest Landscape Restoration in Latin America: Stakeholder Perceptions on Legal Frameworks." *Land Use Policy* 104: 9.

Senganimalunje, T.C., P.E. Chirwa, and F.D. Babalola. 2015. "Potential of Institutional Arrangements for Sustainable Management of Forests under Co-Management with Local Forest Organisations in Mua-Livulezi Forest Reserve, Mtakataka, Malawi." *International Forestry Review* 17, no. 3: 340–354.

Senganimalunje, T.C., P.E. Chirwa, F.D. Babalola, and M.A. Graham. 2016. "Does Participatory Forest Management Program Lead to Efficient Forest Resource Use and Improved Rural Livelihoods? Experiences from Mua-Livulezi Forest Reserve, Malawi." *Agroforestry Systems* 90, no. 4: 691–710.

USAID-Malawi. 2015. "Protecting Ecosystems and Restoring Forests in Malawi (PERFORM): Governance Assessment of Participatory Forest Management in Malawi." Lilongwe, Malawi: Tetra Tech.

USAID-Malawi. 2021. "Modern Cooking for Healthy Forests in Malawi: Fiscal Year 2021 Second Quarter Report (January – March 2021)." Lilongwe, Malawi: Tetra Tech.

Wilson, S.J., and D. Cagalanan. 2016. "Governing Restoration: Strategies, Adaptations and Innovations for Tomorrow's Forest Landscapes." *World Development Perspectives* 4: 11–15.

Zulu, L. 2013. "Bringing People Back into Protected Forests in Developing Countries: Insights from Co-Management in Malawi." *Sustainability* 5: 1917–1943.

Introduction to Chapter 9

This chapter on Uganda, like Chapter 8 on Malawi, examines a formal governmental attempt, sharing features with Adaptive Collaborative Management (ACM), to link communities and the government in forest management. The Collaborative Forest Management (CFM) legislation specifies that Non-Governmental Organizations (NGOs) are the prime supporters of the program, helping communities to develop the requisite formal agreements and strengthen community management capabilities. CFM legislation pre-dates the development of CIFOR's version of ACM but shares some features with it. Here, Egunyu focuses on a particular Ugandan community, where four NGOs have been instrumental in implementing CFM. She strives to better understand the roles the NGOs have played and their influence on the endurance of CFM over time.

The frequency with which NGOs' assistance is required in developing countries makes an examination of their performance of wider interest. Egunyu begins with an introduction to CFM and a comparison of the similarities and differences between idealized CFM and ACM. She then describes CFM implementation in Uganda. This is followed by an account of the potential roles of NGOs in forest management. ACM has also been implemented elsewhere with the aid of NGOs. This section is followed by the methodology which includes a description of a local Ugandan community-based organization implementing CFM as well as data collection and analysis approaches. After presenting the results of her study, Egunyu discusses how the two approaches (ACM and CFM) play out in practice and what scholars and practitioners can learn from the implementation of CFM in Uganda.

As we move forward in our attempts to strengthen cross-scale interactions and influence, these examples of the institutionalization of collaboration are particularly pertinent. Egunyu highlights the difficulties that NGOs have had sustaining their financial solvency, and its importance in maintaining their community-level activities – providing further evidence of the importance of longer-term funding for ACM-like approaches.

Although Egunyu does not expand on this, we suspect that providing training to NGO personnel in the kinds of facilitation skills highlighted in Chapter 6 (Hagmann et al.) may be a valuable way to strengthen (a) links between

DOI: 10.4324/9781003325932-16

communities and outside actors (whether NGOs or government or donors) and (b) buy-in from communities as they implement agreed-upon management strategies.

Another possibility, mentioned only in passing, is the simplification of legislation – so communities have less trouble interpreting and using it (reducing the load on NGOs); and increasing its flexibility, for locally determined visioning, attention to local social structural elements and incorporation of relevant indigenous knowledge.

9 Collaborative Forest Management in Uganda

Policy, implementation, and longevity

Felicitas Egunyu

Introduction

Natural forests are being lost at an alarming rate in Uganda (FAO 2020; Global Forest Watch 2020; Ministry of Water, Land and Environment (MWLE) 2016). For example, in 1990, Uganda had a forest cover of 3,575,000 ha; in 2020, this had dropped to 2,338,000 ha (FAO 2020). Forest cover loss in Uganda is being driven by agricultural expansion, fuelwood harvest, charcoal production, illegal logging, policy failures, and a large rural population that looks to forests for non-timber forest products (NTFPs) to meet their needs and supplement their livelihoods (MWLE 2016). The Ugandan government subscribes to the principles of sustainable forest management, an element that the Conference of Parties (COP) 26 has identified as critical for addressing climate change. Recognizing that forestry resources and forest cover loss cannot be managed apart from the communities that depend on them, the National Forest Authority (NFA) has tried different approaches, including Collaborative Forest Management (CFM), to involve forest-adjacent communities in the management and protection of forestry resources. The fact that these approaches have been tried in Uganda for over 20 years provides us with ready-made experience of relevance from which we can improve our efforts to address climate change.

CFM is viewed as promising for managing forests more sustainably because it provides opportunities for forest-adjacent communities to participate in and benefit from forest management (Mawa, Babweteera, and Tumusiime 2020, 2022; Mawa et al. 2020, McDougall et al. 2013). In addition to Uganda, CFM is also practiced in other countries, including Canada, Mexico, Cameroon, Ghana, India, Nepal, the Philippines, and the United States (Carter and Gronow 2005; Petheram, Stephen, and Gilmour 2004); see Chapter 8 for the similar Participatory Forest Management in Malawi. CFM has been described as

> … a working partnership between the key stakeholders in the management of a given forest — key stakeholders being local forest users and state forest departments, as well as parties such as local governments, civic groups and nongovernmental organizations, and the private sector.
>
> (Carter and Gronow 2005, 1)

DOI: 10.4324/9781003325932-17

Initiated in Uganda in 1998, CFM is designed to benefit forest-adjacent communities, enabling them to access forest resources that were previously out of their reach (Carter and Gronow 2005). It also provides communities with opportunities to participate in forest management. However, Uganda's approach to forestry management, including CFM, is based on regulations and policies that are far removed from the traditional cultural management practices that were employed by forest-adjacent communities. It is therefore not surprising to find that local communities do not have the particular knowledge and skills to participate in CFM, as manifest in government protocols and legislation (Kazoora et al. 2020). NGOs (local, national, and international) work with local communities to equip them to manage forests in ways that work better with governmental protocols; they also guide local communities through the process of developing community-based organizations (CBOs) – another external requirement from government – negotiating a CFM agreement and implementing it.

CFM – which developed independently of Adaptive Collaborative Management (ACM) – is similar to that approach to forest management, as developed and pioneered by CIFOR in the early 2000s (Fisher, Prabhu, and McDougall 2007; Mutimukuru-Maravanyika 2010). Colfer, Prabhu, and Larson (2022, 1) define ACM as "… an approach in which shared learning, experimentation, and adaptation are key principles, as are inclusivity and shared decision making at various levels…". Communication and creation of a shared vision, social learning among stakeholders, and collective action are central tenets of ACM (Fisher, Prabhu, and McDougall 2007). ACM has been practiced in several countries beyond those represented here a number of countries, including Kenya (Kalibo and Medley 2007), Nepal, Indonesia and Philippines (Fisher, Prabhu, and McDougall 2007), Nicaragua (Evans, Larson, and Flores 2020), Uganda (Bomuhangi et al. 2022), Zimbabwe (Mutimukuru-Maravanyika 2010), and others. Although collaborative management and engaging local stakeholders are central tenets of both ACM and CFM, there are also some major differences between the two forest management approaches, as described in the literature. Table 9.1 provides an overview of the similarities and differences between the 'idealized versions' of ACM and CFM.

In Uganda, CFM preceded ACM and is described in the national forest policy (MWLE 2001) – providing the kind of institutionalization ACM proponents would like to see. CFM was operationalized through the signing of CFM agreements with CBOs created by forest-adjacent communities. In both approaches, local communities are pivotal to CFM. By using CBOs, CFM is designed to increase representation, ownership, inclusivity, and diversity in forest management. Studies have also highlighted ACM and CFM's contributions to gender equity. For example, Bomuhangi et al. (2022) found that women's participation and representation were higher in communities that implemented ACM than in communities that did not, while Mukasa et al. (2022) found that ACM enhanced gender equity in forest management.

Although the NFA adopted CFM as an approach to involving forest-based communities in forest management, it lacks the resources, money, and workforce to implement CFM, as ideally envisioned. To get around this, the government

Table 9.1 Idealized similarities and differences between ACM and CFM

Forest management approach	*Adaptive Collaborative Management (ACM)*	*Collaborative Forest Management (CFM)*
Similarities	• Joint or collective action • Social learning among stakeholders • Communication and creation of a shared vision • Involvement of local and relevant stakeholders • Leaders drawn from the local communities • Attention to gender and other social differentiation • Discussion and encouragement of cross-scale linkages	
Differences		
Learning	Seek and apply knowledge actively and deliberately	Seek and apply knowledge incidentally
	learning is institutionalized	learning is not institutionalized
	Learning is at a higher level enabling stakeholders to reframe their perspectives or part of their worldview – transformative learning	Learning is at variable levels.
	Iterative learning process that includes monitoring and reflection and strengthens communities' abilities to learn from their own experience/experiments	Repetitive learning process focused on accumulation of knowledge and acquisition of skills
Vision	Vision serves as a reference point for stakeholders to navigate through decision making and management	Vision is a fixed point to be achieved through management actions
Action	Emphasis on good, equitable bottom-up facilitation	Emphasis on good facilitation
Empowerment	Obvious empowerment of local people	Local people may be empowered

Source: Fisher, Prabhu, and McDougall (2007); Colfer, Prabhu, and Larson (2022).

created a niche for NGOs (non-governmental organizations) and local communities to work collaboratively with the NFA to implement CFM. This niche is enshrined in the national forestry policy (MWLE 2001). In this arrangement, government and NGOs are actors that are external to forest-based communities and contribute to CFM implementation. However, not much is known about the influence of external actors on CFM implementation in Uganda, an issue with implications far beyond Uganda itself. This study seeks to address this gap.

The purpose of this chapter is to examine who is influencing CFM implementation in Uganda by responding to these questions:

1 Who are the main actors in CFM implementation?
2 How is CFM implemented in Uganda?
3 How does CFM differ in practice from ACM?
4 What influence do NGOs have on the longevity of CFM in Uganda?

The next section describes CFM implementation in Uganda. This is followed by a section on the role of NGOs in forest management. The methodology section follows, including a description of Kapeka Integrated Conservation Development Agency (KICODA), the CBO that was investigated in this case study. Results are presented and discussed including a description of both how CFM in practice differs from ACM and what commonalities have evolved. The chapter ends with a conclusions section that highlights how local and international NGOs influence the implementation and longevity of CFM in Uganda.

Literature

Collaborative Forest Management Implementation in Uganda

The goal of Uganda's Forestry Policy is: "An integrated forest sector that achieves sustainable increases in the economic, social and environmental benefits from forests and trees by all the people of Uganda, especially the poor and vulnerable" (MWLE 2001, Glossary). The Forestry Policy further states on page 18 that "Collaborative partnerships with rural communities will be developed for the sustainable management of forests". It was envisioned that CFM would define the Rights, Roles and Responsibilities or three Rs (3Rs), of partners as well as the basis for sharing benefits from forest management. These 3Rs are an important basis for collaborative management as well as address climate change.

CFM is implemented and operationalized through the signing of agreements with communities; these agreements are seen by the NFA as the main way to achieve community involvement in sustainable forest use and management in Uganda. According to the Uganda Forest Policy, "Collaborative forest management means that local communities are genuinely involved in the management of the forest resource through a negotiated process in which rights, roles, responsibilities and returns for the sustainable management of such forest resources are shared" (MWLE 2001, 5).

The Ugandan CFM approach requires communities to set up a CBO and negotiate a CFM agreement. Since local communities lack the legal knowledge and scientific skills to do so, they are assisted by NGOs. The Uganda Forest Policy and the National Forestry and Tree Planting Act (NFTPA 2003) anticipated that local communities would need help to engage in CFM, as defined by the government. Hence, the Forestry Policy and NFTPA make explicit provisions for the involvement of NGOs in the forest sector by establishing the following functions: (1) mobilizing and sensitizing local people; (2) strengthening civil society; and (3) supporting civil society participation in forestry management (MWLE 2001; NFTPA 2003). The next section examines these roles of NGOs further.

Non-Government Organizations in Forest Management

Researchers have investigated the multiple roles that environmental NGOs play, such as NGO involvement in community development and government service delivery (Brass et al. 2018; Fisher 1997; Gupta, Lele, and Sahu 2020; Hearn 2007); the relationship between NGOs, donor agencies, and states (Banks, Hulme, and Edwards 2015; Contu and Girei 2014; Gupta, Lele, and Sahu 2020; Watkins, Swidler, and Hannan 2012); NGO governance (Barr, Fafchamps, and Owens 2005); and the role of NGOs in environmental services delivery (Gupta, Lele, and Sahu 2020; Reed 1997). Other researchers have examined the role of NGOs in forest policy development (Aryal et al. 2021); conservation (Benson 2012); and community forestry implementation (Barsimantov 2010; Gupta, Lele, and Sahu 2020; Shandra 2007; Wright and Anderson 2013). NGOs have been found to play a strong role in forest management (Gupta, Lele, and Sahu 2020; Wright and Anderson 2013).

NGOs may implement projects that produce measurable results; however, researchers point out that these projects may not address the underlying causes of social or environmental problems (Contu and Girei 2014). Furthermore, sometimes conditions attached to funding constrain NGO activities through dictating the focus and scope of work (Benson 2012; Brass et al. 2018; Contu and Girei 2014; Shandra 2007; Watkins, Swidler, and Hannan 2012; see also Chapter 10 on donor effects on the Central African Model Forest Program). NGOs are not businesses; most have to rely on members and/or donors for funding. Some researchers note that the pressure to obtain funds has led NGOs to implement projects that comply with donor agendas (Brass et al. 2018), to the detriment of their attention to issues they or their clients may consider most important. Consequently, projects supported or implemented by NGOs may become limited in scope or temporal scale and/or become detached from local needs (Benson 2012; Hearn 2007; Shandra 2007).

NGOs often have a positive impact on development outcomes, including forestry (Brass et al. 2018; Gupta, Lele, and Sahu 2020). Despite some noted shortcomings of NGO-supported interventions, there is empirical evidence to show that their activity is associated with improved forest management outcomes. For example, some scholars have found a positive relationship between increased NGO activity and decreased deforestation (Shandra 2007; Wright and Andersson 2013). Scholars have also noted that NGOs contributed to forest regeneration in Nepal and India by facilitating the formation of forest user groups and protection committees (Pretty and Ward 2001), while others have shown NGO support for decentralized forest governance (Gupta, Lele, and Sahu 2020). All the Indonesian cases included in this volume worked closely with NGOs. This chapter examined whether and how, by undertaking multiple roles in forest management, NGOs support CFM implementation in Uganda. However, even when NGOs are acknowledged as having a strong role in forest management, researchers studying CFM have not directly considered how they may influence the quality and longevity of CFM implementation.

Studies in forestry, environment, and natural resource management were reviewed to identify possible roles of NGOs in support of CFM that sustains environmental, social, and cultural values. The five roles identified from this earlier period are funding, capacity building, monitoring, advocacy, and research (Banks,

Table 9.2 Possible NGO roles identified from literature

NGO role	*Description*
Advocacy	• Petitioning government for funds to implement community-based projects • Lobbying for government policy changes that support local communities
Funding	• Providing funds for activity implementation or equipment purchase • Providing in-kind supplies • Hiring staff to work with communities
Capacity building	• Training community members in governance, management activities, and policy interpretation/implementation • Raising awareness and/or building support in the community about the natural environment
Research	• Collecting and analyzing data to track effects of CFM interventions
Monitoring	• Monitoring compliance with governmental laws

Source: Banks et al. (2015); Barr et al. (2005); Barsimantov (2010); Contu and Girei (2014); Gupta et al. (2018); Hearn (2007); Reed (1997); Shandra (2007).

Hulme, and Edwards 2015; Barr, Fafchamps, and Owens 2005; Barsimantov 2010; Contu and Girei 2014; Gupta, Lele, and Sahu 2020; Hearn 2007; Reed 1997; Shandra 2007). These roles are listed and described in Table 9.2. The emphasis on empowerment, so important in ACM, came in later studies. I then used a case study approach to examine on site the patterns seen in the literature review as summarized in Table 9.2.

NGOs may take up one or all five roles described in Table 9.2 in supporting CFM implementation in Uganda. The three NGO functions listed in the forest policy are compatible with the five roles listed there. Hence in Uganda, NGOs are important CFM implementation partners (MWLE 2001), as has also been the case with many cases explicitly using an ACM approach.

Methodology

Case Study: Kapeka Integrated Conservation Development Agency

This study investigated KICODA. Registered as a CBO in 2005 by the residents of Kapeka village, KICODA was formed to provide a forum to represent residents in a CFM Agreement with the NFA. Kapeka village has a population of over 3,000 people and is located along the south-eastern border of Budongo Central Forest Reserve (Budongo). Although the study area is located in Bunyoro, the population is ethnically diverse with more than 18 tribes represented. The most dominant tribes are the Banyoro, Alur, and Lugbara; others include the Acholi, Langi, Alur, Bakyiga, and Iteso. There are also people who have migrated from the Democratic Republic of Congo and Southern Sudan. Most people are small-scale subsistence farmers, the majority of whom are sugarcane out-growers for Kinyara Sugar Factory, the largest employer in the area.

Budongo was gazetted as a central forest reserve in 1932. The forest occupies an area of 835 square kilometres making it the largest forest reserve in Uganda. Budongo is an important biodiversity conservation area, with more than 360 bird species, 465 tree species, and 24 mammal species, of which nine are primate species, including chimpanzees (*Pan troglodytes*), which are internationally protected. Budongo is well known for its mahogany (e.g. *Khaya anthotheca*) and ironwood (*Cynometra alexandrii*) species. It also has species that are endemic to Budongo, including two tree species, *Muscicapa sethsmithi* and *Illadopsis puveli*, and two bird species. There are several species of conservation concern, with mahogany and chimpanzees being the most preeminent.

To facilitate management of Budongo Central Forest Reserve, the NFA divided the forest reserve into compartments. KICODA signed a CFM agreement with the NFA to manage compartment W24, which is adjacent to Kapeka village and 767.70 ha in size.

Data Collection and Analysis

Data were collected in 2013 using semi-structured personal interviews and focus group discussions (FGDs). In addition, participant observation was used during forest walks and visits to different parts of Kapeka village. These observations helped to provide context to activities and practices. While the initial interview participants were identified with the assistance of KICODA board members, the subsequent interview participants were identified using the snowball method to ensure diverse input from a variety of study participants. Each person interviewed was requested to propose two names for further interview. New interview participants were sought until data saturation was reached, and for this study, this was at 31 interviews (see Guest, Bunce, and Johnson 2006). Of the people who were interviewed, 14 were women and 17 men. Ten of the 31 interviewees were members of the KICODA Executive Committee, 19 were members, and two were non-members.

An interview guide with open-ended questions was used to provide participants with the opportunity to actively participate and contribute more substantively to the study. Interview questions were designed to obtain information regarding the activities people participated in, which NGOs supported KICODA's activities, and whether participants thought the forest was being effectively managed. All interviews were recorded with the consent of the interviewees. I could understand two of the main languages spoken within the study area, though not fluently. Thus, I hired translators (a man and a woman) for interviewees who were not comfortable speaking in, or did not know, English.

Key person interviews were conducted with six individuals from NFA and NGOs: two each from the NFA and Community Conservation and Development Agency (CODECA), and one each from the Jane Goodall Institute (JGI) and Budongo Conservation Field Station. The interview with key persons explored the role of government and NGOs in CFM implementation and longevity. NGO participants agreed to having the names of their organizations used, although their own names have been omitted to protect their identity.

In addition to individual interviews, five focus group meetings were held in Kapeka village: two all-women groups, two all-men groups, and one mixed group. Various permutations of focus groups were used to meet the needs of participants (Morgan 1997). All focus group meetings were audio-recorded with the consent of participants. The focus group meetings were used to confirm themes found from initial data analysis of interviews with individuals.

All interviews and FGD data were transcribed verbatim, exported to nVivo and coded. Data were analyzed following an inductive approach whereby themes that emerged from data guided analysis in conjunction with variables that were previously identified from reviewed literature (Miles and Huberman 1994). The results are presented in the next section in the form of tables, quotes, and figures. The quotes are used to represent the majority of the views shared by study participants unless stated otherwise.

Results

The results are presented in two parts: the first part identifies the three main types of actors in CFM implementation (NFA, KICODA, and NGOs) and then describes the NGOs in more detail as they support nearly all CFM activities that are implemented. The second part describes how CFM is implemented by KICODA with a focus on activities implemented and the NGOs that support these activities.

Main Actors in Collaborative Forest Management Implementation

The three main types of actors in CFM implementation were government through the NFA, the local community through their CBO KICODA and NGOs. A fourth type of actor, Nyabyeya Forestry College in Budongo Forest, hosted a training session in conjunction with one of the NGOs.

The National Forest Authority (NFA)

The NFA's role was largely passive. They provided the Uganda Forest Policy (MWLE 2001) that guided CFM implementation, and one position at the NFA Headquarters in Kampala in charge of national CFM implementation. During the 2013 interview, the person in charge of CFM stressed financial constraints for CFM implementation, including the limited ability to travel to attend CFM meetings. A recent review of CFM implementation in Uganda also noted that NFA had limited funding and staffing for CFM implementation (Kazoora et al. 2020).

Kapeka Integrated Conservation Development Agency (KICODA)

KICODA is a village- and membership-based organization. Members receive life membership by paying a membership fee once. KICODA is run by an elected executive committee made up of six men and six women. KICODA's annual general meetings are open to the public. However, voting is restricted to members. At the time of fieldwork, the membership fee was a one-time payment of 5,000 Uganda shillings (equivalent to US$1.40 in 2022). KICODA's members were required to donate a percentage of earnings from their tree sales to the organization.

Unlike most CBOs, which have *not* completed negotiating a CFM with the NFA, KICODA signed a CFM agreement in 2006. The objectives of KICODA's CFM agreement are to:

1 Conserve forest biodiversity
2 Protect the forest from illegal activities
3 Improve people's income through forestry activities
4 Contribute to improved livelihoods of the people of Kapeka
5 Ensure regeneration of the forest
6 Sensitize people about forest management

These objectives guided the activities that government and NGOs supported in KICODA.

Interview results showed that KICODA, like other Ugandan CBOs, lacked the resources required to meet its obligations under the CFM agreement they signed with the NFA. Results from the interviews also showed that there were four NGOs that supported the implementation of CFM activities by KICODA.

Non-Government Organizations (NGOs)

The four NGOs that supported the implementation of CFM activities by KICODA were CODECA, Budongo Conservation Field Station, JGI, and Environment Conservation Trust Uganda (Ecotrust). All four NGOs were external to the community. Table 9.3 provides a description of these NGOs. One of the NGOs, JGI is Canadian, while the rest are Ugandan.

Of the four NGOs identified by interview participants, CODECA was well known in the villages around Budongo Forest and had participated in the negotiation and signing of over 15 CFM agreements, including KICODA's. CODECA supported KICODA activities related to capacity building, funding, and monitoring. Of these four, CODECA is the NGO that has supported KICODA for the longest period and provided the most funding for KICODA's activities.

Budongo Conservation Field Station (the Field Station) hosts local and international researchers. The Field Station's original focus was chimpanzee research but this expanded to include community engagement when employees realized that community activities affected chimpanzee conservation. The Field Station supported KICODA through capacity building and funding. According to participants, these activities included holding seminars to sensitize people about chimpanzees, providing training on alternative livelihood sources such as cultivating fast-growing fruit tree varieties; they also provided equipment for tree grafting activities.

The JGI supported CFM activities through three roles: funding, capacity building, and monitoring. Some of the Institute's activities with KICODA included (a)

Table 9.3 Description of environmental NGOs working with KICODA to implement CFM in Kapeka

NGO	*CODECA*	*BCFS*	*JGI*	*Ecotrust*
# of times mentioned by interviewees (N = 31)	10	4	3	4
# of times listed by focus groups (N = 5)	5	4	1	4
Listed by NFA managers	Yes	No	Yes	Yes
Type of NGO	National	National	International	National
Funding source example	CARE	Royal Zoological Society of Scotland	Public donations, IUCN, REDD, CARE	REDD, World Wildlife Fund, IUCN
Location of NGO office	Masindi town	Budongo Forest	Hoima town	Entebbe town
Approximate distance from NGO office to Kapeka village	30 km	16 km	60 km	280 km
NGO mandate	Training, capacity building	Research, conservation	Research, conservation, education	Sustainable environmental management
Status of support to KICODA	Ceased in 2012	Ceased in 2012	Ceased in 2010	Ceased in 2014
Status of NGO	Changed name/ in 2019 Website inactive 18-Oct-21	Still Active 18-Oct-21	Still Active 18-Oct-21	Still Active 18-Oct-21

Source: Interviews by author and NGO websites.

training in nursery operations, tree planting and management, and apiculture; (b) training and providing start-up materials like beehives and protective gear for honey harvesting; (c) creating environmental awareness; (d) training in small business management and record keeping; and (e) supporting joint forest patrols with the NFA.

Ecotrust supported KICODA through capacity building and funding. This support focused on activities that enhance personal livelihoods and generate income to support KICODA as an organization. Ecotrust was the last NGO to start working with KICODA.

All four NGOs received funding from international organizations with headquarters in the global North. At the time of data collection, three of the NGOs had ceased supporting KICODA. CODECA ceased supporting KICODA's activities because it had run out of funding, while the other NGOs had other funding priorities.

One clear difference between these kinds of activities and those developed by ACM is the emphasis on training, or *teaching* the villagers. Within ACM, the emphasis has been more on listening, though training materials were also developed, when requested by villagers, driven by their own interests and expressed needs.

How CFM is Implemented

CFM is implemented through activities that are developed and executed by CBOs such as KICODA. Nearly all CFM activities implemented by KICODA were supported by the four NGOs that were active in the area. Figure 9.1 provides a list of activities supported by these NGOs. Table 9.4 provides a description of these activities and the NGO that supported it.

The most common activities supported by NGOs were sensitizing members about the laws relating to harvesting NTFPs from Budongo. This activity was mentioned by all 31 community participants who were interviewed. The NTFPs harvested included firewood, water, thatching grass/building materials, honey, and materials for making handicrafts such as baskets. The commonly mentioned activities that were supported by NGOs were meetings and seminars. These were mostly held at the Catholic Church compound at the village center. During data collection, the author held some interviews and some of the focus group meetings in the same church compound. The chairperson of KICODA described his experience, which appears to go beyond the simple transfer of legal or scientific information:

> We were all green about the forest, now you can go inside the forest. You learnt things like bee keeping. You are taken for training, and then you are being trained about advocacy, about lobbying politicians, your communities. So, these are the things, the skills we have really learnt. And not only these ones, sometimes even the community members are taken by the NGOs to other districts to go and learn things like poultry, bee keeping, tree planting, piggery, and others. So, these are some skills the communities have learnt through the CFM.
>
> [a participating man]

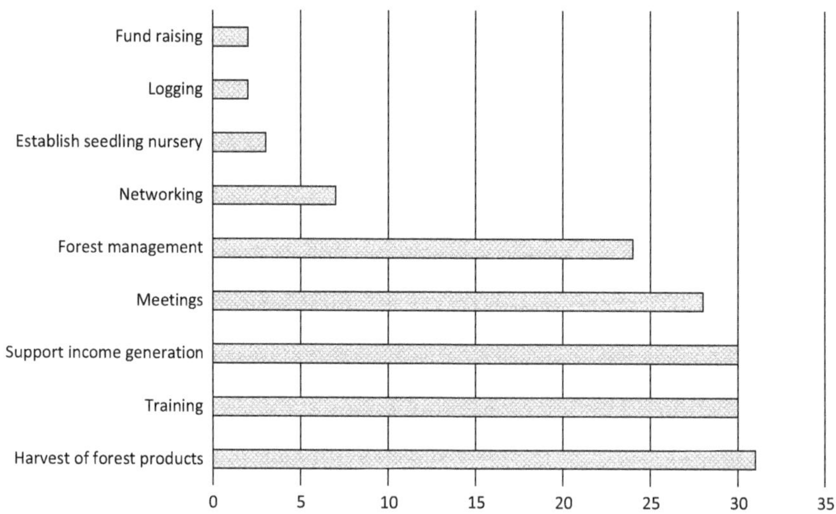

Figure 9.1 List of KICODA activities supported by NGOs with frequency distribution

Table 9.4 Description of some of the activities implemented by KICODA with the support of environmental NGOs

Activity type	*Activity description*	*NGO linked to activities*
Harvest of forest products	Forest products include firewood, thatching grass, vegetables, craft materials, etc.	CODECA
Seminars/ training	On-site and off-site training seminars on forestry by-laws, patrolling, business management, bee keeping, etc.	JGI, BCFS, Ecotrust
Meetings	Executive committee meeting, women's group meeting, workshops, planning meetings.	CODECA, Ecotrust,
Income generation	Personal business e.g. bee keeping, tree planting, goat rearing, chilli growing, Access National Forest Authority contracts, e.g., road construction. KICODA is paid a portion of the contract fees, etc.	Ecotrust, CODECA, JGI
Forest management	Enrichment planting, afforestation, boundary planting, forest patrols, forest boundary maintenance.	CODECA, Ecotrust, JGI
Networking	Visited CBOs in Rakai and Bushenyi districts to learn from them.	CODECA
Timber harvest	Logging, one license ~4 months long.	n/a
Seedling Nursery	Established three seedling nurseries	Ecotrust, CODECA
Fund raising	held one fundraiser, mobilized community to access other funds.	Ecotrust, CODECA

Notes: n/a = not applicable.
Source: Table adapted from Egunyu (2015).

NGO support for meetings and seminars included providing resources for the meetings, including trainers, stationary, and meals. For example, Ecotrust was described as providing training in various areas. Again, these suggest skills of value in empowering local communities.

> Ecotrust came and trained people on how to do bee keeping, and sensitized the community on conflict resolution, and also how to conserve natural resources, and the benefits. Also carrying out activities like planting on the boundary, they taught us.
>
> [a participating man]

The training that was provided to KICODA members and or the people of Kapeka depended on the mandate of the NGO. For example, the Field Station supports chimpanzee conservation and that guided the training that they provided as described by a participant:

> There was Budongo Field Station which sensitized us about problems of animal management.
>
> [a participating man]

This quote illustrates how the training provided aligned with the NGO's mandate. Through training on problem animal management, the Field Station was able to encourage community members to refrain from activities such as trapping, thereby supporting chimpanzee protection and conservation (a more top-down approach).

NGOs sponsored training events. Generally, the training that was provided was to sensitize community members about CFM and Budongo Central Forest Reserve. For example, a participant described:

> There was a time when we were developing by-laws here. Ecotrust sponsored lunch for us. We started at around half past eight, we ended at around four.
>
> [a participating woman]

Other times, the training or seminar was explicit in its intent to build people's capacity to implement CFM and empower them to act in their own interest and become financially self-sustaining.

> When I went out for the tour to Bushenyi to visit another [CFM] group, I learned many things especially how to work with people and I also learned how to talk to people. I used to fear talking to people. But I have now learned that. These days I am a free woman.
>
> [a participating woman]

> CODECA came here and held a seminar for women on emancipation. Women, especially those who have not gone to school, fear to talk in meetings. CODECA trained women how to talk. I also trained some women, how to talk in meetings, although some still fear to talk.
>
> [a participating woman]

The people of Kapeka did not have many livelihood options; the main employment option was Kinyara Sugar Factory. It is not surprising that some of the capacity building seminars included personal business establishment and management. Participants described being trained and supported to run diverse personal and organizational income-generating activities:

> I participated in the seminar on managing businesses. I have a small business here, I make some money and it helps me at home to buy some things and to pay school fees.
>
> [a participating male]

> ... the Jane Goodall Institute came in and brought about four beehives; those were supposed to be sited specifically for the office. Then they brought more [honey] harvesting gear. That's when people started harvesting...
>
> [a participating woman]

From this excerpt, we can see the JGI supporting income-generating activities by providing beehives for individuals as well as for KICODA. The intent was for KICODA to sell honey, become financially self-sustaining and stop relying on external funding. Through supporting a variety of activities, the NGOs enabled KICODA to implement CFM.

Discussion

I set out to examine who was influencing CFM in Uganda by investigating CFM implementation by KICODA, a CBO established by Kapeka, a village adjacent to Budongo, the largest Central Forest Reserve in Uganda. Although CFM provides opportunities for forest-adjacent communities to participate in forest management and benefit from forest access (Larson and Soto 2008; Mawa, Tumusiime, and Babweteera 2021), the implementation of CFM at the local level, particularly in countries in the global South like Uganda, Nepal, India, and Ghana, requires the support of external actors.

The CFM agreement that was signed between the NFA and KICODA required that the latter contribute to the protection, conservation, and management of Budongo Forest. Hence, the NFA was an external actor who defined and guided who is involved in CFM implementation as well as what they do. Interviewees described how the implementation of KICODA's activities was supported by NGOs. NGOs supported three main types of activities: funding, capacity building, and monitoring. These types of activities were similar to what was observed by Gupta, Lele, and Sahu (2020) about NGO support for community forest management in India.

Interview results indicate that NGOs helped KICODA to better understand the Uganda Forestry Policy with specific initiatives to address its three core elements related to public involvement: mobilizing and sensitizing local people, strengthening civil society, and supporting civil society participation in forestry management. These NGOs also helped KICODA by playing the roles identified from the literature (Banks, Hulme, and Edwards 2015; Egunyu 2015; Gupta, Lele, and Sahu 2020; Reed 1997) and further established in the results.

Despite the fact that there are NGOs that carry out academic research (see, for example, Budongo Forest Field Station), there was no NGO support for research and advocacy. In addition, it also seemed as if NGO support was not leading to self-sufficiency in KICODA. However, KICODA is not the only community CFM organization that is struggling to become self-sufficient. A review of CFM in Uganda by Kazoora and others (2020) showed that nearly all communities and CBOs implementing CFM required external support. Although communities taking part in ACM have also required some external support, a central goal of that approach has been to minimize the need for such support in the long term; CFM, however, is *designed* – as part of its institutionalization – to be dependent on external (both NGO and government) support.

I argue that although CFM is defined by the NFA (government) in Uganda, implementation and longevity are considerably influenced by national and international NGOs. What does this say about the future of CFM implementation in Uganda? Pagdee, Kim, and Daugherty (2006) reviewed community forestry worldwide and found that factors that influenced the success of community forests were tenure security, clear ownership, expectation of benefits, shared interests between community members and local authorities, strong leadership, and effective enforcement of rules and regulations. Some of these items, particularly expectation of benefits, shared interests between community members and local authorities, and strong leadership, are evident in the practice of CFM in Uganda. Mawa, Babweteera, and Tumusiime (2022) also reported increased access to forest products from state forestry reserves as well as the ability to access credit using CFM groups. Still, it remained challenging for KICODA to financially support most of its activities.

Comparing CFM Implementation to ACM Description

There were several similarities between CFM implementation and ACM as described in the literature (Colfer, Prabhu, and Larson 2022; Fisher, Prabhu, and McDougall 2007). Both ACM and CFM involved joint or collective action, leaders were drawn from the local community and there was learning during implementation. See Table 9.5 for more details. However, there were some differences between CFM implementation and ACM as described in the literature. For example, empowerment is obvious in ACM, less so in the CFM practiced by KICODA.

Table 9.5 Similarities and differences between CFM implementation by KICODA and ACM as described in the literature

Adaptive Collaborative Management (ACM)	*Collaborative Forest Management (CFM) in practice*
Joint or collective action	CFM was implemented using joint or collective action.
Social learning among stakeholders	There was evidence of social learning among KICODA members. They reported acquiring skills and knowledge.
Communication and creation of a shared vision	KICODA's vision was jointly developed by members over a series of meetings.
Involvement of local and relevant stakeholders	CFM practice involved members who were from the local forest-adjacent community, Kapeeka village.
Leaders are drawn from the local communities	KICODA's leaders are drawn from the local community – Kapeeka village.
Attention to gender and other social differentiation	KICODA's executive committee had 12 members, half were female. The committee reflected attention to gender and other social differentiation as there was a youth representative, an elder representative, a women's representative, and a representative for persons with disabilities.
Discussion and encouragement of cross-scale linkages	KICODA hosted two other CBOs that were practicing CFM, they also visited a CBO that was practicing CFM in the south western part of Uganda. The trip was for motivation and learning purposes.

Seek and apply knowledge actively and deliberately	Sought and applied knowledge deliberately.
Learning is at a higher level enabling stakeholders to reframe their perspectives or part of their worldview – transformative learning	Of the 31 KICODA members interviewed, all reported learning. Most of the learning was at a lower level where people acquired information. A few people reported a reframing of their worldviews.
Iterative learning process that includes monitoring and reflection and strengthens communities' abilities to learn from their own experience/experiments	KICODA members described repetitive learning processes that included accumulation of more knowledge and acquisition of skills. A few members described a change in perspectives (worldviews).
Vision serves as a reference point for stakeholders to navigate through decision making and management	KICODA members described their vision and expressed a desire to achieve the vision through implementation of CFM related activities.
Emphasis on good, equitable bottom-up facilitation	KICODA members reported being facilitated by trained professionals and some NGO employees.
Obvious empowerment of local people	Some KICODA participants reported that they were empowered. But this was not obvious and or widespread among KICODA members.

Source: ACM characteristics obtained from Fisher, Prabhu, and McDougall (2007); Colfer, Prabhu, and Larson (2022). CFM practice from interviews.

As is evidenced in Table 9.5, ACM-like features are more obvious in the CFM implementation than in descriptions of the CFM approach and legislation. These features, as applied while implementing CFM in Uganda however, included learning, some empowerment and developing a vision – much like ACM. These elements were not highlighted as central by the NFA to CFM practice, as they had been among ACM practitioners and theorists. Yet, many of the quotations suggest that learning, empowerment, and shared visioning independently evolved in the Ugandan context.

Conclusions

This study provided an in-depth look at the implementation of CFM by a local CBO and identified three types of actors in CFM implementation in Uganda. Although the Uganda Forest Policy and the NFTPA 2003 identify additional actors, this study focused on those actors working with Kapeka village to illustrate the types of workers and the influence they had on CFM implementation.

The NFA was proactive in providing a niche in the forest policy and NFTPA for CBOs and NGOs to support the implementation of CFM in Uganda. However, KICODA relied heavily on NGOs for CFM implementation and when the NGOs reduced their support, KICODA's activity level dropped. CFM implementation in Uganda as evidenced by KICODA is heavily supported by external actors, local, national, and international NGOs. Without the support of these NGOs, the implementation and practice of CFM in Uganda would be in limbo. This observation is also supported by Nsita et al. (2020) and Kazoora et al. (2020), as well as in the experience of Bomuhangi et al. (2022) and Mukasa et al. (2022).

Here is what scholars and practitioners can learn from CFM implementation in Uganda:

- ***Empowering local forest-adjacent communities***: NGOs and the NFA initiated, designed, and held most training and capacity building seminars for CFM practitioners. National and international NGOs and NFA appear to have approached CFM implementation as if KICODA members had no knowledge to contribute to the process. However, local communities also have local and traditional knowledge that they can contribute. Government agencies and NGOs are not only givers of resources and knowledge but are also learners. If KICODA and other external actors had approached the local communities with greater respect toward local knowledge and experience, they might have had a better success rate and more longitudinal staying power. They might thus have been able to contribute more to the wider national issue of weaning community CFM organizations off external support for local activities.
- ***The role of NGOs***: CFM implementation in Uganda requires the registration of a CBO, knowledge of forestry laws and regulations, as well as the ability to negotiate a CFM agreement with the NFA. Because of these requirements, national and international NGOs continue to have an important role to play in CFM implementation in Uganda, particularly during the community initiation phase. However, local communities should be able to own and continue implementing CFM, and NGOs should support them toward this. Simplification of registration procedures and laws could be another way forward.
- **CFM *definitions***: CFM has an ideal description in the literature. But there are various iterations in practice depending on location. CFM implementation differs from country to country and community to community. Therefore, countries do well to modify CFM implementation according to their own local laws and situations. In Uganda, the 3Rs, that is, clarifying rights, roles, and responsibilities, is the basis for CFM. This approach, which explicitly addresses people's rights, could, in turn, help us to address land use and forestry elements of climate change and other environmental issues.
- **CFM *and* ACM** have similarities and differences. The main difference is that for ACM, learning is intentional, empowerment is an obvious outcome, and visioning is locally determined. While in CFM, learning is a byproduct of activities, empowerment is not an explicitly sought outcome and the project vision is a pre-defined goal to be achieved. Yet in practice, the CFM implementation appears to have taken on some of the learning, empowerment, and localized visioning that ACM also found valuable. The other advantage that CFM has over ACM so far is that it is automatically institutionalized in a way that ACM has not yet succeeded in doing.

CFM has the potential to last for a long time in Uganda as it is meeting a need in providing forest-adjacent communities with an equitable way of managing and accessing forestry resources. In addition to providing benefits to forest-adjacent communities, CFM activities have been linked to healthy forests (see Mawa et al. 2020; Turyahabwe et al. 2012). The forest policy and NFTPA also provide NGOs with legitimate structures to support CFM implementation. However, this study found that KICODA has needed external support to continue to implement CFM. As long as forest-based communities need a forum to participate in forest management and NGOs are interested in supporting CFM activities, CFM will continue to be implemented in Uganda. However, I posit that the lack of resources and local capacity to implement CFM will continue to impact its endurance in any particular site.

Acknowledgments

This study was funded by two Canadian funding agencies, the International Development Research Centre and the Social Sciences and Humanities Research Council. Thanks to the anonymous reviewers (who bear no responsibility for any remaining errors), as well.

References

Aryal, K., H.K. Laudari, P.R. Neupane, and T. Maraeni. 2021. "Who Shapes the Environmental Policy in the Global South? Unpacking the Reality of Nepal." *Environmental Science and Policy* 121: 78–88.

Banks, N., D. Hulme, and M. Edwards. 2015. "NGOs, States, and Donors Revisited: Still too Close for Comfort?" *World Development* 66: 707–718.

Barr, A., M. Fafchamps, and T. Owens. 2005. "The Governance of Non-governmental Organizations in Uganda." *World Development* 33, no. 4: 657–679.

Barsimantov, J.A. 2010. "Vicious and Virtuous Cycles and the Role of External Non-Government Actors in Community Forestry in Oaxaca and Michoacan, Mexico." *Human Ecology* 38: 46–63.

Benson, C. 2012. "Conservation NGOs in Madang, Papua New Guinea: Understanding Community and Donor Expectations." *Society and Natural Resources* 25, no. 1: 71–86.

Bomuhangi, A., A.Y. Banana, J. Bushoborozi, C. Mukasa, A. Tibazalika, and E. Mwangi. 2022. "Gender and Adaptive Collaborative Management in a Forested Ugandan Landscapes." In *Adaptive Collaborative Management in Forest Landscapes: Villagers, Bureaucrats and Civil Society*, edited by C.J.P. Colfer, R. Prabhu, and A. Larson, 87–104. London: Earthscan and Routledge.

Brass, J.N., W. Longhofer, R.S. Robinson, and A. Schnable. 2018. "NGOs and International Development: A Review of Thirty-five Years of Scholarship." *World Development* 112: 136–149.

Carter, J., and J. Gronow. 2005. "Recent Experience in Collaborative Forest Management: A Review Paper." Bogor, Indonesia: Center for International Forestry Research.

Colfer, C.J.P., R. Prabhu, and A. Larson, eds. 2022. *Adaptive Collaborative Management in Forest Landscapes: Villagers, Bureaucrats and Civil Society*. London: Earthscan and Routledge.

Contu, A., and E. Girei. 2014. "NGOs Management and the Value of 'Partnerships' for Equality in International Development: What's in a Name?" *Human Relations* 67, no. 2: 205–232.

Egunyu, F. 2015. "The Contributions of Social Learning to Collaborative Forest Governance in Canada and Uganda: Lessons Learned from Forest-based Communities." PhD Thesis. University of Saskatchewan, Saskatoon, SK, Canada.

Evans, K., A.M. Larson, and S. Flores. 2020. "Learning to Learn in Tropical Forests: Training Field Teams in Adaptive Collaborative Management, Monitoring and Gender." *International Forestry Research* 22, no. 2: 189–198.

FAO. 2020. "Global Forest Resources Assessment 2020: Main Report." Rome: Food and Agriculture Organization. https://doi.org/10.4060/ca9825en (Accessed October 23, 2021).

Fisher, R., R. Prabhu, and C. McDougall. 2007. *Adaptive Collaborative Management and Community Forestry in Nepal, Indonesia and Philippines*. Bogor, Indonesia: CIFOR.

Fisher, W.F. 1997. "Doing Good? The Politics and Antipolitics of NGO Practices." *Annual Review of Anthropology* 26: 439–464.

Global Forest Watch. 2020. Dashboards, country, Uganda. https://www.globalforestwatch.org/dashboards/country/UGA/? (Accessed on October 10, 2021)

Guest, G., A. Bunce, and L. Johnson. 2006. "How Many Interviews are Enough? An Experiment with Data Saturation and Variability." *Field Methods* 18, no. 1: 59–82.

Gupta, D., S. Lele, and G. Sahu. 2020. "Promoting a Responsible State: The Role of NGOs in Decentralized Forest Governance in India." *Forest Policy and Economics* 111: 102066.

Hearn, J. 2007. "African NGOs: The New Comprador?" *Development and Change* 38: 1095–1110.

Kalibo, H.W., and K.W. Medley. 2007. "Participatory Resource Mapping for Adaptive Collaborative Management at Mt. Kasigau, Kenya." *Landscape and Urban Planning* 82: 145–158.

Kazoora, C., D. Irumba, N. Smith, and J. Campese. 2020. "A Review of Collaborative Forest Management in Uganda." Kampala, Uganda: National Forestry Authority, Ministry of Water and Environment; supported by United States Forest Service International Programs and United States Agency for International Development.

Larson, A.M., and F. Soto. 2008. "Decentralization of Natural Resource Governance Regimes." *Annual Review of Environmental Resources* 33: 213–239.

Mawa, C., D.M. Tumusiime, and F. Babweteera. 2021. "Are Community Forests Delivering Livelihood Benefits? Insights from Uganda." *Forest, Trees and Livelihoods* 30, no. 2: 133–150. https://doi.org/10.1080/14728028.2021.1904010.

Mawa, C., F. Babweteera, and D.M. Tumusiime. 2020. "Conservation Outcomes of Collaborative Forest Management in a Medium Altitude Semideciduous Forest in Mid-western Uganda." *Journal of Sustainable Forestry*. https://doi.org/10.1080/10549811.2020.1841006.

Mawa, C., F. Babweteera, and D.M. Tumusiime. 2022. "Livelihood Outcomes after Two Decades of Co-managing a State Forest in Uganda." *Forest Policy and Economics* 135: 102644.

Mawa, C., F. Babweteera, J.R.S. Tabuti, and D.M. Tumusiime. 2020. "Changes in Vegetation Characteristics Following a Decade of Community Forest Management in Mid-western Uganda." *International Forestry Review* 22, no. 3: 323–338.

McDougall, C., J. Jiggins, B.H. Pandit, S.K.T.M. Rana, and C. Leeuwis. 2013. "Does Adaptive Collaborative Forest Governance Affect Poverty? Participatory Action Research in Nepal's Community Forests." *Society and Natural Resources* 26, no. 11: 1235–1251.

Miles, M.B., and A.M. Huberman. 1994. *Qualitative Data Analysis: An Expanded Sourcebook*, 2nd ed. Thousand Oaks, CA: SAGE Publications, Inc.

Morgan, D.L. 1997. *Focus Groups as Qualitative Research*, 2nd ed. Thousand Oaks, CA: SAGE Publications Inc.

Mukasa, C., A. Tibazalika, A.Y. Banana, E. Mwangi, and T. Mutimukuru-Maravanyika. 2022. "Strengthening Women's Tenure Rights and Participation in Community Forestry in Central Uganda." In *Adaptive Collaborative Management in Forest Landscapes: Villagers, Bureaucrats and Civil Society*, edited by C.J.P. Colfer, R. Prabhu, and A. Larson, 106–132. London: Routledge and Earthscan.

Mutimukuru-Maravanyika, T. 2010. "Can we Learn Our Way to Sustainable Management? Adaptive Collaborative Management in Mafungautsi State Forest, Zimbabwe." PhD Thesis. Wageningen University, The Netherlands, p. 251.

MWLE (Ministry of Water Lands and Environment). 2001. "The Uganda Forestry Policy." Kampala, Uganda: Government of Uganda Printery.

MWLE (Ministry of Water Lands and Environment). 2016. "The Uganda Forestry Policy." Kampala, Uganda: Government of Uganda Printery.

NFTPA. 2003. "National Forestry and Tree Planting Act." Kampala, Uganda: Government of Uganda Printery.

Nsita, S.A., B. Nakangu, A.Y. Banana, M. Mshale, E. Mwangi, and D. Ojwang. 2020. "Forest Tenure Reform Implementation in Uganda: Current Challenges and Future Opportunities." Bogor, Indonesia: Center for International Forestry Research.

Pagdee, A., Y.-S. Kim, and P.J. Daugherty. 2006. "What Makes Community Forest Management Successful? A Meta-Study from CFs Throughout the World." *Society and Natural Resources* 19: 33–52.

Petheram, R.J., P. Stephen, and D. Gilmour. 2004. "Collaborative Forest Management: A Review." *Australian Forestry* 67, no. 2: 13.

Pretty, J., and H. Ward. 2001. "Social Capital and the Environment." *World Development* 29, no. 2: 209–227.

Reed, M.G. 1997. "The Provision of Environmental Goods and Services by Local-Governmental Organizations: An Illustration from the Squamish Forest District, Canada." *Journal of Rural Studies* 13, no. 2: 177–196.

Shandra, J.M. 2007. "International Nongovernmental Organizations and Deforestation: Good, Bad, or Irrelevant?" *Social Science Quarterly* 88: 665–689.

Turyahabwe, N., J.G. Agea, M. Tweheyo, and S.B. Tumwebaze. 2012. "Collaborative Forest Management in Uganda: Benefits, Implementation Challenges and Future Directions." In *Sustainable Forest Management: Case Studies*, edited by J.J. Diez and B. Martin-Garcia, 52–74. Rijeka, Croatia: InTech.

Watkins, S.C., A. Swidler, and T. Hannan. 2012. "Outsourcing Social Transformation: Development NGOs as Organizations." *The Annual Review of Sociology* 38: 285–315.

Wright, G., and K. Andersson. 2013. "Non-Governmental Organizations, Rural Communities and Forests: A Comparative Analysis of Community-NGO Interactions." *Small-scale Forestry* 12: 33–50.

Introduction to Chapter 10

In this chapter, we switch from a concentration on specific, Anglophone, East African countries – Zimbabwe, Malawi and Uganda – to a broader overview, from a regional, Francophone, Central African perspective – based on experience in Cameroon and the Democratic Republic of Congo.

Rooted in Central Africa, this chapter takes a global view of forest management, building on the critiques of development processes so powerful in the academic social sciences. It examines the continuing influence of Africa's colonial past and the colonial and capitalist paradigms which have endured – partially due to their continuing benefit to powerful actors ('long-distance claimants'), many, though not all, from the global North. Most of the chapter sets the stage for the examination of participatory approaches as exemplified by the approach the authors took in Central Africa: a combination of ACM and Model Forests.

These authors document a wide range of activities that were carried out collaboratively with Central African partners at various scales. But they also highlight the unpredictability and powerlessness that derive from dependence on external funding – using a sad experience of their own as a demonstration. Unlike the previous two chapters, by Kamoto et al. and Egunyu, Diaw et al. argue that such dependence will not satisfy in the long run. Ultimately, they also conclude that although the participation of local communities and governments is important and necessary, a sole concentration at this level is inadequate to improve the lives of local communities — unless it is solidly grounded in 'blue' and 'social' economies (as discussed in the chapter).

As Diaw et al. point out, "The MF [Model Forests] approach is one of the rare landscape approaches with a systematic method for convening stakeholders at different scales in a well-defined space." Another of the important lessons from this chapter: they describe cases where successes would have been greater had they been able to scale up further. Their dependence on external, unreliable funding played a major role in their inability to do so.

Both Diaw et al. and Prabhu and Colfer (Chapter 11) lead from broader perspectives than have most ACM analysts – suggesting directions: this one emphasizing the need for structural changes in the global economic system, the other for moving toward more substantive incentives for ongoing and potential environmental stewardship.

DOI: 10.4324/9781003325932-18

10 ACM and Model Forests

A new paradigm for Africa

Mariteuw Chimère Diaw, Joachim Nguièbouri, Bruhnel Vambi N'Tambu, Julie Gagoe Tchoko, Marie Françoise Roselle Ngo Baneg and Caroline Bilogui

Introduction

This chapter questions the paradox between (a) the intellectual, political, and participatory vitality of the African forest sector (particularly in Central Africa), and (b) the inversion of the African forest economy. The African forest economy is 'inverted' insofar as it does the reverse of what all other, more successful, world regions do to add value to and grow their economies. In this chapter, we make the point that this 'forest paradox' expresses an even greater oddity: the immense economic value and multifaceted wealth embedded in African natural environments and the extraordinary material poverty of the people living in those resource-rich places. This 'African poverty paradox' has its own intellectual component in the array of policies and discourses promising development, conservation, democracy and other social goods without addressing the relevant transformative conditions for Africa's people. In fact, the whole cognitive, ideological, institutional, and symbolic system of external representation of, and intervention in the African forest universe forms the flip side of this joint forest and poverty paradox, referred to also as more simply, 'the paradox'.

This chapter is not about poverty, but about ground-breaking lessons from an attempt to achieve development in rural/forested Africa through transformative change, using novel participatory governance and innovation tools and approaches. The chapter's central objective is to question and rethink the roles and ambitions of participatory engagements for local democratic governance in Africa, considering the gravity of the poverty paradox and its momentous significance for the viability of future African and world sustainability. The chapter explains why Africa is critical to current and future population, climate, environment and development trajectories, and opens a window into how future options could eventually look from the ground up. Relying on a critique of the colonial legacy of mercantilism and environmentalism in Africa, the chapter further argues that, for the participatory paradigm to remain relevant and avoid acting as a smokescreen for externally driven conservation and financial interests, it must contribute to addressing the poverty paradox in a way that is truly transformative.

DOI: 10.4324/9781003325932-19

The next two sections contextualize and define the conceptual frame for this transformative change, which we see, following the Intergovernmental Science Policy Platform on Biodiversity and Ecosystem Services (IPBES 2019: 18), as

> ...fundamental changes in development paradigms, entailing changes in society, including much more efficient use of land, water, energy and materials, rethinking of consumption habits and major transformations of food systems.

In the African context, we also see it as a pathway to structural transformation, defined in Section 2. In Section 1, we start with the forest side of the poverty paradox to show how value creation is structurally truncated in this sector and what this tells us of poverty and structural un-development in Africa. This section also discusses the informal economy because of its neglect in the forest sector and its hidden economic and social value in light of future ecological economic options for Africa. This plants the seeds of further discussion of the informal economy from historical, ideological, and strategic angles in the following section. These discussions provide a broad context for the debates that took place in the African Model Forests Network (AMFN) when the time came to define its economic program (the *One Programme*; see Section 3).

Section 2 frames our understanding of the complex strategic problem underlying the forest poverty paradox and sets up the chapter's conceptual building blocks. We define the key concepts of structural transformation, which we see as a system shift in the way we create economic value and incorporate ecological and social values in the African forest economy. The system of value creation is at the core of the poverty paradox and other hidden consequences of the 'Western bias' in the framing of Africa's environmental problems. The section highlights the flip side – paradigmatic and ideological – of the paradox and analyzes the relationship between some of the concepts and common paradigms influencing the structure of external interventions in the sector. We look briefly at the convergence of climate, demographic, environmental, economic and social issues, including agriculture, food, and deforestation, that demand a fresh outlook. At the heart of our exploration is the type of global system shift that needs to happen in the coming decades and the reason Africa has such an important role to play in the process.

The third and last section draws on the trajectory of Adaptive Collaborative Management (ACM) in Central Africa and its expression in the Model Forests (MF) network. It considers what links them, how they differ and complete each other, and how they came to face the poverty paradox. We see how fully engaged actors, resourced with convening platforms and open, problem-solving learning tools as offered by embedded MF and ACM concepts, could not fail to be confronted with the poverty paradox. In doing so, we show the fertilizing, eye-opening content of this joint ACM-MF approach, as well as the formative and transformative value of the tools, concepts and practical innovations developed to address the paradox. We draw lessons from these concepts and practices regarding the emergence of communities of practice, the transformative power of people-centered innovation, and the construction of territorial identities. But we also consider the dire limitations of such processes when they lack long-term investment and reinvestment capability in order to project and take root durably in the economy.

Finally, the chapter draws both broad and specific lessons from these trajectories in Central Africa. We document the nature of the successes obtained and the failures sustained and provide insights on some fundamental limitations to transformative change at the landscape level. In doing so, we explore, from experience, the conditions needed for grassroots and collaborative governance, like ACM, to contribute to these processes and endure in concrete places.

The African forest economy and the poverty paradox in Central Africa

Only over the past seven years have some reports begun to shed light on the structural disabilities of Africa's forest economy. These are fundamental impairments that could be 'fixed' only by changing the very structure of the economy. They are at the heart of the sad position and performance of Africa in global forest value chains, as documented by two recent African Development Bank (AfDB) publications (ANRC 2021a, 2021b). These papers highlight the weak intra-African wood trade and the dismal external trade balance deficit of 12 of the biggest African forest economies, confirming previous studies, particularly FAO's *State of the World Forests* (2014) and Diaw (2015). Additionally, CIFOR studies have shown that the forest wealth produced in Central Africa comes mostly from the informal sector (Cerutti et al. 2014), and that agro-plantations in Cameroon and Congo can be six times more profitable than forest concessions (Lescuyer et al. 2014).

Value creation in the African wood economy

The World Food Organization (FAO 2011) estimates that there were nearly 675 million hectares of forests in Africa in 2010, representing about 17% of the world's forest area.[1] FAO's (2014) State of the World Forests (or FAO 2014) gives the gross value added (GVA, i.e., the sum of all sector revenues minus the cost of purchases from other sectors) of the formal timber sector in different regions of the world in 2011. With an added value of US$ 606 billion, the forest economy represented a little less than 1% of the world economy, a ratio that was roughly the same for all world regions. All had a forestry sector adding more or less 1% to the value of the general economy.[2] With a GVA of US$ 17 billion, Africa was worth 2.8% of the global forest economy. All other regions were doing better. With about the same amount of forests (18%), the Asia and Oceania region was doing 15 times better and accounted for 43% of the global forest economy. It was followed by Europe (27%) and North America (17%). Latin America and the Caribbean, the second weakest economy, did almost three times better than Africa (8%), with only 5% more forests (22%). It is thus not just that Africa is losing forest biodiversity at unprecedented rates, it is also not making any significant economic gain from the use of its forest resources (Archer et al. 2018) (Table 10.1 and Figure 10.1).

Table 10.1 Global value added (GVA) in the world forest sector (Billion USD at 2011 prices)

Regions	*Forestry and logging*	*Sawnwood and wood-based panel*	*Pulp and paper production*	*Total*	*Share of the national economy (GDP)*	*Share of the global forest economy*
Asia and Oceania	**84**	**66**	**111**	**260**	**1.1**	**42.9**
%	*32*	*25*	*43*	*100*		
Europe	**35**	**61**	**68**	**164**	**0.9**	**27.1**
%	*21*	*37*	*41*	*100*		
North America	**26**	**29**	**61**	**115**	**0.7**	**19.0**
%	*23*	*25*	*53*	*100*		
Latin America and Caribbean	**14**	**12**	**24**	**49**	**0.9**	**8.1**
%	*29*	*24*	*49*	*100*		
Africa	**11**	**3**	**3**	**17**	**0.9**	**2.8**
%	*65*	*18*	*18*	*100*		
World	170	171	267	**606**	**(0.9)**	**99.8**
	28	*28*	*44*	*100*		

Data source: FAO (2014)

The internal structure of forestry activities explains this poor performance. Africa is, by far, the worst performing forestry region in the world because it exports its wood raw or barely processed, instead of transforming it locally into high-end wood products that create domestic jobs and significant local added value. Two-thirds of Africa's GVA in FAO (2014) came from primary activities. It was the only world region to present such a profile. All the other regions concentrate on value-adding processing activities from which they derive between two-thirds and three-quarters of their forest value (FAO 2014, Figure 10.3). In fact, in 2014, Africa's GVA share of global processing activities represented a tiny 1.8% for lumber industries and 1.1% for the pulp and paper sector. Even the primary sector's performance is disappointing with only 6.5% of the value produced worldwide. This situation still prevails today. Africa barely exists in high-end value segments of the forest value chain and its share of global aggregate income sharply falls in inverse ratio to the degree of transformation (Figures 10.1c,d and 10.2).

This failure to create high-grade processing added value is confirmed by the latest data from the AfDB. Using a selection of 12 of the most forest-endowed African countries[3] in ITTO and FAOSTAT databases, the AfDB studies (notably, ANRC 2021a) show that from 2011–2020 these countries had a combined positive trade balance of US$ 26.7 billion for primary wood products (logs, sawnwood, plywood and veneers) but sharply negative ones for more processed products. The combined deficit of the 12 countries over that period was over US$ 20.4 billion in the trade of 3 selected secondary processed wood products, where, for the same products, Asian countries made a trade surplus of US$ 250 billion

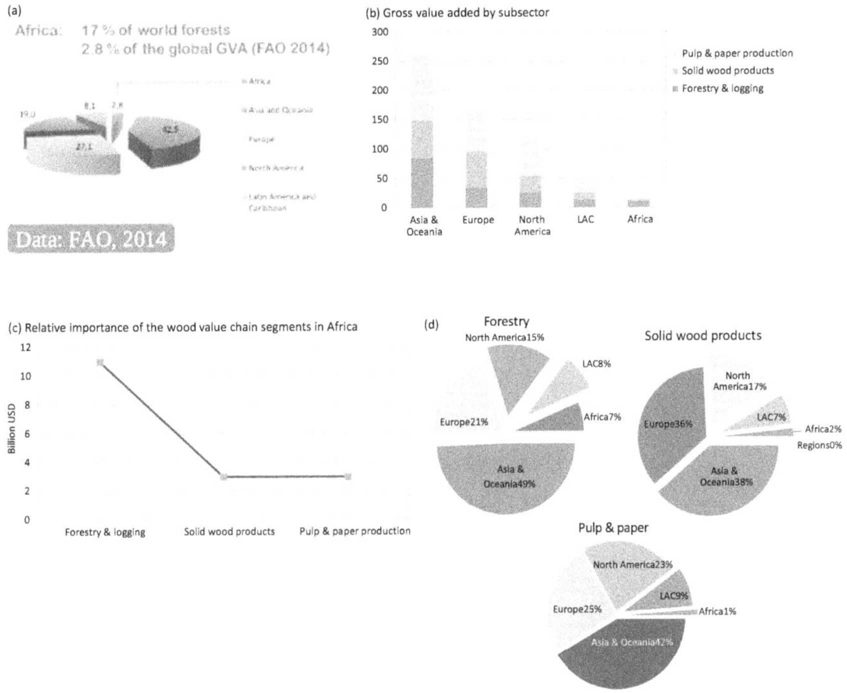

Figure 10.1 Africa in forest value chains

(see Table 10.2). With respect to seven selected tertiary wood products, these African countries registered a trade deficit of US$ 6.98 billion between 2010 and 2019, where Indonesia alone made a surplus of US$ 41.2 billion. In order to feed its exports of wooden furniture to the US and other countries, Vietnam imports 250 different species of wood annually, 100 from Africa (MARD 2022; TTM 2020). Vietnam made US$ 7 billion in 2016 from such exports (ANRC 2021a), whereas, in the whole 2011–2020 decade, the seven West African countries cited by AfDB barely made US$ 2.01 billion from the cumulative export value of wooden furniture (Dokua 2021). By 2021, Vietnam wood export turnover to the US market alone had grown to US$ 8.8 billion[4] (MARD 2022), while the government aims to hit US$ 20 billion worth of exports to the US, Japan, South Korea, Western Europe, and China, by 2025 (Dalheim 2022).[5] In a nutshell and following the famous quip, Africa exports wealth (in the form of raw logs and slightly processed wood) and imports poverty (as jobs and forest revenue losses) through that process. We are reminded that ten years ago GEF (2013) was forecasting that Africa would be importing wood by 2030. What will we do then? (Figures 10.3 and 10.4).

Table 10.2 Selected trade and export figures comparing Africa with Asia and Latin America (from the African Development Bank, ANRC 2021a)

Countries	*Primary WP – 2011–2020 Logs, sawnwood, veneer, plywood*	*Secondary WP – 2011–2020 Wooden furniture, builder woodwork (doors, floorings, etc.), Cane and bamboo*		*Tertiary processed WP – 2010–2019 Writing, printing & sanitary papers, newsprint, chemical wood pulp, case material, cartonboard*
5 Central Africa	US$ 19.18 billion	US$ −941.2 million		US$ −785 million
7 West Africa	US$ 7.5 billion	US$ −2.807 billion		US$ −6.2 billion
Indonesia	–	–		US$ 41.2 billion
Vietnam	–	US$ 7 billion (2016)	Furniture exports only	–
Malaysia	–	US$ 2.6 billion (2017)		–
Africa overall	–	US$ −20.4 billion		–
Asia overall	–	US$ 250 billion		–
Latin America	–	US$ −48 million (down from a surplus of US$ 6.8 billion in 2006–2013)		–

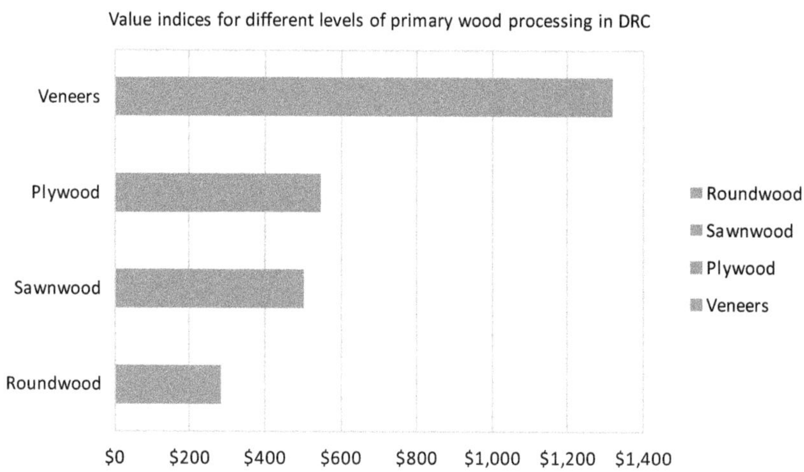

Figure 10.2 Value indices for different levels of wood processing in Africa

Wasted potential: the neglect and under-representation of the informal economy

The poor performance of the wood industry still gives an incomplete picture, as it ignores non-timber forest products (NTFPs) and other informal components

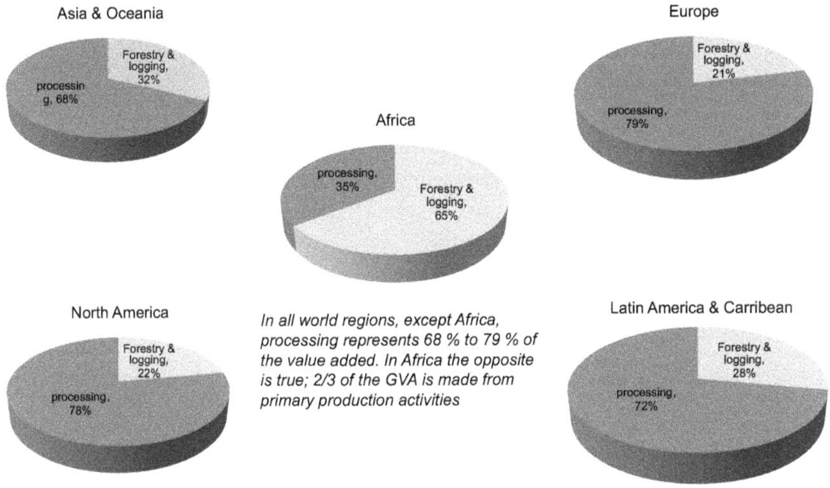

Figure 10.3 Africa as the world's inverse

of the forest economy.[6] The estimated global NTFP income in 2011 was US$ 88 billion (FAO 2014), 77% of it captured by the Asia-Oceania region. Africa represented only US$ 5.3 billion, that is, 6% of that total; though that amount is still worth a third of the formal African timber sector GVA. However according to FAO (2014), approximately 43% of the global income from the informal production of fuelwood and construction materials (US$ 33 billion) was generated in Africa. This is the only domain where Africa predominates and the amounts concerned are almost equal to the GVA of the formal wood sector (US$ 17 billion). In addition, anecdotal evidence suggests that both estimations of the incomes from NTFPs as well as fuelwood and construction activities may be considerably underestimated.

The relative weight of NTFPs and other informal value chains, combined with their serious underestimation, reflect both their central place in the culture, health, and livelihoods of millions of Africans (see Ingram et al. 2011) and the fact that they routinely escape official statistics and strategic planning in Africa. The series of studies CIFOR conducted between 2010 and 2013 in Cameroon, Democratic Republic of Congo, Gabon, Ecuador, and Indonesia (Cerutti et al. 2014) shows that employment in the informal domestic timber market can be several times higher than in the formal sector, ranging from 45,000 people in Cameroon to 1.5 million in Indonesia. In Cameroon, this sector accounts for the greater part of the wood economy and employs many more people than export-oriented industrial logging.

Thus, the inversion of the formal timber sector in Africa is aggravated by the lack of understanding and strategic integration of the informal timber and non-timber forest economy. The way Africa consumes its wood, by favoring fuelwood and the export of primary products, logs and lightly processed sawnwood and plywood, is a vector of underdevelopment. The incentive for foreign operators to invest in local processing is weak, while private African actors are, for the most part, trapped in informality and illegality, for lack of financial strategy and capacity to invest in the higher segments of the wood chain. NTFPs are equally neglected.

The development of cutting-edge industries in food and beverages, health, cosmetics, flavors, fibers and other natural products is within reach. But the lack of political attention and strategic investment to develop this immense potential is daunting. This contrasts with high investments, whether political, legislative, financial, and/or intellectual, in much less consequential projects, while the people's economy is going nowhere.

The flip side: framing a response to the forest poverty paradox

We started this chapter by highlighting the inverted nature of the African forestry sector and its role in maintaining poverty conditions. Yet, 20 years ago, the *Forestry Outlook Study for Africa* (FAO 2003) had already identified the continuing focus on fuel wood and low value-added exports, such as industrial roundwood and sawn wood, as a trap for the region. The report maintained this was keeping Africa in a sluggish segment of trade, the antipode of 'the phenomenal growth in the trade of processed items' and a corresponding 'decline in the share of unprocessed items' (FAO 2003). So why is this unique phenomenon still the norm in Africa, and what might be the options for changing the status quo? We argue in the following two subsections that this un-development, epitomized by the forest poverty paradox, persists because of deep colonial and ideological roots, each with a role of its own. We then critique the scientific discourse on NTFPs and tropical deforestation, to serve as the basis for critical propositions on how Africa should address and integrate its nature management and sustainable development. We then present new models and principles of collaborative and ecological economy, which can open credible alternative pathways to viable African futures. Such analyses and perspectives greatly influenced African MF in our attempts to develop an economic framework of transformative change at the landscape level.

Colonial and ideological continuities in the construction of the poverty paradox

The current structure of the African forest sector, notably, its parceling into myriad parks, reserves, and concessions, is a colonial legacy. Systemic inertia alone, however, was insufficient to maintain colonial influence, no matter how deep, for so long. Active systems of thought and values endowed with their own political economy, that is, their own rationale in the global interplay of powers and interests, proved essential in order to preserve some of the colonial fundamentals in the evolving status quo. We briefly examine and illustrate these two – historical and ideological – components of the problem.

The colonial historicity of the concessionary model

Historically, concessionary politics and fortress conservation (Hulme and Murphree 2001) were key drivers that shaped the architecture of forested landscapes. A cornerstone of feudal tenures in Europe, the concessionary system evolved in the 17th century to become a tool of European colonial expansion with British and Dutch charter companies (Karsenty 2010). Portugal established huge trading concessions in the 18th century to attract Portuguese settlers to Mozambique; the King of Belgium, the infamous Leopold II (Hochschild 1998), inaugurated the system in Central Africa at the end of the 19th century. France copied the system and established 40 massive territorial concessions over 700,000 km^2 covering about 78% of the present-day Central African Republic, Chad, Congo and Gabon. Until 1929, private companies had state power of justice and policing in these giant concessions that encompassed all types of resources (except mines), along with native villages and towns (Coquery-Vidrovitch 1972).

Around 1910, the concessionaires' exorbitant privileges were curtailed. They had to give back parts of 'their' territories to the state and to limit themselves to one product, either timber or rubber. In exchange, they were given full ownership of the land they chose (Karsenty 2010). The first forest concessions were born around that time, granting huge powers despite official restrictions. In 1910, European commercial traders began organizing large-scale commercial hunting in Central Africa (Roulet and Hardin 2010). As part of the concession system, vast expanses of forestlands were withdrawn from traditional communal controls (Diaw and Njomkap 1998). Meanwhile, protected areas were expanded, initially as forest or game reserves (Diaw and Tiani 2010).

Originally inspired by American theological romanticism, the first modern parks were created in settler territories at the end of the 19th century (Diaw and Tiani 2010). This was a time when policies could be imposed by force, without true negotiation with the natives. Sabie (Kruger) in South Africa and Amboseli in Kenya were established as early as 1892 and 1899, respectively. Other reserves followed in the 1920s and later within the framework of land ordinances, forest acts, laws and decrees targeting broader aspects of the relationship between natives and Europeans (Diaw and Njomkap 1998). The reclassification of reserves into national parks accelerated after WWII (Adams and Hulme 2001), continuing to the present.

Thus, in Africa, concessionary and conservation policies were linked early in the 20th century, along with a certain form of 'participation' (Joiris and Bigombe 2010). Conflicting public demands from Europe on colonial authorities to develop the colonies and, at the same time, to intervene to avert the loss of a disappearing natural Eden, found an outlet in the coupling of extractive reserves with 'virgin', 'inviolate' natural parks. Taken together, parks and productive forest reserves thus reconciled the tensions in the West between utilitarian and moral demands on nature to become the twin udders of the 'commercial and aesthetic dreams of colonialism' (Neumann 1995, p. 153). This basic structure, and its estrangement from the deepest transformational aspirations of Africans is still very much the norm in forested African landscapes.

Long-distance claimants in the fate of African forests: advocates, merchants, and scientists

The twin mercantile and conservation functions assigned African forests in colonial times remain the global environmental benchmark for Africa. How is it that the colonial forest infrastructure and its flawed structure of value creation could remain the norm for so long, with all the resulting un-development? Africa as a place of raw material production and export has benefited the developed world for centuries, which surely has some relevance, along with the system's ideological functions. In classical social theory, ideology is a mask, an obfuscation of reality that gives people a truncated picture of how the world works in order to help those in power to maintain control. It is, in fact, all too common to see papers on the very structure of the forest industry falling into that trap. For instance, the critique of the 'inherently anti-poor' model of large-scale industrial timber concessions and its failure 'to contribute to the sustainable development of (African) countries' by Counsell, Long, and Wilson (2007, p. 133) falls short. In none of its 24 chapters, including on economic sustainability (Counsell 2007; Van Dorp 2007), do they mention the low processing, low value-added trap in the sector and the sea change in jobs and prosperity that Africans investing in the advanced processing of their own forest products would induce. Instead, we see a propensity to address legitimate but soft impact topics, e.g., on rights and Indigenous people, or false leads such as 'pro-poor forestry initiatives' – as if the 'poor' were to remain a fixture in a never-ending African forestry status quo.

In fact, as Diaw (2005a) noted, there is a fierce, ongoing struggle for control of African forests, involving three key groups of actors: (1) an inner circle of 'forest owners', in particular African States and local communities; (2) 'Second-tier claimants' holding state-granted subsidiary rights; this group includes protected area managers, industrial loggers, and other industries and projects; and (3) 'long-distance claimants' who presume to have far-reaching rights to nature. This includes wealthy, powerful and influential donors, scientists and conservation advocates who draw on moral and aesthetic ideologies, widely relayed in the global North, to legitimize their demand for rights over distant natural spaces. Their discourse is dual: it posits that biodiversity is a global issue that must benefit all humankind yet refuses to see their national wealth as a global benefit. Thus, while granting themselves rights over forests of developing countries, they do not in return assume their duty of solidarity, for instance, by sharing ecologically relevant technological advances or by allowing freedom of movement to industrialized countries. This amounts to saying: open your parks to us, which are also ours, those which God has given us all as heritage, while we keep our technologies to ourselves and close our borders to you (Diaw 2005a).

This group of long-distance claimants has had the most influence on our forest-related paradigms, which sometimes resemble discursive innovation – that is, the constant producing of new paradigms that do not address core local development issues, yet have deep, intrusive impacts on the organization, language and focus of forest actors and institutions. Looking closely, with sensitivity, we realize that these paradigmatic solutions are inadequate, often stopgap measures that rarely address the fundamental problems highlighted in this chapter. These paradigms typically subordinate the development interests of the global South to environmental objectives as conceived in industrialized countries.

This is most obvious in the struggles surrounding industrial logging, which European extractive interests dominated until their recent progressive displacement by Asian, notably Chinese, interests. Two poles are at loggerheads: foreign industrial loggers vs. long-distance claimants – international advocacy NGOs and their African, science and media allies. The latter are waging a virulent long-term campaign to phase out logging, without ever addressing its ecologically benign transformation to the benefit of Africans. Scientists are actively involved in these struggles (Gray 2002; Legault and Cochrane 2021; Lescuyer et al. 2012), no matter their cautionary approaches and their search for objectivity. In fact, logging, occurring within an ocean of poverty, having done little to improve local people's livelihoods, is an easy strawman to destroy (Fuller et al. 2018). This article, linking Chinese wood imports from Central Africa to deforestation and, anecdotally, to Chinese furniture exports to the US, incited a mini storm of offended Western media laments about *'How Furniture Demand in America Thins Forests Across Central Africa'* (ALERT 2018; Cannon 2018). An analysis of this discourse abundantly shows the bias and presumptive position of the protagonists who consider themselves entitled to call for US consumer boycotts of Chinese imports without ever considering the position and interests of Africans. This is typical. In spite of all the brouhaha, the debate never touches the poverty paradox, the inverted structure of the African forest industry, the very low value additions, the job opportunities lost, or the remedies offered by structural transformation. By structural transformation, we mean the transition from a low productivity economy to a highly productive, imaginative, diversified, low carbon, social economy (Diaw and Franks 2019). As in other cases, the conversation focuses on issues that remain peripheral to our core problem, which is left unspoken and unaddressed, despite being at the core of all true sustainability prospects for Africa.

Is an ecological development future for Africa possible? A critical proposition

To carve out a bold future for themselves, Africans must explore transformative pathways by which to invest their creative imagination, intelligence, and immense natural wealth. They must take responsibility for the current situation, without blaming everything on dominant, Western-inspired paradigms and the mind-eroding, stifling impact of post-colonial institutions and their mimetic organization of States, languages and education systems. Africa's official languages of instruction and government are not the languages of the people but foreign languages – an interesting parallel with the forest paradox. So an effort must be made by Africans to free their minds and exercise their creative freedom to strengthen their position in the world and move into the future they want. We examine two cases: 'sustainable use' of NTFPs, on the one hand, and the conflict between forest conservation and food production, on the other, to problematize this challenge and illustrate possible creative pathways toward unapologetic African futures. We continue with climatic and demographic data showing that Africa really has no choice and must act. We then describe some transformation options based on emerging principles of ecological and collaborative economy. These are not the only possible pathways to a sustainable future, as envisioned scenarios (Durán et al. forthcoming; Pereira et al. 2020) indicate, but they are the closest to AMFN's ecological and collaborative economic strategies and provide an appropriate background to the discussion in Section 3.

Hidden values: a subsistence use paradigm that needs changing

Sills et al. (2011) conducted an important review of international attitudes toward NTFPs, making the case that many had been historically mainstream trade commodities,

> …driving the fabled spice trade between Asia and Europe, expanding in the colonial period with products such as shea butter (*Vitellaria paradoxa*) and gum Arabic (*Acacia* spp.) from Africa, and feeding the industrial revolution with products such as rubber from the Amazon (*Hevea brasilenses*).
>
> (p. 24)

After WWII, the focus on timber exports from colonial concessions coupled with the rise of cheap synthetic substitutes led to the decline of forest products such as gums, resins, fibers and medicines in both international trade and international policy discourse.

FAO stopped collecting and publishing data on NTFPs in 1971, while NTFP research became essentially descriptive, unconnected to strategic management or economic value options (Sills et al. 2011). Though forest products had remained central to the culture, knowledge systems and economies of rural people in Africa and the tropics, they were increasingly seen as 'minor forest products'. They were neglected by governments and, at worse, considered a nuisance, interfering with rational timber management and biodiversity conservation in 'properly administered' concessions and parks.

There have been positive changes since those earlier times. Riding the wave of the sustainability discourse, the late 1980s brought renewed interests in NTFPs. Sills et al. (2011) give a detailed and captivating account of the massive efforts and literature dedicated to this rehabilitation. From development and analytical perspectives, a major flaw underlies the renewed interest in NTFPs. Mostly framed through the lenses of traditional Western interests in the conservation of tropical forests, NTFPs were seen as a way to reduce deforestation through subsistence use and trade. Some scientists debated that they were important for poverty prevention but 'poor instruments' for poverty reduction, which led to speculation about whether they were a 'safety net' or a 'poverty trap' for poor forest people (Angelsen and Wunder 2003). Overall, anyway, NTFPs were cast in the role of an alleviating mechanism – a crutch – for 'poor people using *simple technologies*' that do not require forest destruction.

Thus, the structural transformational potential, neglected in the case of wood processing, was likewise ignored in the new NTFP discourse. Africans were absent from these discussions in the early years. This, of course, did not facilitate any strategic rethink or policy uptake allowing for the development of technologically advanced NTFP-based industries for food, medicine, cosmetics, fibers, and nutraceuticals. The research and debates were restricted to the search for a 'silver bullet' by which 'productive conservation' would save forests from logging and deforestation.

Deforestation models, food, and development: overturning causal analytics in the DRC

Several studies looked at the extent to which food production and forest conservation could be on a collision course in Ethiopia, Tanzania and Ghana (Franks and Hou Jones 2016; Franks et al. 2017). The studies were extended to Mai Ndombe province in the Democratic Republic of Congo, DRC (Diaw and Bisimwa 2017; Diaw and Franks 2019) and showed, as expected, that the production of staple crops involves expansion of agricultural land at the expense of forests in all those countries, and at an increasing pace (Diaw and Franks 2019). This initial conclusion perfectly fits Geist and Lambin's (2002) seminal paper on tropical deforestation, which found that agricultural expansion is, 'by far, the leading land-use change associated with nearly all deforestation cases (96%) worldwide' (p. 145). Though a valid conclusion, this correlation is also a source of complications when we need to devise responses to the mechanistic relation between more people leading to more land to clear and more deforestation. The fact is that Geist and Lambin's causality framework, though seductive in the way it distinguishes direct from indirect drivers, proximate from underlying causes, is also fundamentally mechanistic. Population growth is mechanically correlated with agricultural expansion (Diaw and Franks 2019). The framework is cast in the past and cannot account for the array of future decisions and intervening factors that can transform this correlation. Similarly, viewing poverty as an 'underlying cause' is deeply ideological and of little use in the search for effective strategies. How can poverty, a consequence of something, become the ultimate underlying cause of anything (Diaw and Franks 2019)?

In DRC, Geist and Lambin's causal framework, has had considerable influence in deforestation thinking and related policies. Unlike in Ethiopia, Tanzania and Ghana where the political and economic forces in the food system are far stronger than those in the forest sector, DRC policies are dominated by the externally funded Reduction of Emissions due to Deforestation and Degradation of forests (REDD+) program, which aligned agricultural policies to forest conservation priorities: a unique case in Africa of agricultural policies being defined through an environmental window. The State's weak policy practice in both forest and agriculture sectors led to powerful external actors stepping into this vacuum to set the agenda (Diaw and Franks 2019).

Indeed, some momentous decisions were made, such as pushing for a policy of population displacement to ease the pressure on forests. The DRI framework – Displace, Replace and Intensify – as Diaw and Bisimwa (2017) coined it, sums up this strategy in the REDD+ jurisdictional pilot in Mai Ndombe Province. The DRI triptych expressed the intention to:

1 gradually shift (*displace*) the production of food crops to degraded forests and savanna environments, using a combination of negative and positive incentives, such as preventing funding to food crops in Mai Ndombe forests and paying farmers for 'results-based' environmental services
2 phase out (and *replace*) shifting cultivation (slash and burn)
3 *Intensify* the farming system by introducing perennials, particularly in savannas and degraded forestland.

This strategy's good points – more balanced intensification combining food crops and perennials in savanna areas and developing tree crops and community forests in forest districts – are thwarted by very flawed objectives in the intent to discourage the cultivation of food crops in forest areas and to phase out slash and burn agriculture. Targeting the livelihood systems of the poor as a change strategy is fundamentally wrong, both morally and on realistic strategic bases. It denotes both a common dramatic lack of empathy for African farmers and tragic miscomprehension of the shifting cultivation system, which actually integrates food crops, perennials, and forests into a unified system of tenure and social redistribution of land over time (Diaw 2005b). Such tactics could only be envisioned in a context where powerful external actors, further empowered by longstanding blame-the-poor deforestation narratives, were left free to set the agenda.

This 'framework of causality' – in both its construction of a scientific object and its attribution of blame to the poor – dominates contemporary deforestation analysis, negatively impacting African policymakers' ability to address the disconnect between food production and forest conservation in novel ways. It places a conceptual straight-jacket on our ability to open up to a wider range of future options. The causality framework defines the framework of solutions; this is why most efforts to address deforestation are focused on the production systems of the poor, while perspectives for simultaneously addressing deforestation, food supply, and development run into a wall (Diaw and Bisimwa 2017).

The concept of 'forest transition', originally developed by Mather (1992), opens more interesting possibilities. Forest transition describes the evolution of the forest cover of a country that has developed in three phases: (1) a phase of forest abundance and low deforestation; (2) a phase of rapid deforestation and reduction of forest cover; (3) a phase of stabilization and eventual increase in forest cover. This last phase contains 'the turning point', meaning 'the end of the country's deforestation episode' (Wolfersberger, Delacote, and Garcia 2016). Rudel, Schneider, and Uriarte (2010) observe that forest transitions can unfold over very variable periods of time, 500 years in Scotland, but only about 30 years in Vietnam where, like China and India, the forest transition is in progress, showing that reducing the deforestation episode is possible. GDP is the key factor, while faster economic development allows for a shorter deforestation episode in a country. Other variables, such as the quality of institutions (governance), the decline in corruption and the marginal value of forests positively affect forest recovery. Population density, demographic growth, agricultural exports, and the initial extent of forest cover, also have negative influence and need to be overcome to hasten a turning point. These findings come from a study of 57 developing countries, 11 of which had reached a turning point between 1990 and 2010 (Wolfersberger, Delacote, and Garcia 2016). With regards to the forest paradox, these results tell, among other things, that a country with abundant forest resources tends to valorize them less, and that this lower valorization induces a longer phase of deforestation. The increase in the marginal value of forests, that is, investments adding significant value to standing forests, is then crucial to reducing the period of total deforestation (Ibid.).

Considering these findings, the absence of any serious discussion or study of forest transition in Africa is incomprehensible. As an enlightened conservation manager noted in a conversation (Pers. Com., Kinshasa, June 26, 2017):

> You never hear about forest transition; this dismisses the whole question of the relationship between reforestation and development. We are dealing with short-term explanatory cycles that take away the means to think about the transformation of socio-environmental systems in the long term.

The Congo Basin forest policy space is suffocating under the weight of dystopian narratives promising dauntingly bleak futures if people do not renounce their development aspirations in order to save the forest for 'future generations'. This, of course, is not an option. The real option for African governments concerns the levels of strategic deforestation they will have to accept in exchange for the increased quality of life and improved domestic commodity and food supply needed by their people. It will also be about their engagement to work resolutely on all aspects of the governance, resource coordination, investments, innovations and smart ecological development needed to get to their country's turning point faster. Other regions are doing it. Why not Africa?

Social and ecological economy: a field to invest

Four major facts deserve the attention of those involved in facilitating change through participatory processes in forested African landscapes. The first is that the perpetuation of the poverty paradox is intimately linked via the blindness of external policy and project protagonists to the inversion of the African forest economy. This concerns all actors and approaches and it exposes a collective failure to truly serve the people. The failure is collective and thus starts with African actors and their governments. As to the participatory paradigm, it must be part of the solution, if it does not want to be (or remain) part of the problem. As this chapter, and perhaps, this whole book, illustrates, it has the ingredients and capabilities needed to embed itself in the social economy of the people. It can uniquely contribute to stimulating a truly value-creating economic process in which the constitutive values of governance will be able to express themselves within shared spaces of economic prosperity. Figure 10.4 conceptualizes this integration of governance with the economy.

Why Africa? A new frontier for population, climate, and development

The second major fact comes from prospective data: one in four people will be African in 2050 and 40% of the children will be born in Africa, according to the latest United Nations (UN) estimates. By 2040, the approximate horizon of African emergence plans, the continent will have a labor force of over 1.1 billion people, more than either China or India. According to the UN, it is the only region in which population may still be growing after 2100.[7] All other regions will have stagnant or regressive demographics. In fact, Africa is the new world frontier for both population and climate, as we show below.

The third point is that the continent has become, over the past 20 years, the second fastest growing region in the world; but African economic growth, even when strong and sustained, creates few jobs. Africa's job creation has been driven by sectors like agriculture and forests, whose extroversion has hitherto prevented the necessary internal transformations and synergies between agriculture and industry. Yet, the potential for transformative value addition is immense. Africa's rapid economic and population growth will stimulate future demands for both wood and non-wood forest products. In addition, most wood products are intermediate goods that enter as raw material into other industries, such as construction, furniture, packaging, printing, and textile manufacture (ANRC 2021a). Each value-added segment will fetch up to four times more

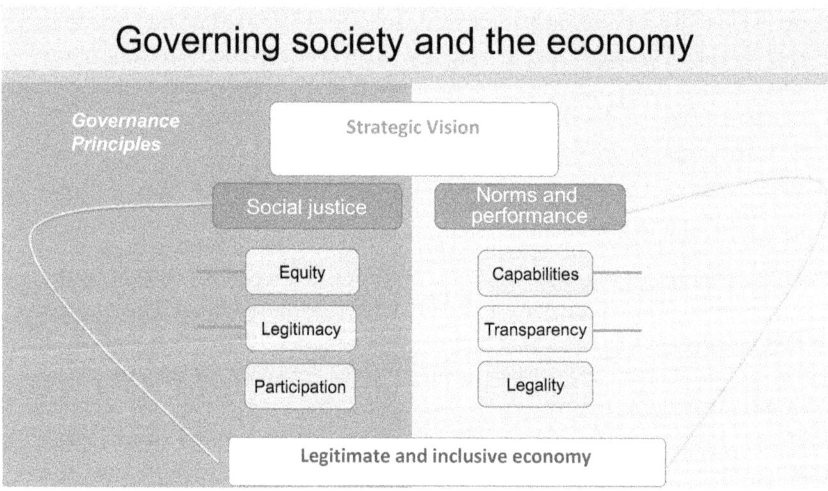

Figure 10.4 Inclusive governance principles and the economy

income than exporting in the raw form; the furniture industry, for instance, will generate as many as 4–12 times more jobs than primary sawmills (Hierold 2010, cited by ANRC 2021a).

This is equally true for NTFPs, with staggering conversion potential into medicinal, natural health, nutraceutical, cosmetic, and food industries. In 2020, the global wellness industry, spreading across 11 sectors, most of them depending on natural products, was valued at US$ 4.4 trillion, a fall from US$ 4.9 trillion in 2019, due to the COVID-19 pandemic.[8] This is almost 8 times the global wood market estimated at $631.11 billion in 2021.[9] Just the markets for Personal Care and Beauty ($955 billion) and for Healthy Eating, Nutrition, and Weight Loss ($946 billion) were each 50% bigger than the wood sector. Together with Traditional & Complementary Medicine ($413 billion)[10] they are nearly four times bigger than the wood industry. How much will Africa gain in market shares, jobs and environmental, cultural and social benefits by entering these markets and bringing out the multiple values of its standing forests and indigenous knowledge?

The fourth and last point concerns the relative coevolution of carbon emissions per capita and GDP per capita (Figure 10.5). The data trend is clear: there has been a growing divergence since the 1990s between global South emissions, which are increasing, and emissions from the North, which are declining. China's emissions exceeded those of the United States in 2006. As for Africa, it accounted for only 3.6% of emissions in 2006, but these have increased 12-fold since the 1950s (Olubusoye, Musa, and Ercolano 2020), and the continent still represents only 2.84% of world GDP for 16.72% of the population.[11] Some scenarios project that Africa will represent 20% of global emissions by the end of the century (Ayompe, Davis, and Egoh 2020). This would quintuple 2006 emissions in the span of a century, although it took China just 30 years to more than triple its 1990 emissions (Larsen et al. 2021). Much will depend on the direction Africa takes for its development.

Africa's growth is still based on the use of fossil fuels, as has been the case in other emerging and industrialized countries. Its population has been by far the most resilient to the COVID-19 pandemic and has shown the most apocalyptic and malicious forecasts of the COVID carnage there to be false. What will happen when Africa's population has doubled and adopted the consumption patterns practiced today in China, Europe or America? Something needs to change. Our goal of a prosperous Africa, almost twice as populous as China in 2050, is inconceivable without an entirely new economy – a challenge for Africa and for the world. It will be necessary to create and share great wealth, while dissociating it from the production of CO2 and the destruction of the environment. It is a strategy of decoupling, around which there is no unanimity but without which it is difficult to envision a viable future for the 10 billion people – many of them Africans – who will populate the planet in a quarter of a century. These questions are behind the global objectives launched in 2015 – the Sustainable Development Goals (SDGs) and the COP21 climate agreement.

Rank in global emissions	Countries	Emissions 2013	Emissions per capita	GDP per capita 2013
2-	United States	[illegible]	[illegible]	[illegible]
9-	Canada	550	15.7	52,939
7-	Germany	[illegible]	[illegible]	[illegible]
6-	Japan	1360	10.7	38,633
8-	South Korea	[illegible]	[illegible]	[illegible]
3-	EU 28 (inc. Germany)	3740	7.3	25500
5-	Russian Federation	[illegible]	[illegible]	[illegible]
1-	China	10330	5.4	6,959
10-	Indonesia	[illegible]	[illegible]	[illegible]
4-	India	2070	1.7	1,508

Data sources: World GDP: Knoema, http://knoema.com/nwnfkne/world-gdp-ranking-2015-data-and-charts; EU 28 GDP: Eurostat, https://www.stat.ee/29958; Emissions and emissions per capita: Netherland Environmental Agency, 2014, http://edgar.jrc.ec.europa.eu/news_docs/jrc-2014-trends-in-global-co2-emissions-2014-report-93171.pdf

Figure 10.5 The coevolution of carbon emissions and the wealth of nations

Ecological and collaborative trends for a new economy

A new economy needs to be defined. It is essential but does not yet exist anywhere, except in embryonic forms. The paths to its future can be many.[12] From our perspective, this economy must be not only green and blue, but also a social solidarity economy. By blue economy, we include but go beyond the model of an ecological maritime economy promoted by the Seychelles and the European Union, and we refer primarily to the concept developed by Gunter Pauli (2009; discussed below). One shortcoming of the green economy is its lack of systems thinking. It aims to be more ecological and sustainable, but its products are expensive and inaccessible to a significant portion of the population. Investment in a green image in one place can cause ecological and social disasters elsewhere. We are still in the linear economy that 'produces, uses, and throws away' without worrying about waste – one reason for so much industrial wood waste in Central Africa.

The blue economy, however, relates to a family of new concepts that engage people and value local resources differently. It is a 'whole system' regenerative economy, built from the bottom up, with constant transformative innovations. It includes principles of the renowned circular economy, under the concept of ecosystem 'nutrient and energy cascades' where the spent substrate from one natural or productive process becomes the raw material of another. Principles of symbiotic economy and ecological intensification, such as the one MF put in practice with their biofertilizer innovation (see next section), mesh well with the blue economy. All use biodiversity and refined understanding of natural, physical, and biological interactions, to reduce waste and harness the productivity and efficiency of natural processes as inputs into a more productive but also more sustainable economic process. The terms blue or ecological in reference to the economy fundamentally mean all these ways of understanding and integrating natural processes and principles into the economic process. Barbault and Weber (2010) describe it beautifully in their plea for an 'ecological revolution of the economy'.

The collaborative economy is a close kin to the blue economy. It refers to the social side of this economic revolution, a social and generative economy that is in the process of emerging. Its borders are fluid in their interconnections among themselves and with their ecological counterpart. Therefore, we use the terms collaborative, social, solidarity and/or generative economy as loose conceptual umbrellas for the range of emergent, socially oriented models of economic organization that create value differently than the current ultra-extractive capitalism. This dominant economic model has been the principal force behind our growing climate and biodiversity crises. Although clearly several types of economies will coexist in the world for some time to come, this new economy is gradually gaining momentum.

By 'new economy', we refer to both the ecological blue economy and the collaborative social economy. We acknowledge the connection as well as the differences among their various components, but also posit that they are all in the process of *becoming* and that attempting to differentiate strictly would be futile and unnecessary. The collaborative economy, for example, was estimated to grow from about US$ 14–15 billion in 2014 to US$ 335 billion by 2025 (Yaraghi and Shamika 2017),[13] although evaluating it has been quite complicated. This is because of continuous growth and evolving differences with various expressions of the sharing economy, from the most altruistic to the most capitalistic, including the Internet of Things, MOOCs (Massive Open Online Courses), crowdfunding, commons-based peer productions, such as FabLabs, coworking, and the design and manufacture of 3D printed objects, as well as peer-to-peer lending, couch-surfing, ride-sharing and car-sharing, knowledge and talent sharing, etc. It is an economy of sharing as much as of functionality, in which we trade services rather than goods, as in carpooling and in paid exchanges of homes, meals, tools or libraries. Companies like Airbnb and Uber extract and accumulate value from social relationships; others, in the movement for generative commons and peer production economy (Bauwens and Niaros 2016) strive to create value through open, voluntary contributions, mutual coordination, and the sharing of resources.

For its part, the ecological economy, blue, circular, and symbiotic, emphasizes local solutions and skills, and short value chains; it is closer to people and less expensive in transport. It uses gravity and the sun as a source of energy, physical rather than chemical processes; it reuses waste as nutrients or as raw material for other uses and integrates the five kingdoms (algae, fungi, plants, bacteria and animals) in productive processes. This imaginative and inventive economy, adaptive and based on intimate knowledge of natural resources and interactions, already exists in Africa. This is the case, for example, of the Songhai Centres, which, after 30 years of experimentation in Porto Novo, offer a convincing prototype of agro-sylvo-pastoral integration, based on complex nutrient cascades, productive integration, and value adding innovations. Songhai Centres use animal, plant and human waste, microorganisms, solar and biomass energy, as well as machines made from recovered waste, while also applying structural transformation principles in medium-sized food and beverage industries.

The mycorrhizal biofertilizers, produced by the AMFN (see the next section) are part of this movement and are the subject of increasing applications in West and Central Africa. African soils are immensely rich in the symbiotic microorganisms – fungi and bacteria (rhizobia) – needed to produce such biofertilizers, which capture nitrogen from the air or colonize the root system of plants to collect phosphorus, water and nutrients from the deep soil (Davet 1996; Jeffries et al. 2003; Kittiworawat, Youpensuk, and Rerkasem 2010; Mousain, Matumoto-Pintro, and Quiquampoix 1997). These biofertilizers are widely used in India and Brazil. Bio-pesticides and other fertilizers of organic origin (compost, manure, green manures, bio-char, and uri-char) are also used in Africa, as are crop rotations, improved fallows, agroforestry and physical arrangements (stone bunds, grass strips, Zaï). The production of lumber (breeze blocks, pens) based on recovered materials is part of the same logic, while the renewable energies, of which the African continent is rich, can reduce the costs of transforming the nutritional, cosmetic and therapeutic sectors of agriculture and forests.

Africans are increasingly making innovations based on circular economy principles and inventive adaptations of the smartphone to the needs of small producers (Diaw 2014). In Central Africa, rural development projects have been experimenting for years in agroforestry and the domestication of wild species, the development of NTFPs, and unconventional breeding (snails, grasscutters, honey, etc.). Young social enterprises are emerging and national NGOs developing new products. MF partners such as *Santé-Nature Congo,* in the development and transformation of *Moringa olifeira*, and Agro-PME Cameroon, in the market expansion of Penja pepper as a Protected Designation of Origin (PDO), are regional leaders in their domains. Basic elements therefore exist for scaling up and building on these micro-experiments.

The experience and know-how developed in participatory frameworks in Africa can contribute to this change, but this, though necessary, will be insufficient, as the MF experience shows in the next section (see also Chapter 11, this volume). The seeds of change must be part of a deeper movement to reform the African economy, and this will require strategic vision as well as investment and innovation capabilities to support the high value-added products and novel social-ecological creations coming from private start-ups as well as grassroots social enterprises. Africa has a 'latecomer advantage' because it has not yet industrialized and can, thus, leapfrog or skip-stage to renewables, to nature-based technologies and eco-friendly industrial models without having to deal with the infrastructural inertia of the old economy (Lee, Juma, and Mathews 2014). Grassroots innovation will be insufficient; sustained backing by governments in terms of R&D support and financing of business innovators, whether privately or socially oriented, will be needed to harness emerging opportunities. Asia did it (Tseng, Lin, and Tuong Vy 2012) and Africa can learn from it and act accordingly.

Model Forest trajectories: a social economy pathway

MF[14] trajectories in Africa illustrate the ways in which the participatory paradigm can be inscribed in the social economy. It is by no means a unique or exhaustive pathway and, even less, a model to be followed to the letter. Rather, its lessons provide information on the strengths available to players in the sector as they work for the emergence of a new economy, one which truly creates value for the people.

From ACM to Model Forests

In Africa, MF first emerged as an ACM project. ACM's initial aim was to look for the conditions of self-sustaining forest management systems; in that line, it examined a confluence of questions related to adaptation and collaboration in community-based management, complex multifunctional landscapes, conservation projects, and environmental governance. When we began developing MF in Cameroon, circa 2003, ACM practitioners soon realized that they had found the perfect 'hardware' for the long-term transformative work they had started as a simpler action research program. MF are built as a place, a partnership and a process, intended to last several generations, well beyond the lifetime of a single project. The place is a landscape covering all land uses, from community forests, to parks, logging concessions, farms and cities; the partnership is voluntary and inclusive of all stakeholders; and the process is a journey of dialogue, experimentation, and innovation designed to work with the partnership to achieve a common vision of 'sustainability'. Each MF is unique but all share this common framework, underpinned by six core principles: partnerships, landscapes, sustainability, governance, program of work, and networking (Bonnell et al. 2012).

Stemming from the participatory paradigm, both ACM and MF are rooted in research and learning approaches, particularly learning by doing. MFs were therefore built in Africa as learning platforms and open systems, capable of, and committed to transforming themselves. This gave them the ability to persist in the face of difficulties and to reinvent themselves at various times by incorporating lessons learned from experience. MF practitioners were thus able to work on the participatory paradigm from within, constantly questioning its viability from the perspective of communities' sustainability. This constant questioning was a lesson integral to the ACM process, wherein practitioners aim to 'walk the talk', learning as they proceed and altering their actions and goals as experience, learning and collaboration dictate. The result is a bumpy and uneven process. In this case, it led the MF teams, step by step, to recognize that participatory governance alone had little for the people, unless it could manifest in the economy, as a means to truly meet the local demand for a better life.

Landscape construction as experiential transformation

The concept of landscapes, applied to the planning or management of territories, dates back at least half a century (see, e.g., Butler 2014; Colvin 1970; Crowe 1958).

Only in the last decade, however, have landscape approaches gained ground – in response to the failure of conventional conservation approaches, the need to build a more resilient ecological agriculture, and adapt to, and mitigate climate change.

The MF approach is one of the rare landscape approaches with a systematic method for convening stakeholders at different scales in a well-defined space (Diaw 2017). In the African experience, this convening occurs at three levels, via (i) political entry at the highest possible level of government and with full knowledge and integration of a country's priorities and transformation policies; (ii) landscape-level entry through the MF territorial convention; and (iii) lateral solidarity expressed at various territorial, national, regional and international scales through the MF Network – each further explained below. The economy has proven, with experience, to be a fourth, essential, level of landscape integration.

The political entry is based on the systems theory of interactive feedback loops, as formulated within the framework of ACM research (Diaw, Prabhu, and Aseh 2009; Mandondo, Prabhu, and Matose 2008). It starts from the idea that policies must incorporate local experience to ensure their social relevance and to be able to make the necessary adjustments and corrections over time. In turn, local processes need political decision-making levels to resolve some of their problems, whether they be conflicts, regulatory bottlenecks, or access to certain resources. The construction of this interactivity facilitates the scaling up of solutions resulting from local initiative and innovation. This aspect of the MF method has been one of the most fruitful in the African experience (Diaw 2017).

Landscape governance is the second key to a successful landscape approach. It relates to ACM's 'mid-level entry', as formulated by Diaw, Prabhu, and Aseh (2009), and takes the form of a two- to three- day landscape convention. This initial gathering of landscape actors is followed by a confirmation process that lasts several months and that includes 'common vision' meetings or workshop(s), the development of a strategic plan and a plan of work by the actors. This longer process can be thought of as the overall 'Model Forest landscape convention', even though the initial three-day convention is the triggering event and real starting point of the MF's landscape governance process.

To convene landscape stakeholders, AMFN developed a family of tools at the center of which is FELA – Freely Explored Landscape Agreements. The methodology empowers stakeholder groups to freely explore the ins and outs of the MF concept and decide on its relevance to addressing their issues. The actors decide whether or not to join the MF as a social project or a 'life project' and how they would structure it locally, before electing their representatives in provisional governance structures. FELA can be thought of, in retrospect, as a multi-stakeholder FPIC – Free, Prior and Informed Consent; FPIC being the internationally agreed standard for protecting Indigenous people's rights regarding projects planned on their lands. However, by putting all landscape actors on the same plane, FELA goes further to level the playing field between Indigenous peoples, local communities, local governments and other actors; thus, laying the ground for a common vision and a partnership for shared governance of the landscape.

Landscapes are socio-ecological constructions. Their distinct geographies and ecologies, socially interpreted and constructed in time, encompass senses of self and belonging that are creators of social identity. The places we have lived and the meaningful experiences we have had are integral to our self-image and how we feel about ourselves. This identity is therefore never fixed. It is constantly transformed and reconstructed by social practice. Because the initial landscape convocation is a mid-level entry involving a limited selection of landscape actors, it necessarily lacks 'depth'. Indeed, not all people in all places across the landscape are involved or aware of the convention. This is something that happens later as part of the longer term landscape building and information process. At the start, the process also lacks 'density', i.e., cumulative and synergistic achievements that impact people's lives, such as new infrastructure, impact projects, enterprises, or cultural events. The MF process still lacks the 'experiential contents' that eventually give new meanings to local identities. It takes time, practice, rigor, and courage for the structural foundations put in place by the landscape conventions to gain in depth and density.

MF communities in Africa have experienced plenty and scarcity, they have known moments of communion and flamboyance, and have gone through conflicts over process control or resource allocation; they have persisted on the land, but remain challenged in terms of self-possession and financial autonomy. As a new social construct, the MF brought together pre-existing organizations, without denying them but rather by integrating them more or less imperfectly into its collaborative matrix.[15] This allowed for the emergence of an open identity that partially deconstructs former divisions to strengthen the solidarity ties on the land. People thus experience, in conviviality or constructive conflict, their belonging to a community of practice that is expressed through lateral solidarity in concrete progress in the field.

MFs are therefore built in a retro-prospective logic where the actors recognize themselves as agents of economic and social transformations. This transformative project is credible in their eyes, because they have in-depth knowledge of the landscape resources and their potential for value. They understand the need to mobilize this potential for the benefit of a shared vision of the future. It is this set of visions and experiences, as well as their transformation into results and concrete impacts in people's lives, that give density and credibility to the MF's social project. This allows the structure to keep its organizational foundations as well as its existential cohesion, and to transform itself into a 'social infrastructure'. Landscape resilience is intrinsically linked to such shared identities as social constructions and as projections into the future.

New mechanisms for social governance of the economy

This co-construction aimed at transforming the landscape has brought the MFs to squarely face the question of the economy. From the MFs' elaboration of their first strategic plans in Cameroon in 2005, it appeared that, to succeed in Africa, MFs had to 'walk on their own two feet', that is to say with governance, from one side (as a container), and the economy, on the other (as content). Only some five years later, when the AMFN first received significant funding from Canada, was it able to launch and begin to shape its economic transformation program. Initiated in the form of classic development projects, this movement very quickly encountered the limits of such an approach, at the same time as it validated the extraordinary untapped potential of the territories.

The manufacture of pens from Cameroonian forests' wood wastes was the first concrete demonstration of this potential. Only about 30% of logs cut by the forest industry are actually used; the remaining 70% is abandoned to rot or serve as firewood. Set up by the Dja and Mpomo Model Forest (FOMOD) with technical input from the Lac St. Jean Model Forest in Quebec, this low-cost project revealed hidden economic value by transforming 10-centimetre wood scraps into high-end products offered at a price of 12,000–15,000 CFA (~US$ 24–30) per unit on local markets and up to US$ 100 in international wood fairs (Figure 10.6).

Dozens of such projects have been proposed in MF Assemblies, Councils and Platforms. They focused on agriculture (plantain, cassava, macabo, or pistachio) and on many untapped forest resources. In the first year, nearly 700 people (particularly, women and Indigenous peoples) engaged in giant snail breeding, oyster and mushroom cultivation, honey production, fruit trees, coconut groves, *Moringa olifeira* cultivation, and wildlife domestication. Nurseries were set up, as well as projects for the collection, storage and marketing of NTFPs and workshops for manufacturing rattan and wood bark fabric decorations and furniture. All was done with little money. For example, with only 10 million CFA francs (< US$ 20,000), women's platforms set up 87 micro-projects in 2010. Over time, they went on to concentrate mostly in food and cosmetics production, using NTFPs such as *Moringa olifeira*, *Allanblackia* floribunda, moabi (*Baillonella toxisperma*), cocoa, *mbalaka (Pentaclethra macrophylla)*, and *njansang (Ricinodendron heudelotii)*.

However, this very proliferation and the structural difficulties in carrying out and sustaining these projects illustrate the limits of the approach. In particular, the lack of funds to truly launch value chains has proven to be a critical handicap. More and more readily available investment is needed to achieve significant productivity gains, purchase machinery for processing, treatment and packaging, and establish relays in storage, transport and markets. But classic donor funding notably lacks the reliability needed to sustain long-term community transformations,

Figure 10.6 The Dja and Mpomo wooden pens

while microfinance penetration is low in rural areas and investment projects rare, as are social investors. The conditions for supporting the transformation of the economy by small producers are definitely not yet in place, whether financially or in terms of the mastery of innovation and technical know-how. These deficiencies have yielded mixed returns on the governance of the nascent economy. In local assemblies and platforms, allocation conflicts arose, as well as debates and disputes over priority projects: should people invest the limited resources available in agriculture, timber or NTFP sectors? Should they favor *Moringa*, *njansang* or Allanblackia? These choices were not just technical; they also determined who, among the diverse groups of actors, would be funded; not everyone could be served.

Within the management of the Network, these limits have fostered an acute awareness of the paradox of poverty. This led us to revisit Amartya Sen's capability theory, so relevant to the problem, how to understand it, and formulate an answer. The root of the paradox is indeed the immensity of African natural resources and indigenous knowledge in the face of the lack of financial sovereignty and the inability of actors to transform their natural endowments into goods and services over which they have 'legitimate control' (Sen 1984). Beyond the important issues of governance and deliberative democracy, another essential question relates to the investment and innovation capabilities of the actors (Lee, Juma, and Mathews 2014; Sen 1985). People must be able to convert their assets (resources and rights) into goods and services they can use. Capabilities, that is to say the bundles of means, knowledge, know-how, ideas and freedoms available to people for using nature and developing its potential are the key to this conversion process (Sen 1985; see Figure 10.7). This failure of capability is still, unfortunately, until today and despite all our efforts, the major problem that continues to confront African landscapes and the network of MF within them.

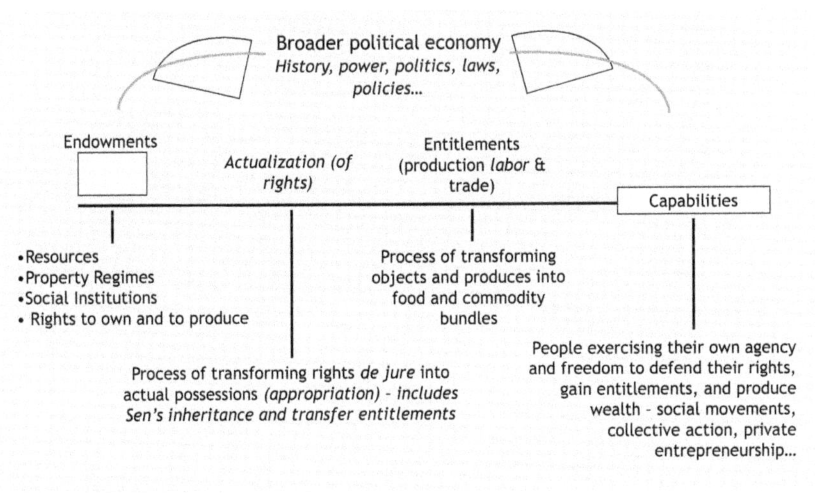

Figure 10.7 Entitlements, endowments and capabilities

The One Programme: emancipating tools for transformation

The objective of this subsection is not to showcase AMFN or market its 'exceptional tools'. That would not be very useful. The point, rather, is to provide a limited sample of the array of empowering capabilities people need to face the poverty paradox in their lives, and show that that is still not enough. The AMFN One Program was born in 2012 from the troubled but exciting deliberative process described in the preceding subsection; it was conceived as an integrative economic framework, a matrix that tried to match 'capacitation' tools with transformative change objectives, while being pragmatic in trying to attract the external partnerships and funding needed for investment and innovation in the economy of small producers. The program is based on social entrepreneurship, that is, a way of doing business that is driven by a cooperative social agenda rather than the search for maximum profit. The prime motivation was to create real value in the local economy and to reduce local organizations' massive dependency on external funding. It was also a way to reward local entrepreneurship, after observing the rent seeking behavior of a handful of influential project 'beneficiaries'. This type of behavior had been the source of uncounted local plots, allocation conflicts and inequalities. Breaking away from the negative consequences of donor dependency was seen as a basis for the emergence of an innovative model that would fully develop local value chains and capabilities. There was no illusion that this would not take time – at least a couple project cycles – and that it would require turning classic projects into investment projects, if the possibility arose. By investment project,[16] we meant funding that allowed for the purchase of agriculture and processing equipment, trucks and other productive machinery (see Figure 10.8).

Figure 10.8 The AMFN One Programme

To develop a social business culture in the MF, the One Program relied on the training of Local Expert Facilitators (LEFs), along with MF focal points, state extension agents, and rural leaders, particularly women. This was done through the MF Practical Itinerant School (PRAIS), which is defined by its content, not a fixed place or a building. As a mobile rural business school, PRAIS is oriented toward practical, quickly actionable learning.[17] It has enhanced the entrepreneurial competencies of a couple hundred LEFs, who, in turn, trained other farmers in the development of local agro-forest value chains. The MF also developed a system of farm schools facilitated by PRAIS trainees, particularly LEFs and extension agents, as part of a multi-stakeholder extension system. Each farm school associated a 'model producer' to nine 'partner producers' for scaling purposes, notably to speed up the adoption of innovations, once accepted by farm school participants.

Other tools, such as participatory monitoring and evaluation (PME), a label in construction under the name of *Model Forest Green Business* and the *Africa Living Land* (ALL) trademark, are part of the package developed by the AMFN to meet the requirements of developing local value chains and businesses. PME puts local actors in the foreground and gives them a means of controlling the dynamics set in motion in their living space. The MF Green Business label and the ALL brand give them greater visibility in markets. All contribute to making local actors more autonomous and resilient, in resonance with the values and ethics of the MF concept.

This practical philosophy contributed to effectiveness in the One Programme innovations and extensions in the field. In a span of 13 months, between 2013 and 2014, NTFP and biofertilizer innovations were disseminated to 2,700 producers and micro-entrepreneurs in Cameroon through the Business for Adaptation to climate change (B-Adapt) project. This was the MF's first investment project in ten years of existence in Africa. The project was extremely successful and seemed to provide an optimistic blueprint for scaling its many lessons, innovations, and social business model to Central Africa and the larger African region. But we had not accounted for the unexpected. In this case, an exacerbated power struggle between the African network and a powerful partner, an international NGO; the latter preferred to renege on a previous agreement and derail a highly successful generation project covering five Congo Basin countries, rather than carry it out in co-responsibility under the leadership of the African partner.

This second generation cycle was the foundation of the AMFN 2014–2024 Strategic Plan presented to the AMFN Board a few weeks earlier with its partners. The program's basic ambition was to complete the social entrepreneurship strategy that had emerged from the first phase and been refined using this second, much more ambitious, investment program in five countries, intended then to scale up again to other countries. This was to benefit more than 1.2 million farmers, facilitate the development of 100 social enterprises in key natural product areas, and generate a projected US$ 300 million in the 'real' economy by 2024, and US$ 1 billion by 2034. The program's fundamental flaw was its short-term dependency on a single dominant donor and its resulting vulnerability to external pressure or to a decisive partner reneging on a key agreement. This is a tangible reality in the political economy of projects, which MF 'ACMers' had to learn the hard way.

Conclusion

This chapter documents the AMFN attempt to establish the germs of a collaborative ecological economy in Central African landscapes, as a response to fundamental social and economic problems encountered in the field. Those problems reflect a condition found across most if not all African forested landscapes, which we summarize as the forest poverty paradox. As such, this chapter contributes to illustrating the exploratory principles and considerable transformative power of a participatory paradigm such as ACM when it embeds itself into real-life movements of change in society; it also shows its limitations, which correspond to broader limitations of the African rural/forest environment, in terms, particularly, of robust and independent investment and innovation capability.

The chapter looks at the conditions of landscape transformation from that perspective and analyzes how ACM contributed to this rich experience of establishing a novel form of local governance and trying to inject it into the local economy. The analysis may be unique in the way it documents this attempt to transform an African economy from the bottom up, with its richness as well as its challenges and limitations. The first take home lesson from this journey, then, is the role of ACM as a 'software' for the 'hardware' of MF. As an action research paradigm in Central Africa, ACM successfully embedded itself into the MF Network to better address the concerns of the people. The two joined to become a learning, evolving platform for transformative change and thus became inseparable in the African MF experience.

The contradiction between the importance of the investments made in Central Africa's participatory engagements and conservation policies, on the one hand, and the structural inversion of its forest economy, on the other, has long gone unnoticed by scientists and environment actors. The participatory paradigm has grown at the periphery of the economy, with a focus on questions of law, governance, and democracy that did not question the way the economy works. It took some time for scientists in the ACM program to take the full measure of the forest paradox – and only after a few years of working to establish the MF program. MFs, thus, did not start by implementing a social economy strategy in the African landscapes but remained for a time in a classical 'project mode'. It is only progressively that the complex strategic problem posed by the poverty paradox was perceived, understood, conceptualized, responded to, and learned from, through recursive cycles of deliberation and action. Applying these recursive learning principles is one of the distinctive marks of ACM.

In this chapter we tried to show the progression of this awareness and its gradual integration in the MF approach of transformative change in Africa. Through that process, we learned that it is not just a matter of improving one aspect here and there, contributing to small successes in some localities, and counting how many people got better off by the end of a project to strengthen applications for additional funding. New awareness of the paradox forced us to go beyond this vision of incremental change. We could no longer separate from the unspoken – maybe unintended – effect of external interventions in contributing to maintaining conditions of 'bearable poverty' in African landscapes. The scope was just no longer acceptable given the scope of the problem and the scope of the engagements and investments necessary. This is what justified the One Programme's ambitious social business and scaling strategies and the diversified set of tools and innovations the AMFN designed to initiate a transfiguration of local landscapes capabilities to meet the challenge of overcoming the paradox.

At this stage, this strategy has, so far, failed to trigger the scale of transformations needed for a gradual system shift in the African landscapes. It is difficult, in retrospect, not to see the discrepancy between the scale of ambition and the politically narrow and insecure basis of funding and investment. This leads to the most essential question: who is going to mobilize and secure the investments for a transformative program of the magnitude envisioned? Is it possible or even viable to mainly rely on external donors? Private corporations and donors have their own priorities and limitations. The corporations' priority is to develop their own supply chains, not whole landscape innovation and investment strategies, even when they are willing to support local landscape efforts. However, donors have their own country's competitive advantages and foreign policy interests, which do not always fit with the idea of investing durably in another country's industrial development, be it at the regional, national, or local level. Funding is conditional on the funder's strategies and interests, and not just based on the excellence or intrinsic value of a project. Other considerations come into play, which can be geopolitical in nature. So, who is going to fund the innovation capability for local value chains and industrial development and all the investments needed for innovative ecological-economic transformation in Africa? The roles of African governments, private actors and multilateral organizations and banks are fundamental and must be brought to bear in this critical area of structural transformation, the African forest sector. This is a key question in need of an answer.

The world is on the cusp of fundamental upheavals in the way it produces wealth. The evolving COP debates confirm this trend and the need for decoupling emissions production and the wealth of nations. In the future, an economy freed from carbon and poverty will have to be green and blue, social and collaborative, and will require governance mechanisms that will empower people to be full players in the economy. Africa, which has yet to industrialize, has a major latecomer and collaborative advantage to assert, since it can move directly to post-carbon infrastructure. For this, it will have to rely on innovations led by its people, particularly at grassroots, in the landscapes and the villages, in order to make use of its vast reservoirs of hidden potential for value and achieve the structural ecological transformation of its economy. The participatory paradigm can and must be part of the solution. The desire for emancipation and democratic deliberation that it expresses is a requirement of contemporary societies, rural and urban, African or otherwise, who want to participate in all awareness in societal choices and decisions. This demand, this internal democratic thrust, cannot be underestimated, as shown by the ongoing story of the African MF network.

Notes

1 According to SOFO 2020, Africa lost 3.94 million ha of forests/year in 2010–2020, which would amount to a loss of about 40 million ha, the highest in the world after South America's 2.60 million ha lost/year.
2 It is worth noting the higher added value of the forestry sector in some Central African countries. It represented up to 2.7% of Cameroon's GDP in the early 2010s (Eba'a Atyi et al. 2013), a rate higher than that of the non-oil mining sector and three times higher than the African forestry average (0.9%).
3 Those include seven countries in West Africa: Côte d'Ivoire, Ghana, Liberia, Nigeria, Benin, Mali, and Togo: and five in Central Africa: Cameroon, Central African Republic, Gabon, Congo, and DR Congo.
4 In a dramatic industry shift, Vietnam actually surpassed longtime giant China in 2020, to become the largest furniture exporter to the United States.
5 Though less reliant on African log imports, Malaysia's annual furniture exports to over 160 countries are worth mentioning as they amounted to US$ 2.74 billion in 2020 and were targeted to reach US$ 5.47 billion by 2022 (MIDA 2022).
6 For instance, payments for environmental services represented global annual revenues of US$ 1.9 to 2.5 billion between 2005 and 2011, but Africa received only 0.9% of these amounts in 2011.
7 Editors note: The demographic transition may affect these projections, if encouraged and supported by populations and governments.
8 https://globalwellnessinstitute.org/press-room/statistics-and-facts/.
9 https://www.thebusinessresearchcompany.com/report/wood-products-global-market-report.
10 https://www.thebusinessresearchcompany.com/report/wood-products-global-market-report.
11 According to IMF 2021 World Economic Outlook data cited by https://statisticstimes.com/economy/africa-gdp.php and United Nations population estimates consulted on Worldmeters on June 9, 2022 https://www.worldometers.info/world-population/africa-population/.
12 For instance, IPBES scientists working with the Nature's Future Framework (NFF) have identified at least six sustainable future narratives based on variations within the broad outlines of the NFF, some close to the principles described in this subsection, others quite distant (Durán et al. forthcoming).
13 See also, e.g., https://www.forbes.com/sites/forbeslacouncil/2019/03/04/the-sharing-economy-is-still-growing-and-businesses-should-take-note/?sh=3e4edcba4c33; https://spendmenot.com/blog/sharing-economy-statistics/
14 Model Forests refers to the program initiated in Cameroon as part of the African and global Model Forests Networks. The program strives to create innovative landscape partnerships that improve people's livelihoods, well-being, and environments.
15 This approach to institutional plurality shares some features with the US-based Collaborative Forest Landscape Restoration Program described by Butler and Schultz (2019), among others.
16 Projects of this type are rare in forestry R&D, which tends to favor the 'soft', conservation-oriented side of development (research and awareness-building, along with limited 'poverty alleviation', 'income generation' activities).
17 Currently four components make up PRAIS: ***STAR** (Start and Run your business); ***TOP** (Techniques of the One Program, focused on NTFPs' domestication and transformation, *mycorrhizal biofertilizers*, drip solar irrigation, etc.); ***AWARE** (environment and climate awareness, corporate social responsibility, landscape trade-offs, conflict management, facilitation and mediation, etc.); and ***FIELD** (a hands-on, practical component of PRAIS, direct trainees' engagement in Model Forests extension activities).

Bibliography

Adams, W. and D. Hulme. 2001. "Conservation and Community: Changing Narratives, Policies and Practices in African Conservation." In *African Wildlife and Livelihoods: The Promise and Performance of Community Conservation*, edited by D. Hulme and M. Murphree, 9–23. Oxford, UK: James Currey, Ltd.

ALERT. 2018. "Turning Rainforest to Furniture – Global Markets Gobble Up the Congo Basin." *Alliance of Leading Environmental Researchers and Thinkers*, 28 September 2018. https://us9.campaign-archive.com/?e=a973de8aa7&u=228c18608b6eb3bd8fc42327e&id=17585fa683.

Angelsen, A. and S. Wunder. 2003. "Exploring the Forest – Poverty Link: Key Concepts, Issues and Research Implications." *CIFOR Occasional Paper* 40: 104.

ANRC (African Natural Resources Centre). 2021a. "Intra-African Trade in Wood Products: Case Studies from Benin, Cameroon, Ghana and the Republic of Congo." Abidjan, Côte d'Ivoire: African Development Bank.

ANRC (African Natural Resources Centre). 2021b. "Wood Processing and Trade of Wood Products in Africa." Abidjan, Côte d'Ivoire: African Development Bank.

Archer, E., L. Dziba, K.J. Mulongoy, M.A. Maoela and M. Walters, eds. 2018. *The IPBES Regional Assessment Report on Biodiversity and Ecosystem Services for Africa*. Bonn, Germany: Secretariat of the Intergovernmental Science-Policy Platform on Biodiversity and Ecosystem Services.

Ayompe, L.M., S.J. Davis and B.N. Egoh. 2020. "Trends and Drivers of African Fossil Fuel CO_2 Emissions, 1990–2017." *Environmental Research Letters* 15: 124039.

Barbault, R. and J. Weber. 2010. *La Vie, Quelle Entreprise! Pour une Révolution Ecologique de L'économie*. Paris: Seuil.

Bauwens, M. and V. Niaros. 2016. *Value in the Commons Economy: Developments in Open and Contributory Value Accounting*. Amsterdam: P2P Foundation.

Bonnell, R., R. de Camino, C. Diaw, M. Johnston, P. Majewski, I. Montejo, M. Segur and J. Svensson. 2012. "From Rio to Rwanda: Impacts of the IMFN over the Past 20 Years." *The Forestry Chronicle* 88, no. 3: 245–253 (May/June).

Butler, A. 2014. "Developing Theory of Public Involvement in Landscape Planning: Democratising Landscape." PhD Diss., Swedish University of Agricultural Sciences, Uppsala.

Butler, William H. and Courtney A. Schultz, eds. 2019. *A New Era for Collaborative Forest Management: Policy and Practice Insights from the Collaborative Forest Landscape Restoration Program*. London: Earthscan and Routledge.

Cannon, J.C. 2018. "How Furniture Demand in America Thins Forests Across Central Africa." *Pacific Standard*, 10 October 2018. https://psmag.com/environment/african-deforestation-is-affected-by-american-furniture-demand.

Cerutti, P.O., Y. Artati, A. Dermawan, A. Kelly, G. Lescuyer, E. Mejía, K. Obidzinski, P. Pacheco, L. Putzel, R. Tsanga and A. Wardell. 2014. "Possibilités d'Action Publique pour une Meilleure Prise en Compte des Marchés Domestiques du Bois dans le Cadre des Accords de Partenariat Volontaire (APV): Synthèse des Enseignements Dégagés au Cameroun, en République Démocratique du Congo, en Équateur, au Gabon et en Indonésie." *Infobrief No.* 88, September. https://www.cifor.org/publications/pdf_files/infobrief/5231-infobrief.pdf

Colvin, B. 1970. *Land and Landscape: Evolution, Design and Control*. London: John Murray.

Coquery-Vidrovitch, C. 1972. *Le Congo au Temps des Grandes Compagnies Concessionnaires: 1898–1930*. Paris: Mouton and Co. Editions, Ecole Pratique des Hautes Etudes.

Counsell, S. 2007. “For Richer, for Poorer: Logging and Livelihoods in the Congo Basin.” In *Concessions to Poverty: The Environmental, Social and Economic Impacts of Industrial Logging Concessions in Africa’s Rainforests*, edited by S. Counsell, C. Long and S. Wilson, 107–114. London, Cambridge: Rainforest Foundation UK and Forests Monitor.

Counsell, S., C. Long and S. Wilson, eds. 2007. *Concessions to Poverty: The Environmental, Social and Economic Impacts of Industrial Logging Concessions in Africa’s Rainforests*. London and Cambridge: Rainforest Foundation UK and Forests Monitor.

Crowe, S. 1958. *The Landscape of Power*. London: Architectural Press.

Dalheim, R. 2022. “Vietnam Hopes to Boost its Wood Furniture Exports by $9 Billion.” *Furniture Today*, 25 March. https://www.furnituretoday.com/international/vietnam-hopes-to-boost-its-wood-furniture-exports-by-9-billion/.

Davet, P. 1996. *Vie Microbienne du Sol et Production Végétale*. Paris: Ed. INRA.

Di Méo, G. 2002.” L’Identité: Une Médiation Essentielle du Rapport Espace/Société.” *Géocarrefour* 77, no. 2: 175–184.

Diaw, M.C. 2005a. “Un Plaidoyer Humaniste. Conserver pour Qui?” *Courrier de la Planète* 75: 44–47.

Diaw, M.C. 2005b. “Modern Economic Theory and the Challenge of Embedded Tenure Institutions: African Attempts to Reform Local Forest Policies.” In *Sustainability, Institutions and Natural Resources: Institutions for Sustainable Forest Management*, edited by S. Kant and A. Berry, 43–81. Amsterdam: Springer.

Diaw, M.C. 2014. “Post-2015 Sustainable Development Goals and Forests: Prospects, Integration, Priorities and Experience of Africa.” London: IIED.

Diaw, M.C. 2015. “Forests Must Create Jobs: Evidence for the Structural Transformation of the African Forest Economy.” Conference Paper: African Forest Forum, XIV World Forestry Congress, Durban, 4–5 September.

Diaw, M.C. 2017. “Learning from Each Other: A Convening Method for Integrated Landscape Management in Africa.” Ottawa, Canada: IMFN. https://imfn.net/integrated-landscape-management-in-africa-learning-from-each-other/

Diaw, M.C. and A.M. Tiani. 2010. “Fences in our Heads: A Discourse Analysis of the Korup Resettlement Stalemate.” *Journal of Sustainable Forestry* 29: 221–251.

Diaw, M.C. and E. Bisimwa. 2017. “Food Production, Agricultural Expansion and Deforestation in Mai Ndombe Province of DRC.” London: IIED Report.

Diaw, M.C. and J.C. Njomkap. 1998. *La Terre et le Droit. Une Anthropologie Institutionnelle de la Tenure Coutumière au Sud Cameroun*. Yaounde, Cameroon: Inades-Formation.

Diaw, M.C. and P. Franks. 2019. “Production Alimentaire, Expansion Agricole et Déforestation au Mai-Ndombe, RDC.” London: IIED Working document. http://pubs.iied.org/17652FIIED.

Diaw, M.C., R. Prabhu and T. Aseh 2009. “Assembling the ACM Mosaic: Reflections and Conclusions.” In *In Search of Common Ground: Adaptive Collaborative Management in Cameroon*, edited by M.C. Diaw, T. Aseh and R. Prabhu, 443–474. Bogor, Indonesia: CIFOR.

Dokua, S. 2021. “Wooden Furniture Export Value in West Africa 2011–2020, by Country.” *Statista* (June). https://www.statista.com/statistics/1254913/wooden-furniture-export-value-by-country-in-west-africa/.

Durán, A.P., J. Kuiper, A.P.D. Aguiar, W.W.L. Cheung, M.C. Diaw, et al., Forthcoming. “Bringing the Nature Futures Framework to Life: Creating a Set of Illustrative Narratives of Nature Futures.” *Sustainability Science* (forthcoming).

Eba’a Atyi, R., G. Lescuyer, J.N. Poufoun and T.M. Fouda, eds. 2013. *Etude de l’Importance Economique et Sociale du Secteur Forestier et Faunique au Cameroun*. Bogor: CIFOR.

FAO (Food and Agriculture Organization). 2003. *African Forests: A View to 2020*. Rome: European Commission, African Development Bank, UN Food and Agriculture Organization.

FAO. 2011. *State of World's Forests 2011*. Rome: Food and Agriculture Organization.

FAO. 2014. *State of the World Forests 2014: Enhancing the Socioeconomic Benefits from Forests*. Rome: Food and Agriculture Organization.

Franks, P. and X. Hou Jones. 2016. "Managing Trade-offs between Growing Food and Conserving Forests in Sub-Saharan Africa." *IIED Briefing* (December).

Franks, P., X. Hou-Jones, D. Firkreyesus, M. Sintayehu, S. Mamuye, E. Yaw Danso, C.K. Meshack, I. McNicol and A. van Soesbergen. 2017. *Reconciling Forest Conservation with Food Production in Sub-Saharan Africa: Case Studies from Ethiopia, Ghana and Tanzania*. London: IIED Research Report.

Fuller, T.L., Narins, T.P., Nackoney, J. et al. 2018. Assessing the Impact of China's Timber Industry on Congo Basin Land Use Change. *Area* 00:1–10. https://doi.org/10.1111/area.1246910. Available from: https://www.researchgate.net/publication/326036300_Assessing_the_impact_of_China%27s_timber_industry_on_Congo_Basin_land_use_change [accessed Jun 03 2022].

GEF (Global Environment Fund). 2013. "Africa will Import, not Export Wood." Chevy Chase, Bethesda, MD: GEF. https://www.criterionafrica.com/wp-content/uploads/2017/06/Africa-will-Import-not-Export-Wood.pdf.

Geist, H.J. and E.F. Lambin. 2002. "Proximate Causes and Underlying Driving Forces of Tropical Deforestation." *BioScience* 52, no. 2: 143.

Gray, John A. 2002. "Forest Concession Policies and Revenue Systems: Country Experience and Policy Changes for Sustainable Tropical Forestry." World Bank Publications – Books (14094). Washington, DC: The World Bank Group (December).

Hierold, J. 2010. Value Addition in the Wood Processing Sector – Options for Production Waste Minimization along the Wood Value Chain. PTC/AGR/AIT. June.

Hochschild, Adam. 1998. *King Leopold's Ghost: A Story of Greed, Terror, and Heroism in Colonial Africa*. London: Pan Macmillan.

Hulme, D. and M. Murphree, eds. 2001. *African Wildlife and Livelihoods: The Promise and Performance of Community Conservation*. Oxford, UK: James Currey Ltd.

Ingram, V., S. Asaha, T. Sunderland, and A. Tojoacha. 2011. "Governance and NTFP Chains in Takamanda-Mone Landscape, Cameroon." In *Collaborative Governance of Tropical Landscapes*, edited by C.J.P. Colfer and J.L. Pfund, 183–216. London: Earthscan, CIFOR.

IPBES. 2019. "IPBES/7/6, Next Work Program of the Platform." Plenary of the Intergovernmental Science-Policy Platform on Biodiversity and Ecosystem Services, Seventh Session, Item 9, 5 March.

Jeffries P., S. Gianinazzi, S. Perotto, K. Turnau and J.-M. Barea. 2003. "The Contribution of Arbuscular Mycorrhizal Fungi in Sustainable Maintenance of Plant Health and Soil Fertility." *Biology and Fertility of Soils* 37: 1–16.

Joiris, V., and P. Bigombe Logo, eds. 2010. *La Gestion Participative des Forêts d'Afrique Centrale*. Paris: Editions QUAE.

Karsenty, A. 2010. "La Responsabilité Sociale et Environnementale des Entreprises Concessionnaires." In *La Gestion Participative des Forêts d'Afrique Centrale*, edited by V. Joiris and P. Bigombe Logo, 35–55. Paris: Editions QUAE.

Kittiworawat, S., S. Youpensuk and B. Rerkasem. 2010. "Diversity of Arbuscular Mycorrhizal Fungi in *Mimosa invisa* and Effect of the Soil pH on the Symbiosis." *Chiang Mai Journal of Science* 37, no. 3: 517–527. https://www.thaiscience.info/Journals/Article/CMJS/10905476.pdf.

Larsen, K., H. Pitt, M. Grant and T. Houser. 2021. "China's Greenhouse Gas Emissions Exceeded the Developed World for the First Time in 2019." Rhodium Group (6 May). https://rhg.com/research/chinas-emissions-surpass-developed-countries/.

Lee, K., C. Juma and J. Mathews. 2014. "Innovation Capabilities for Sustainable Development in Africa." WIDER Working Paper 2014/062, World Institute for Development Economics Research, United Nations University, Tokyo.

Legault, D.D. and L. Cochrane. 2021. "Forests to the Foreigners: Large-Scale Land Acquisitions in Gabon." *Land* 10: 420. https://doi.org/10.3390/land10040420.

Lescuyer, G., J.N. Poufoun, A. Collin and R.I. Yembe Yembe. 2014. Le REDD+ à la Rescousse des Concessions Forestières? Analyse Financière des Principaux Modes de Valorisation des Terres dans le Bassin du Congo." *CIFOR Working Paper* 160.

Lescuyer, G., S. Assembe Mvondo, J.N. Essoungou, V. Toison, J.-F. Trébuchon and N. Fauvet. 2012. "Logging Concessions and Local Livelihoods in Cameroon: From Indifference to Alliance?" *Ecology and Society* 17, no. 1: 7. http://dx.doi.org/10.5751/ES-04507-170107.

Mandondo, A., R. Prabhu and F. Matose, eds. 2008. *Coping Amidst Chaos: Studies on Adaptive Collaborative Management from Zimbabwe*. Bogor, Indonesia: CIFOR.

MARD. 2022. "Vietnam's Wood Exports to the US Towards the Goal of 10 Billion USD." Ministry of Agriculture and Rural Development, Socialist Republic of Vietnam (January 25). https://www.mard.gov.vn/en/Pages/vietnam%E2%80%99s-wood-exports-to-the-us-towards-the-goal-of-10-billion-usd.aspx.

Mather, A.S. 1992. "The Forest Transition." *Area* 24, no. 4: 367–379.

MIDA (Malaysian Investment Development Authority). 2022. Malaysian Investment Development Authority Official Website, consulted 26 May 2022. https://www.mida.gov.my/securing-the-future-of-malaysias-furniture-industry/.

Mousain, D., P. Matumoto-Pintro and H. Quiquampoix. 1997. "Le Rôle des Mycorhizes dans la Nutrition Phosphatée des Arbres Forestiers." *Revue Forestiere Francaise* 49: 67–81.

Neumann, R.P. 1995. "Ways of Seeing Africa: Colonial Recasting of African Society and Landscape in Serengeti National Park." *Ecumene* 2, no. 2: 149–169.

Olubusoye, O.E., D. Musa and S. Ercolano. 2020. "Carbon Emission and Economic Growth in Africa. Are They Related?" *Cogent Economics and Finance* 8: 1.

Pauli, G. 2009. *10 Ans, 100 Innovations, 100 Millions d'Emplois*. Paris: Caillade.

Pereira, L.M., K.K. Davies, E. Belder, et al., 2020. "Developing Multiscale and Integrative Nature–People Scenarios Using the Nature Futures Framework." *People and Nature* 2: 1172–1195. https://doi.org/10.1002/pan3.10146.

Roulet, P-A and R. Hardin, 2010. "Des Domaines de Chasse aux Zones d'Intérêt Cynégétique à Gestion Communautaire." In *La Gestion Participative des Forêts d'Afrique Centrale*, edited by V. Joiris and P. Bigombe Logo, 123–142. Paris: Editions QUAE.

Rudel, T.K., L. Schneider and M. Uriarte. 2010. "Forest Transitions: An Introduction." *Land Use Policy* 27: 95–97.

Sen, A.K. 1984. *Resources, Values and Development*. Oxford: Blackwell.

Sen, A.K. 1985. *Commodities and Capabilities*. Amsterdam: North-Holland (re-published, Oxford, 1999).

Sills, E., P. Shanley, F. Paumgarten, J. de Beer and A. Pierce. 2011. "Evolving Perspectives on Non-Timber Forest Products." In *Non-Timber Forest Products in the Global Context*, edited by S. Shackleton, C. Shackleton and P. Shanley, 23–51. Heidelberg: Springer-Verlag.

Tseng, M.L., S.H. Lin and T.N. Tuong Vy. 2012. "Mediate Effect of Technology Innovation Capabilities, Investment Capability and Firm Performance in Vietnam Procedia." *Social and Behavioral Sciences* 40: 817–829.

TTM (Tropical Timber Market). 2020. *Tropical Timber Market Report* 24, no. 22: 16th – 30th November. https://www.itto.int/files/user/mis/MIS_16-30_Nov2020.pdf.

van Dorp, M. 2007. "Economic Impacts of Industrial Logging Concessions: What are the Benefits?" In *Concessions to Poverty: The Environmental, Social and Economic Impacts of Industrial Logging Concessions in Africa's Rainforests*, edited by S. Counsell, C. Long and S. Wilson, 82–95. London, Cambridge: Rainforest Foundation UK and Forests Monitor.

Wolfersberger, J., P. Delacote and S. Garcia. 2016. "Une Analyse Empirique de la Transition Forestière et des Changements d'Usage des Sols dans les Pays en Développement." *INRA Sciences Sociales* 3.

Yaraghi, N. and R. Shamika, 2017. "The Current and Future State of the Sharing Economy." *Brookings India IMPACT Series* No. 032017. March 2017.

Introduction to Chapter 11 "Changing the game

An economy built around stewardship"

In the previous chapter, Diaw et al. called for a change in the global economic system, which has so penalized African forest dwellers (and others). In this final chapter, Prabhu and Colfer call for a perhaps more dramatic change in the economic system, such that genuine stewardship – something many (not all) forest groups have practiced traditionally – be explicitly acknowledged and fairly compensated, in what Prabhu and some of his previous co-authors have called 'stewardship economics'. Although this concluding chapter is not the place to flesh out the concept fully, we consider some of the vital elements in the proposed approach and how it might function as (part of) a logical 'next step' for ACM proponents.

The chapter also pulls together some thoughts that emerge from the previous ten chapters. One of the most important and consistent insights from this longitudinal compilation is the critical nature of the surrounding context (whether economic, political, environmental, or sociocultural) as it interacts with what happens locally. No one is suggesting returning to a world where local realities and local people are ignored (as was the case for a long time); rather, these authors see an important improvement to ACM being the strengthening of links between these now more fully understood local systems and the broader systems in which they function and with which they interact.

Another key insight is the likely utility of ACM's approaches – its emphases on collaboration and on learning, as we try to adapt to and communicate with each other in a changing world – at broader scales.

DOI: 10.4324/9781003325932-20

11 Changing the game

An economy built around stewardship

Ravi Prabhu and Carol J. Pierce Colfer

Setting the scene

It is 30 years since the world first formally agreed in Rio de Janeiro, Brazil, that climate change threatened our planet because of humans. After 27 years of global negotiations on reducing greenhouse gas concentrations in the atmosphere, all we can bear witness to is the constant rise of their atmospheric concentrations and decades of broken promises (IPCC 2022). Let's therefore agree that we are digging ourselves into a metaphoric hole and what we need to do is stop digging it deeper (Czech 2002), a sentiment also observed in 1911: "Nor would a wise man, seeing that he was in a hole, go to work and blindly dig it deeper…" (Washington Post 25, also referred to as the 'First law of holes').

The most recent IPCC assessment (IPCC 2021) reports on the state of global biodiversity (IPBES 2019) and land degradation (Critchley, Harari and Mekdaschi-Studer 2021), and an assessment of progress against the Sustainable Development Goals (Katila et al. 2019) all bear witness to our predicament. While these reports provide evidence of our inability to deal with wicked problems, with their cryptic causalities (as Kusumanto et al. have also elaborated in Chapter 4), there are also problems festering in plain sight, with extreme power differentials, inequities, selfish ambitions and information perversion as the apparent causes. The planet, as we know it, is threatened because of all of these and a future we are beginning to fear.

We began this book by stressing the need to change direction and finding encouragement in the Glasgow Leaders' Declaration on Forests and Land Use,[1] which was a call for just such a change in direction and a call for stewardship at the last conference of the parties to the United Nations Framework Convention on Climate Change (UNFCCC – CoP26). Making good on their call will require a radical transformation; we explore a pathway to this here.

In the preceding chapters we have been offered insights into how ACM and similar or related processes can help, through equitable, collaborative and evidence-based learning and decision making. We will return here to address some of these insights as to how ACM could contribute to the transformative change we need. Managing forested landscapes is a game where the rules – incentives, power disparities, access to information, etc. – are unfortunately stacked against

DOI: 10.4324/9781003325932-21

local stakeholders and communities, preventing them from reaching the kind of sustainable, resilient and productive outcomes they and the planet need (FAO, IFAD, UNICEF, WFP and WHO 2021; Kristjanson 2020; Laird et al. 2022; Shepherd, Warner and Hogarth 2020). Local people in forested landscapes are often trapped within a system they have little influence in shaping – regulations, restrictions to rights, incentives, i.e., the 'rules of the game' – that often locks them into pathways to unsustainable outcomes, even when they do not want to go down those tracks. This is by no means a novel insight, as several examples in this book and its companion volume (Colfer, Prabhu and Larson 2022; see also, e.g., Davis and Ruddle 2012; Li 2014; Li and Semedi 2021; Paulson 2017) show. While most reform efforts are focused on the 'rules' that determine how the game is played, we contend that changing the game instead of just focusing on better rules is what is needed. Failing this, the fate of forests, and the people and planetary processes that depend on them, will likely continue to succumb to global challenges such as climate change and other wicked problems, some of which were described in preceding chapters. We explore this further here.

But first let's pause and explain why it makes sense to see local communities (and others!) as 'trapped' in a game. In 1950, working at the Rand Corporation, Merrill Flood and Melvin Dasher proposed what would later come to be known as the Prisoner's Dilemma and that eventually gave rise to a branch of mathematics called game theory. The Prisoner's Dilemma explores the impacts of choices that are selfish or cooperative and it has been used to explain our current lack of progress in the climate change negotiations (Madani 2013). The benefits of cooperation are clear: every nation (and every individual) benefits if others reduce their greenhouse gas emissions but the selfishness baked into the system means that no nation wants to risk jeopardizing its own prosperity by taking the first step. The Prisoner's Dilemma has also been used as an explanation for the tragedy of the commons that has befallen our other shared environmental spaces – beyond what has happened to the air we breathe, our atmosphere (Carrozzo Magli, Della Posta and Manfredi 2021).

In this book, Yuliani et al. (Chapter 2) provide evidence of the choices that reflect the Prisoner's Dilemma when they note that … *people said quite realistically, "It will be cleared anyway by the company. But if we clear it first, we can get some money, and we can also become the company's partner"*. So, despite recognizing that the actions were not in accordance with their ACM goals – which are by their design collaborative, strategic and systemic – some succumbed instead to narrower, more selfish, if still rational objectives as predicted in the Prisoner's Dilemma.

We have seen that ACM and allied approaches do help communities, and local governments to adapt, deal with challenges and take advantage of opportunities. But we have not seen evidence that these efforts made a significant, enduring contribution to the kinds of lasting transformative changes we need. This is not surprising, because local communities and other stakeholders are prisoners in the wrong game, one where selfish, short-term choices are a feature, not a bug, of the system. We will return to this and explore how the term 'stewardship' in the Glasgow Leaders' statement may provide the avenue to changing the game and direct us out of the hole we presently find ourselves in. Let us start, however, by exploring our first question: In what ways has ACM helped with achieving transformative change? Following that, we explore the degree to which the communities we worked in exhibit stewardship and what opportunities this provides, to break out of unsustainable games. We close by exploring how we might achieve outcomes that are more equitable and offer a change of direction towards a more resilient and sustainable future, in other words how we might begin to change the game in its entirety towards a caring economy built around stewardship.

ACM and learning across scales

At the outset, we suggested, based on our experience, that to get lasting change we must have active engagement of local communities, forest users, managers and policymakers. It is now manifestly clear that all the relevant stakeholders (Colfer 1995) must collectively want to save, sustainably use and restore forests as a means also of safeguarding and improving their own livelihoods. This cannot be a burden that rests solely on the shoulders of local communities, even if they have a principal part to play. The intervening chapters have shown how ACM can support self-empowerment through mobilization of collaboration, building of trust, social learning and adaptive improvements. But they also demonstrate ACM's limitations in achieving this.

As Yuliani et al. note, "(O)ne of the main outcomes of the ACM project … was the strengthened capacity of local institutions … to perform participatory, transparent and democratic decision making at the village level". This is important because it speaks to a key feature of ACM – the process of self-empowerment. If a change in direction is to take place, people must not only desire it but feel capable of attaining it. Kamoto et al. (Chapter 9) note relatedly that the transformation of the relationship between many communities and the Department of Forests was the precursor to a reduction in deforestation and degradation. Kozanayi et al. (Chapter 7) also remark on the improved ability of local communities to engage in decision making, testimony to their self-empowerment. They also note that learning took place not only among local communities, but also within the Forestry Commission and among more distant stakeholders. Indeed, in many cases, this kind of multi-scale learning did emerge, but often with quite long lag times. Prabhu, Larson and Colfer (2022) describe a more recent ACM-inspired intervention that was designed to address the need for multi-scale learning from the outset.

In Chapter 10, Diaw et al. devote considerable attention to the contributions of ACM to acquiring power and the role of power within the struggle to achieve equitable economic outcomes in Central Africa. Working across scales and groups during processes of learning and self-empowerment necessitates the building of trust and requires facilitation as Liswanti et al. (Chapter 3), for Sumatra, and Hagmann et al. (Chapter 6), more broadly, have pointed out. But it is very clear that this is not sufficient – there are often greater economic forces at play, as Kozanayi et al. have documented in the case of Zimbabwe and as Diaw et al. go to considerable length to explain. Diaw et al. recognize that ACM as part of a strategy of transformation had "…failed to trigger the scale of transformations needed for a gradual system shift in the African landscapes". In their discussion of what they call the 'poverty paradox', they attribute several reasons to this failure to transform, despite apparent benefits of ACM; and they assert that

> …it is not just a matter of improving one aspect here and there, contributing to small successes in some localities, and counting how many people got better off by the end of a project to strengthen applications for additional funding.

Indeed, they argue that transformative change based on ACM alone has not been possible because there was a need to "question the way the economy works". We agree. Clearly, we can no longer hope to get out of our present predicament simply by changing the tools we have, to return to our metaphor: we need to stop digging and change direction. This means questioning some of the fundamental roles, power structures and framing conditions that determine the economy and political economy within which local people manage forests and their own livelihoods.

Yuliani et al. acknowledge that the ACM process helped researchers to learn to recognize and anticipate signs of conflict, manage unrealistic expectations and how to build "self-reliance, self-motivation and self-efficacy to achieve their goals". In many cases, researchers and policymakers initially found themselves conceptually hemmed in by their own unrealistic expectations and assumptions. They had to learn to set those aside as they came to understand local conditions and the ACM process more fully.

Despite good intentions, the success of any intervention is conditional on a number of factors such as building of trust and the bringing of the right people to the table, in the right way, as Liswanti et al. note. On the whole, ACM has given us tools and provided invaluable experience on the 'how' of building self-empowerment and enabling collective learning processes in the efforts to improve and sustain livelihoods and forests. It has contributed to the generation of improved material benefits as well in some cases, but many of these promising gains have eroded with time. So, the questions around lasting benefits and changes remain incompletely answered, despite promising beginnings and some lasting benefits, such as those described in multiple chapters. ACM has also contributed to our understanding that 'softer' parts of any system are as, perhaps more, important than 'harder' institutional elements in the pursuit of better livelihood and sustainability outcomes. Thus, trust, collaboration, communication, sustaining engagement, facilitation and learning are as important as institutional rules, regulations, laws and bylaws. Having the right people in the space is as important or perhaps more important than any efforts to equitably change the formal distribution of power. As Kozanayi et al. observe, when the "…ACM approach is recognized through policy, resources (financial and otherwise) are allocated towards its implementation, and … ACM becomes institutionalized within the government system [and it] has more chances of success in the long term". The needs for resources and capacity are echoed once more by Kamoto et al. and Egunyu.

Kozanayi et al. go on to observe that "... beyond the latent processes of engagement and adaptation that are hallmarks of the ACM approach, other subtle power and interest configurations and design of adaptive strategies need to be embraced". Could this embrace include or be circumscribed by stewardship, at least partly? What benefits might accrue from a stronger view of local people as stewards? These are questions we intend to pursue, even as we begin by exploring what stewardship means.

The importance of a culture of care: why stewardship matters

Clearly, calls for stewardship are being echoed up to the highest levels of international governance (cf. the EPA Innovation Council 2005; Glasgow Leaders' Declaration; Saner and Wilson 2003); however, the actual stewards of forests and lands continue to either be marginalized or forced by circumstances to abandon principles of stewardship. Aldo Leopold (1949) defined stewardship as "the conducting, supervising, or managing of something, especially the careful and responsible management of something entrusted to one's care". In its essence, stewardship is about caring for what we value (Berry 2006). The focus on stewardship has grown rapidly in the last five years as Mathevet, Bousquet and Raymond (2018) note: about 75% of the citations and 62% of publications they analysed have used stewardship as a key concept or a pathway for action in conservation and environmental science.

Prabhu, Lawry and Colmey (2021) have suggested that stewardship is "a deliberate and informed combination of solicitude, foresight and skill – a marriage of practice and ethics – that has tangible impacts in landscapes" (no page numbers). In their view, land stewards, then, are not simply owners or producers of commodities (food, timber, fibre, etc.), as water stewards are not just those making use of water resources. Stewards are engaged in their landscapes, but in ways that uphold a 'duty of care' – an ethos of responsibility for all the human and ecosystem services the land currently provides, as well as the integrity of its history and, importantly, its future (see, e.g., Singh 2015). Such landscapes are multifunctional in that they provide a diversity of ecosystem functions or services that underpin social and economic functioning for a range of beneficiaries or landscape stakeholders (Minang et al. 2014).

Stewards – individuals or communities – can exercise their stewardship best when they hold secure rights to their land and waters, giving them the legal assurance to invest in the longevity of their natural resources. Throughout this book and indeed in the preceding volume, we have seen how ACM practitioners have sought to facilitate collaboration in multistakeholder, multiple-objective, contested landscapes. The need to support collective action and learning from both the bottom-up and the top-down has been emphasized. But in all of these studies, despite clear evidence of its frequent, rather subterranean existence, we have stopped short of calling for stewardship and the emergence of a culture of care. We rectify that here.

As Johnson, Campbell and Svendsen (2020) note and we concur, "collaboration among individuals and groups is a critical aspect of stewardship capacity and social innovation". Enqvist et al. (2018), citing multiple authors, point out that,

> …recently, stewardship has been used to indicate a broad shift away from techno-managerial, control-oriented approaches to landscape and environmental management, policy and planning, towards those that prioritize participatory, cross-scale, and trans-disciplinary engagements rooted in shared values.
>
> (p. 18)

Rather than forcing a single definition of stewardship upon policymakers and practitioners, this approach offers them tools to understand what its advantages and disadvantages might be. We agree that there is a need for this conceptual flexibility – this, in our view, is a form of constructive ambiguity that allows us to find iterative pathways to better outcomes when dealing with complex adaptive systems.

Johnson, Campbell and Svendsen (2020) also assert that there are a number of questions that still need to be answered about stewardship; and Mathevet, Bousquet and Raymond (2018) have pointed out that the term is still contested. Despite this, it is a term that lends itself to communicative action, emerging as it does as a property of (a) care, along with creativity and values, (b) knowledge and know-how, and (c) agency, power and resources, in a novel framework for connecting the multiple meanings of stewardship proposed by Enqvist et al. (2018) and Andersson, Enqvist and Tengö (2017).

While we have dealt extensively with knowledge and its collaborative generation and use in decision making in the preceding chapters – indeed, we have also dealt with agency, power and resources – we have not yet addressed the need for care sufficiently. This is equally true of ACM writings, despite many having quietly acknowledged local stewardship as one factor in their grounded research. Like population (Crist, Mora and Engelman 2017) or domestic abuse (Arora-Jonsson, Colfer and González-Hidalgo 2021), the human value of caring has been subject to academic taboos that render many researchers reluctant to discuss such topics publicly, let alone address them as serious issues in forests and elsewhere (see Kashwan and Ribot's 2021 'violent silence', discussed below).

While we have established that stewardship encompasses care, why is a *culture* of care necessary? For this, we go back to Aldo Leopold, Wendell Berry, Rachel Carson and others who have emphasized the role of ethics and our sense, echoed by many local communities, that forests and land are not simply objects to be exploited, but are values to be cherished and nurtured in their own right. It is no surprise then that forests, streams, trees and even rocks in a landscape are often considered sacred, that culture and nature are, among many groups, explicitly intertwined inextricably. The commodification of nature has led us to lose sight of its values, beyond the returns squeezed out of narrow and often disadvantageous market relationships. It is in this narrow over-commodification of nature that the Prisoner's Dilemma finds space to play out, where sustainability goals are not met, and the future of our planet is jeopardized. All this despite many local communities recognizing what they lose when they eschew the culture of care that they know, value and likely have practised. Thus, when the duty of care becomes embedded in attitudes, norms and behaviours – beyond policies, rules, regulations and laws – we have the emergence of a culture of care, where caring becomes an intuitive and reflexive posture towards nature and people. We recognize that this is likely more an ideal to strive for than one we can hope to achieve in the near term, but one we must recognize as key to a sustainable and equitable future.

Cockburn et al. (2020) assert that systems are relationally constituted, i.e., they are what they are by virtue of the multiple, dynamic relations or interconnections which link the elements of a system together. In their view, the nature and functioning of social-ecological systems are strongly shaped by the nature of the web of relationships in that system. This is not to say that the nature of the elements is not relevant but rather that the system emerges out of both the elements and the relations among them. We agree, and it is in maintaining and setting the tone for those relationships that a duty of care must express itself within a striving for a culture of care towards nature and those who nurture it. If a global shift to a culture of care and a nurturing of stewardship is to take place in forested landscapes, then a good starting place would be in the communities that are actively caring for their environment and each other. It will, of course, take much more – changes to incentives, rewards, regulations and policies, along with changes to knowledge, attitudes and behaviours. We take this forward in the next section.

Where might ACM find a place in all this? Cockburn et al. (2020) provide the signposts: "Collaboration for landscape stewardship requires people to build a shared sense of identity and belonging, in spite of these differences.... Growing together by interacting regularly and building common knowledge ... Learning and adapting together with humility and empathy". Caretta (2020) sees collaboration among individuals and groups as a critical aspect of stewardship capacity and social innovation.

We find examples of a duty or a culture of care, linked to knowledge and action, in the work on ACM in this book and in its companion volume. We offer these examples here fully cognizant of the fact that this is a post-hoc analysis – our focus on stewardship succeeds our work on ACM and as such the authors did not include it in their analytical frames.

Yuliani et al. document the care taken for the mountain in Baru Pelepat as a result of a mixture of hopes for long-term sustainability of products and a sense of stewardship. The fact that it is recognized as a 'customary forest' carries within it a sense of its links to custom, tradition, some sense of care – this we recognize as embedding care in culture, and at the very least a precursor to a culture of care. Here, we cite some examples from Sumatra:

> ...their goals, include[ed] maintaining their natural resources, in particular timber, NTFP and clean water from the HA, and fish from *lubuk larangan* (a protected river area) sustainably.
>
> (p. 28)

> Local people might want to protect their forests for many reasons, e.g., for cultural reasons or as a reserve for later use. National and regional policies, on the other hand, are prioritizing economic development over other objectives.
> (Maryani et al. 2021, p. 17)

> Such isolated small patches of forest protected by traditional belief systems contribute to upholding some ecosystem services such as crop pollination and seed dispersal.
> (p. 18)

In Chapter 7, Kozanayi et al. document how traditional and modern institutions have at times worked together to help local Zimbabwean communities exercise a duty of care:

> For Batanai, the village head is at the forefront of facilitating these processes at his village court. He oversees the granting of harvesting permits to resource harvesters as well. Two other RMCs [Resource Management Commitee], around Mafungautsi are reportedly still operational and using elements of the ACM approach which they had copied from CIFOR's three original ACM sites.
> (p. 180)

> …the Gababe RMC chairperson indicated that they still use results of foregoing trials to convince people not to harvest broom grass through the ecologically more damaging method of uprooting the grass.
> (p. 183)

Fisher et al. (Chapter 5) document how a culture of care influences the stewardship of the forests these people of Sulawesi consider sacred:

> First, it was a small community, and crucially, there was no question about the extent of local adherence to Kajang cultural norms given the clear outward appearances of customary practice (Maarif 2012). The inner zones, *ilalang embaya*, were particularly strict in their adherence to the moral code (*pasang*), with residents wearing all black to symbolize values of modesty. While there are several thousand people that identify ethnically as Kajang across the Subdistrict of Kajang in Bulukumba, strict adherents of the *pasang* are largely confined to the village of Tanah Toa. *Ilalang Embaya* spatially overlaps with much of Tanah Toa village, which also extends into the Kajang sacred forest, an area that according to state administrative documents had been enclosed under the designation of *limited production forests.*
> (p. 117–118)

> …the sacred forest was very much still under the control of the Kajang, where rights of entry are closely guarded, various ritual sites are located, and rituals continue to be performed.
> (p. 120)

> With the village planning initiatives, the water resources discussions helped to identify mutual interests for protecting water, given key concerns over irrigation resources, the loss of drinking water resources, and the flooding that has taken place due to land use change.
>
> (p. 125)

Kamoto, Missanjo and Djenontin offer this from Malawi:

> The 'tangible' benefits that are often referred to as a necessary pre-condition for communities to enter into PFM do not have to be direct cash. Intangibles are also important to people.
>
> (p. 212)

We return to intangible benefits in the next section.

A culture of care requires supporting stewardship and stewards at all levels, beginning with those directly engaged. This means that larger players, such as governments, businesses, educational and research institutions, nonprofit organizations, and the slew of interested others must begin to recognize and internalize in their own cultures of relationships, the societal benefits of environmental stewardship. Their relationship with landscapes must begin to connect mind and action within a culture of care, which may look different in different places. Indeed, there are already longstanding examples of stewardship and a culture of care, this is not something new. What is new is the recognition that without stewardship and a duty of care, we cannot find the balance between nurture and use of nature. The fact that stewardship is constructively ambiguous leaves room for hyper-local solutions to emerge and evolve, within a larger framework of a culture of care. We can thus respect and cherish, without artificially constraining the diversity in nature and culture that has been the bedrock of our resilience as a species and as a planetary system, especially now that we are firmly in the Anthropocene. We need to see stewardship as part of a land ethic that accepts the need for "… a mode of guidance for meeting ecological situations so new or intricate, or involving such deferred reactions, that the path of social expediency is not discernible to the average individual" (Meine 2020, citing Leopold 1949, p. 203), which to us sums up a culture of care.

Breaking out: towards a new game

Describing the *kukiya kiya* or 'making do' economy in Zimbabwe, Kozanayi et al. inform us that communities were forced to make do by felling trees and cutting forests in order to meet their livelihood needs; that 'low value' products such as thatch and broom grass were insufficient in a failing economy to help them secure their livelihoods. There is a lot to unpack here.

As they acknowledge, ACM had worked well in Mafungautsi: local communities, in collaboration with the Forestry Commission, indeed under its umbrella, had developed resource management institutions that helped them to meet their livelihood needs while protecting the forest from degradation. There was learning, collaboration, collective decision making and action. Stewardship could be seen as emerging out of care, creativity, shared values, shared knowledge and collective action. But, as they point out, it is not enough for this to happen at one scale, when there are disruptions at higher scales. In fact, ignoring the need to address other scales simultaneously could lead to a false sense of complacency. Yuliani et al. make very similar points about their experience in Sumatra. We cannot carve out small chunks of larger systems and then aspire to develop stewardship and a culture of care there alone. A sandcastle on the beach exists only for the hours between high tides. These, as we have shown, are clearly not lasting solutions. In this, our concluding section we ask ourselves how do we get to lasting or sustainable outcomes?

We have dealt with many of the elements in this and preceding chapters, but we return to two conclusions here: (1) the need to account for multiple scales with the associated disparities in power, and (2) how changing the game, not just tweaking the rules, appears as our only course of action. This requires us to consider the planet as a system, and understand the roles that ACM, stewardship and other elements of a culture of care might play in changing the way we express our relationships with each other, including within economic and political systems, and with nature. We follow Prabhu, Lawry and Colmey (2021) and Prabhu (2022) in calling this new game a Stewardship Economy. For now, the Stewardship Economy remains an emerging set of ideas, which we briefly set out below, but it also includes sets of as yet unconnected and likely imperfect institutions, attitudes, cultures and actions that also exist in the real world. We hope that in describing the Stewardship Economy we will facilitate its emergence as the 'game' that takes us out of the Prisoner's Dilemma and towards a more resilient and adaptive future framed within stewardship and a culture of care.

Diaw et al. have rightly called for the definition of a 'new economy'. They require it to "…be not only green and blue, but also a social solidarity economy". As these authors recognize, the dominant ultra-extractive economic model has been "…the principal force behind our growing climate and biodiversity crises". However, Diaw et al. focus their attention on the lack of power and capability of African stakeholders to convert the assets they have at their disposal for their own well-being. This is necessary, but, like ACM alone, far from sufficient, as removing the power and capability deficits would simply be improving (significantly) the rules in the current game of 'over commodification of nature' (Prabhu, Lawry and Colmey 2021). We need an entirely new game.

Prabhu (2022) has suggested that adopting a 'duty of care' relationship towards nature would reduce or reverse the commodification of nature without jeopardizing our ability to feed and sustain humanity. It is this over-commodification of nature that Prabhu, Lawry, and Colmey (2021) see as the primary problem. They define a Stewardship Economy as

> an equitable system of exchange that rewards those managing land sustainably for the goods and services we derive from those landscapes without disrupting the rights of people to food, nutrition, health, voice and decent livelihoods.

Such an economy would operate both within and outside markets as we know them, connecting stewards, in nested and interacting landscapes. In a Stewardship Economy, as they understand it (and as we mean it here), there is a need to step beyond rewarding good management of produced commodities, to also considering the needs and welfare of the stewards. This implies, for instance, that the welfare of stewards in landscapes providing important but lower-priced commodities (e.g., rice) are not disadvantaged when compared to those producing higher-priced commodities (e.g., vanilla or coffee). This protects both the people in those landscapes and the ecosystem services produced in those landscapes. Prabhu (2022) explores the example of Indian agricultural landscapes to show how a Stewardship Economy might emerge and be supported. His focus on low value but essential bulk commodities shows the limitations of 'market only' solutions. Prabhu, Lawry and Colmey (2021) postulated a 'stewardship dividend' as one of the cornerstones of a Stewardship Economy in order to: "...make up the difference between market-based income and the total income stewards need to pursue their duty of care toward essential non-commodified products and services and simultaneously achieve equitable welfare outcomes" (no page numbers). The assumption underpinning this is that the Stewardship Dividend is sufficient to ensure that local communities and other stakeholders have no incentive to reduce the capacity of their managed ecosystems to deliver services. The focus is on incentives rather than disincentives, though clearly there must be some thought given to the latter as well. This suggests that while there may be scope to improve the current rules of the game, and that this is necessary (as Diaw et al. note), sufficiency conditions will only be met if we also look beyond markets to the system of rewards and benefits as a whole, of which markets are only a part. In determining the Stewardship Dividend, Prabhu (2022) calls for it to be based on (1) the needs and aspirations of the stewards, i.e., the incentives needed to support current behaviour or catalyse changes to it if necessary; (2) any economic value attributed to non-commodified products and services, from an environmental goals perspective; and (3) the fair[2] price of a commodity in the market. One notable difference between a Payment for an Ecosystem Service (PES) and the proposed Stewardship Dividend is that the former sets out a payment on the basis of a valuation of an ecosystem service, which can be very problematic as we have seen with the carbon and REDD+ discussions, because values of the service tend to be arbitrary and often determined in imperfect markets. The Stewardship Dividend, with its three prongs of focus on the welfare needs of communities, economic value information (of a bundle of ecosystem services) and a consideration of a fair or equitable price for commodities, is a more systemic approach to dealing with rewards for stewardship.

A simple way of thinking of a Stewardship Economy may be to consider it as being based around an expanded universal basic income (UBI) platform (Gentilini et al. 2020) that is tied to collectively agreed performance expectations and monitored through distributed and participatory efforts. A UBI according to Gentilini et al. (2020) is a programme that delivers in cash, unconditionally, to everyone. In this way, it would seek to meet social and environmental goals at multiple scales simultaneously, with stewardship as the crucial axis and iterative improvement driven by social learning, inclusivity, intersectionality, equity, multistakeholder engagement and trust: the spokes of the wheel of our collective aspirations for a sustainable and decent future.

Note the difference to a UBI in Figure 11.1. This is not to exclude that within a given landscape the Stewardship Economy could become unconditional, if threshold conditions for that landscape/community are met, in order to lower transaction costs. If there is free riding, this would be left to the collective to deal with.

If we look around, we can already see elements of a Stewardship Economy in action, from payments for ecosystem services (which however still seek to operate through markets and commodification of nature), through to the management of sacred groves, the historical sacrifices of the Bishnoi community in India (https://en.wikipedia.org/wiki/Khejarli_massacre), etc. We would include in this list efforts to repurpose subsidies and incentives towards sustainable and equitable outcomes. Hitherto, efforts to reduce pressure on nature and improve sustainability have generally focused on the value of nature, insufficiently on the welfare and well-being of the stewards of that nature. Only by recognizing that the value of nature is more than what is realized in markets (Finlayson 2022; IPBES 2022) and that the value of stewardship is not reflected in the rewards stewards receive for their care, knowledge and action, can we truly move forward to sustainable outcomes. Fundamentally, this is the correction that the Stewardship Economy seeks to apply.

While the Stewardship Economy is very much a work in progress, consider that already more than half the global population is urban and by 2050 roughly 68% of

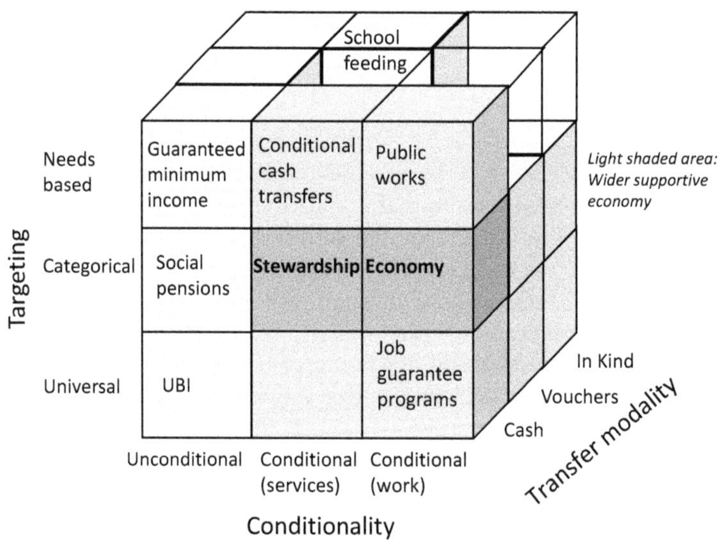

Figure 11.1 Stewardship Economy within a social assistance cube

that population will be urban (Ritchie and Roser 2018). The need to improve the kind and total amounts of rural investment (e.g., away from farm subsidies) has been recognized (OECD 2006) to ensure that the gulf between the development and economic benefits urban populations enjoy compared to rural populations does not continue to widen. The Stewardship Economy, by linking duty of care, knowledge, know-how and action to economic rewards and incentives, would add a valuable dimension to guide such rural investments and transfers towards more equitable, productive and sustainable outcomes in future.

We want to turn to the likely contributions of ACM to the institutional elements of a Stewardship Economy now. Kusumanto et al., in describing the needs of a multilayer, nested governance structure, lay out the institutional elements for what Minang et al. (2014) have called landscape democracy, and which ACM practitioners recognize as necessary for multi-scale social learning. Kusumanto et al. see such a "*…learning platform as consciously constructed opportunities for multiple stakeholders to jointly learn*". In determining what is a commodity and what is not, who should be rewarded, how, for what and how much, such learning platforms, processes of social learning, collaborative monitoring and collective action are essential. Moving to a Stewardship Economy is therefore not just about equitable economic rewards and welfare; it is about leaving space for adaptation, creativity, care and innovation and the new knowledge that needs to be developed as we deal with rapidly changing manifestations of our multiple global crises. As Kamoto et al. recognize, "…we need to continuously apply the ACM concept which is aimed at catalyzing change while continuously monitoring performance and consciously learning from it" (p. 212).

> There is of course much more to say about institutions. Kashwan and Ribot (2021), for instance, identify the powerful concept of 'violent silence':
>
> … the institutionalized assumptions—such as those incorporated in the models used by the IPCC, the policy instruments of the UNFCCC, and the Paris Agreement—that avoid history, spurred by the desire (or compulsion) of some parties to avoid responsibility and blame.
>
> (p. 331)

Surely more such subtle, equally 'hidden' concepts will emerge as we move forward with the Stewardship Economy. We are only at the beginning but we must pursue this path with urgency and energy.

A Stewardship Economy, in whichever form it finally emerges, must provide a pathway to a new game, where the over-commodification of nature, a shortage of resources (based on related over-consumption and population growth) and power differentials do not force stewards into perverse and competitive behaviours. We must develop and learn to play by the rules of this new game, if we are to have any chance of averting the catastrophic crises that are upon us. ACM has a key role to play, and that role is likely more important where a culture of care and stewardship prevails.

Notes

1 https://ukcop26.org/glasgow-leaders-declaration-on-forests-and-land-use/.

2 Prabhu, Lawry, and Colmey (2021) distinguish between an equitable 'fair price' and the 'true price' of an agricultural commodity, which would reflect all the costs of stewardship, but may not be affordable for poor people.

References

Andersson, E., J. Enqvist, and M. Tengö. 2017. "Stewardship in Urban Landscapes." In *The Science and Practice of Landscape Stewardship*, edited by C. Bieling and T. Plieninger, 222–238. Cambridge, UK: Cambridge University Press. https://doi.org/10.1017/9781316499016.023.

Arora-Jonsson, S., C.J.P. Colfer, and M. González-Hidalgo. 2021. "Seeing the Quiet Politics in Unquiet Woods: A Different Vantage Point for a Future Forest Agenda." *Human Ecology* 49, no. 3: 297–308. https://doi.org/10.100.7/s10745-021-00233-0.

Berry, R.J. 2006. *Environmental Stewardship*. London: A and C Black.

Caretta, M.A. 2020. "Homosocial Stewardship: The Opposed and Unpaid Care Work of Women Water Stewards in West Virginia, USA." *Ecology and Society* 25, no. 2: 29. https://doi.org/10.5751/ES-11555-250229.

Carrozzo Magli, A., P. Della Posta, and P. Manfredi. 2021. "The Tragedy of the Commons as a Prisoner's Dilemma. Its Relevance for Sustainability Games." *Sustainability* 13: 8125. https://doi.org/10.3390/su13158125.

Cockburn, J., E. Rosenberg, A. Copteros, S. Francina (Ancia) Cornelius, N. Libala, L. Metcalfe, and B. van der Waal. 2020. "A Relational Approach to Landscape Stewardship: Towards a New Perspective for Multi-Actor Collaboration." *Land* 9: 224. https://doi.org/10.3390/land9070224.

Colfer, C.J.P. 1995. "Who Counts Most in Sustainable Forest Management?" CIFOR Working Paper No. 7, p. 16.

Colfer, C.J.P., R. Prabhu, and A. Larson, eds. 2022. *Adaptive Collaborative Management in Forest Landscapes: Villagers, Bureaucrats and Civil Society*. London: Earthscan and Routledge.

Crist, E., C. Mora, and R. Engelman. 2017. "The Interaction of Human Population, Food Production, and Biodiversity Protection." *Science* 356, no. 6335: 260–264. https://doi.org/10.1126/science.aal2011.

Critchley, W., N. Harari, and R. Mekdaschi-Studer. 2021. "Restoring Life to the Land: The Role of Sustainable Land Management in Ecosystem Restoration." UNCCD and WOCAT. https://www.unccd.int/sites/default/files/documents/2021-10/211018_RestoringLifetotheLand_Report%20%282%29.pdf.

Czech, B. 2002. *Shoveling Coal for a Runaway Train*. Oakland: University of California Press.

Davis, A., and K. Ruddle. 2012. "Massaging the Misery: Recent Approaches to Fisheries Governance and the Betrayal of Small-Scale Fisheries." *Human Organization* 71, no. 2: 244–254.

Enqvist, J.P., S. West, V.A. Masterson, L.J. Haider, U. Svedin, and M. Tengö. 2018. "Stewardship as a Boundary Object for Sustainability Research: Linking Care, Knowledge and Agency." *Landscape and Urban Planning* 179: 17–37. https://doi.org/10.1016/j.landurbplan.2018.07.005.

EPA Innovation Council. 2005. *Everyday Choices: Opportunities for Environmental Stewardship*. Washington, DC: US Environmental Protection Agency Archive Document.

FAO, IFAD, UNICEF, WFP and WHO. 2021. *The State of Food Security and Nutrition in the World 2021. Transforming Food Systems for Food Security, Improved Nutrition and Affordable Healthy Diets for All.* Rome, Italy: FAO. https://doi.org/10.4060/cb4474en.

Finlayson, R. 2022. "Biodiversity Science–Policy Panel Calls for Broadening Value-of-Nature Concepts in Sustainable Development." *CIFOR Forest News* (11 July). https://forestsnews.cifor.org/78341/biodiversity-science-policy-panel-calls-for-broadening-value-of-nature-concepts-in-sustainable-development?fnl=en.

Gentilini, U., M. Grosh, J. Rigolini, and R. Yemtsov, eds. 2020. *Exploring Universal Basic Income: A Guide to Navigating Concepts, Evidence, and Practices.* Washington, DC: World Bank. https://doi.org/10.1596/978-1-4648-1458-7.

IPBES. 2019. *Global Assessment Report on Biodiversity and Ecosystem Services of the Intergovernmental Science-Policy Platform on Biodiversity and Ecosystem Services*, edited by E.S. Brondizio, J. Settele, S. Díaz, and H.T. Ngo. Bonn, Germany: IPBES Secretariat. https://doi.org/10.5281/zenodo.3831673.

IPBES. 2022. *Summary for Policymakers of the Methodological Assessment of the Diverse Values and Valuation of Nature of the Intergovernmental Science-Policy Platform on Biodiversity and Ecosystem Services*, edited by U. Pascual, P. Balvanera, M. Christie, B. Baptiste, D. González-Jiménez, C.B. Anderson, S. Athayde, R. Chaplin-Kramer, S. Jacobs, E. Kelemen, R. Kumar, E. Lazos, A. Martin, T.H. Mwampamba, B. Nakangu, P. O'Farrell, C.M. Raymond, S.M. Subramanian, M. Termansen, M. Van Noordwijk, and A. Vatn. Bonn: IPBES Secretariat. https://doi.org/10.5281/zenodo.6522392.

IPCC. 2021. "Summary for Policymakers." In *Climate Change 2021: The Physical Science Basis* (Contribution of Working Group I to the Sixth Assessment Report of the Intergovernmental Panel on Climate Change), edited by V. Masson-Delmotte, P. Zhai, A. Pirani, S.L. Connors, C. Péan, S. Berger, N. Caud, Y. Chen, L. Goldfarb, M.I. Gomis, M. Huang, K. Leitzell, E. Lonnoy, J.B.R. Matthews, T.K. Maycock, T. Waterfield, O. Yelekçi, R. Yu, and B. Zhou, 3–32. Cambridge, UK: Cambridge University Press.

IPCC. 2022. "Climate Change 2022. Mitigation of Climate Change." Working Group III contribution to the Sixth Assessment Report of the Intergovernmental Panel on Climate Change. Approved Summary for Policy Makers. https://www.ipcc.ch/report/ar6/wg3/.

Johnson, M.L., L.K. Campbell, and E.S. Svendsen. 2020. "Conceptualizing, Analyzing, and Supporting Stewardship: Examining the Role of Civil Society in Environmental Governance." *Ecology and Society* 25, no. 4: 14. https://doi.org/10.5751/ES-11970-250414.

Kashwan, P., and J. Ribot. 2021. "Violent Silence: The Erasure of History and Justice in Global Climate Policy." *Current History* 120, no. 829: 326–331.

Katila, P., C.J.P. Colfer, W. Dejong, G. Galloway, P. Pacheco, and G. Winkel, eds. 2019. *Sustainable Development Goals: Their Impact on Forests and People.* Cambridge, UK: Cambridge University Press.

Kristjanson, P. 2020. "Closing Gender Gaps in Forest Landscape Initiatives." *International Forestry Research* 22: 44.

Laird, S., A. Awono, C. Okia, G.A. Anaya, V. Ingram, P. Sola, C. Watson, C. Muthuri, P. Gilruth, R. Mendum, and M. Njenga. 2022. "Social and Environmental Transformation of Refugee and Hosting Community Landscapes in Central and Eastern Africa." *CIFOR Occasional Paper* 229.

Leopold, A. 1949. *A Sand County Almanac and Sketches Here and There.* New York: Oxford University Press.

Li, T. 2014. *Lands End: Capitalist Relations on an Indigenous Frontier.* Durham, NC: Duke University Press.

Li, T.M., and P. Semedi. 2021. *Plantation Life: Corporate Occupation in Indonesia's Oil Palm Zone*. Durham, NC: Duke University Press.

Maarif, Samsul. 2012. *Dimensions of Religious Practice: The Ammatoans of Sulawesi, Indonesia*. Tempe: Arizona State University.

Madani, K. 2013. "Modeling International Climate Change Negotiations More Responsibly: Can Highly Simplified Game Theory Models Provide Reliable Policy Insights?" *Ecological Economics* 90: 68–76.

Maryani, R., D. Djaenudin, M. Lugina, Indartik, M. Moeliono, and E.L. Yuliani. 2021. "Perspectives on Landscape Approaches in Indonesian Policies." In *P3SEKPI Infobrief 2*: 1–16. Bogor: Center for Research and Development for Social, Economy, Policy and Climate Change of the Forestry Research Development and Innovation, Ministry of Environment and Forestry (P3SEKPI) and CIFOR.

Mathevet, R., F. Bousquet, and C.M. Raymond. 2018. "The Concept of Stewardship in Sustainability Science and Conservation Biology." *Biological Conservation* 217: 363–370.

Meine, C. 2020. "From the Land to Socio-Ecological Systems: The Continuing Influence of Aldo Leopold." *Socio-Ecological Practice Research* 2: 31–38. https://doi.org/10.1007/s42532-020-00044-5.

Minang, P.A., M. van Noordwijk, O.E. Freeman, C. Mbow, J. de Leeuw, and D. Catacutan. 2014. *Climate-Smart Landscapes: Multifunctionality in Practice*. Nairobi, Kenya: World Agroforestry Centre (ICRAF).

OECD. 2006. "Investment Priorities for Rural Development." OECD Rural Policy Conference. Edinburgh, Scotland: OECD. https://www.oecd.org/cfe/regionaldevelopment/Investment-Priorities-for-Rural-Development.pdf

Paulson, S. 2017. "Changing Masculinities and Femininities in the (Re)Production of Andean Silvopastoral Systems." In *The Earthscan Reader on Gender and Forests*, edited by C.J.P. Colfer, M. Elias, B. Sijapati Basnett, and S. Stevens Hummel, 163–280. London: Earthscan and Routledge.

Prabhu, R. 2022. "Would Indian Agriculture Benefit from a Stewardship Model?" *Ecology, Economy and Society–the INSEE Journal* 5, no. 1: 5–14. https://ecoinsee.org/journal/ojs/index.php/ees/article/download/612/201

Prabhu, R., A.M. Larson, and C.J.P. Colfer. 2022. "Circles and Spirals." In *Adaptive Collaborative Management of Forest Landscapes: Villagers, Bureaucrats and Civil Society*, edited by C.J.P. Colfer, R. Prabhu, and A. Larson, 31–61. London: Earthscan and Routledge.

Prabhu, R., S. Lawry, and J. Colmey. 2021. "Creating a New Relationship with Nature Through a 'Stewardship Economy.'" *Forests News* (October).

Ritchie, H., and M. Roser. 2018. "Urbanization." Published online at OurWorldInData.org. https://ourworldindata.org/urbanization [Online Resource].

Saner, M., and J. Wilson. 2003. "Stewardship, Good Governance and Ethics." *Institute on Governance Policy Brief* 19.

Shepherd, G., K. Warner, and N. Hogarth. 2020. "Forests and Poverty: How has our Understanding of the Relationship been Changed by Experience?" *International Forestry Review* 22, no. 1: 29–43.

Singh, N.M. 2015. "Payments for Ecosystem Services and the Gift Paradigm: Sharing the Burden and Joy of Environmental Care." *Ecological Economics* 117: 53–61.

Index

Note: **Bold** page numbers refer to tables; *italic* page numbers refer to figures and page numbers followed by "n" denote endnotes.

For Product Safety Concerns and Information please contact our EU representative GPSR@taylorandfrancis.com
Taylor & Francis Verlag GmbH, Kaufingerstraße 24, 80331 München, Germany

www.ingramcontent.com/pod-product-compliance
Ingram Content Group UK Ltd.
Pitfield, Milton Keynes, MK11 3LW, UK
UKHW022318250926
475702UK00046B/924
9781032352282